ANALYSIS OF INTERPLANETARY DUST

NASA/LPI WORKSHOP

AIP CONFERENCE PROCEEDINGS 310

ANALYSIS OF INTERPLANETARY DUST

NASA/LPI WORKSHOP
HOUSTON, TX MAY 1993

EDITORS: **M.E. ZOLENSKY**
T.L. WILSON
NASA, JOHNSON SPACE CENTER
F.J.M. RIETMEIJER
UNIVERSITY OF NEW MEXICO
G.J. FLYNN
SUNY, PLATTSBURGH

American Institute of Physics   New York

Authorization to photocopy items for internal or personal use, beyond the free copying permitted under the 1978 U.S. Copyright Law (see statement below), is granted by the American Institute of Physics for users registered with the Copyright Clearance Center (CCC) Transactional Reporting Service, provided that the base fee of $2.00 per copy is paid directly to CCC, 27 Congress St., Salem, MA 01970. For those organizations that have been granted a photocopy license by CCC, a separate system of payment has been arranged. The fee code for users of the Transactional Reporting Service is: 0094-243X/87 $2.00.

© 1994 American Institute of Physics.

Individual readers of this volume and nonprofit libraries, acting for them, are permitted to make fair use of the material in it, such as copying an article for use in teaching or research. Permission is granted to quote from this volume in scientific work with the customary acknowledgment of the source. To reprint a figure, table, or other excerpt requires the consent of one of the original authors and notification to AIP. Republication or systematic or multiple reproduction of any material in this volume is permitted only under license from AIP. Address inquiries to Series Editor, AIP Conference Proceedings, AIP Press, American Institute of Physics, 500 Sunnyside Boulevard, Woodbury, NY 11797-2999.

L.C. Catalog Card No. 94-71292
ISBN 1-56396-341-8
DOE CONF-9305328

Printed in the United States of America.

CONTENTS

Preface ix
Workshop Summary xi
Organizing Committee and Sponsors xiii

INTRODUCTION

Introduction 3
The Origin and Role of Dust in the Early Solar System 5
    D. Brownlee

REMOTE SENSING AND DYNAMICS OF INTERPLANETARY DUST

Topical Summary 11
Detection of Asteroidal Dust Particles from Known Families in Near-Earth
Orbits 13
    S. F. Dermott and J. C. Liou
Remote Sensing of Cometary Dust and Comparisons to IDPs 23
    M. S. Hanner
Volcanism and Levitation: Unusual Origins of Interplanetary Dust 33
    T. L. Wilson

MINERALOGY AND COMPOSITION OF INTERPLANETARY DUST

Topical Summary 47
Mineralogical and Chemical Relationships of Interplanetary Dust
Particles, Micrometeorites and Meteorites 51
    W. Klöck and F. J. Stadermann
Mechanisms of Grain Formation, Post-accretional Alteration, and Likely Parent
Body Environments of Interplanetary Dust Particles (IDPs) 89
    J. Bradley
Olivine and Pyroxene Compositions of Chondritic Interplanetary Dust
Particles 105
    M. E. Zolensky and R. Barrett
Helium and Neon in Interplanetary Dust Particles 115
    A. O. Nier
Changes to the Composition and Mineralogy of Interplanetary Dust Particles
by Terrestrial Encounters 127
    G. J. Flynn
Chemical Compositions of Primitive Solar System Particles 145
    S. R. Sutton

Carbon in Primitive Interplanetary Dust Particles 159
    L. P. Keller, K. L. Thomas, and D. S. McKay
Quantitative Analyses of Carbon in Anhydrous and Hydrated Interplanetary
Dust Particles 165
    K. L. Thomas, L. P. Keller, G. E. Blanford, and D. S. McKay
Volatiles in Interplanetary Dust Particles: A Comparison with Volatile-Rich
Meteorites 173
    E. K. Gibson, Jr. and R. Bustin
Origin of the Hydrocarbon Component of Interplanetary Dust Particles 185
    T. J. Wdowiak and W. Lee
Carbon in Comet Halley Dust Particles 193
    M. Fomenkova and S. Chang
Isotopic Constraints on Interstellar Material in Chondritic IDPs 203
    R. M. Walker
$^{6}Li/^{7}Li$, $^{10}B/^{11}B$ and $^{7}Li/^{11}B/^{28}Si$ in Individual Interplanetary Dust Particles 211
    Y.-L. Xu, L.-G. Song, Y.-X. Zhang, and C. Y. Fan
Cometary Dust: A Thermal Criterion to Identify Cometary Samples Among the
Interplanetary Dust Collected from the Stratosphere 223
    G. J. Flynn
A Proposal for a Petrological Classification Scheme of Carbonaceous Chondritic
Micrometeorites 231
    F. J. M. Rietmeijer

COLLECTION OF INTERPLANETARY DUST

Topical Summary 243
Collection and Curation of Interplanetary Dust Particles Recovered from the
Stratosphere by NASA 245
    J. L. Warren and M. E. Zolensky
Windows of Opportunity in the NASA Johnson Space Center Cosmic Dust
Collection 255
    F. J. M. Rietmeijer and J. L. Warren
Collection and Curation of IDPs from the Greenland and Antarctic
Ice Sheets 277
    M. Maurette, G. Immel, C. Hammer, R. Harvey, G. Kurat, and S. Taylor
Meteoroid Investigations Using the Long Duration Exposure Facility 291
    M. E. Zolensky, F. Hörz, T. H. See, R. P. Bernhard, C. Dardano, R. A. Barrett,
    K. Mack, J. L. Warren, and W. H. Kinard
Description of the COMRADE Experiment 305
    J. Borg, J-P. Bibring, C. Maag, W. Tanner, and M. Alexander
Analysis of Remnants Found in LDEF and Mir Impact Craters 313
    L. Berthoud and J. C. Mandeville

Penetration Experiments in Aluminum and Teflon Targets of Widely Variable Thickness 329
    F. Hörz, M. Cintala, R. P. Bernhard, and T. H. See
List of Participants 345
Author Index 349
Subject Index 351

PREFACE

Space research is a very broad discipline of international scope that focuses upon those fundamental aspects of space science and space technology which will improve our knowledge and understanding of the Universe in which we live. This necessarily involves a delicate balance of scientific, technological, and social issues which must be weighed against the pressing fundamental questions of international and national space programs—including the National Aeronautics and Space Administration (NASA), the European Space Agency (ESA), the Russian space program, the Japanese space program, or any of the multitudes of others.

The issue of priorities becomes even more acute when promising new fields of research appear within the scientific subdisciplines, and compromises have to be made at the very time when space exploration is entering a significant and exciting phase with the approach of the 21st century. NASA has attempted to maintain a balance between national space policy and scientific priorities in a number of fashions, two of which come to mind. One is the non-advocate, peer-reviewed research process, and the other is the use of the recommendations of the Space Science Board of the National Academy of Sciences, which independently sets forth periodic guidelines for space science in this and the next century. From this procedure has evolved the present compliment of supported space research in such diverse areas as high-energy astrophysics, astronomy, life science, planetary science, space physics, solar system exploration, and so forth.

The exploration of the solar system has been and continues to be one of the central themes of NASA's space research program because it is intimately tied to the question of our own origin, our ultimate survival, and our irresistible quest for wonder as we search the unknown. So, while the Hubble Space Telescope (HST) is providing marvelous images of active nuclei in other galaxies, while the Cosmic Background Explorer (COBE) has apparently measured slight density variations in the primordial origin of our Universe, and while the Compton Gamma-Ray Observatory (GRO) has possibly discovered new physics in the isotropy of gamma-ray bursts, other man-made names loom about through the orbits of our neighbor planets. These are names like Ulysses, Giotto, Vega, and Voyager, the names of instrumented "flyby's" which have attempted to analyze *in situ* the structure of comets and planets, and the origin of interplanetary dust. To this add high-flying aircraft collecting samples in the Earth's stratosphere as well as NASA's repository of material samples returned from the Moon during the Apollo program and meteorites collected in Earth's polar regions, and one finds a systematic scientific and multidisciplinary analysis (involving plasma physics, geophysics, chemistry, and astronomy) in our own terrestrial laboratories of actual pieces of asteroids, meteoroids, comets, and even interstellar dust.

It is this latter aspect of NASA's origins programs to which the present workshop, held at the Lunar and Planetary Institute (LPI) in Houston on May 15–17, 1993, was dedicated.

Michael E. Zolensky
Thomas L. Wilson
NASA, Houston
February, 1994

WORKSHOP SUMMARY

Michael Zolensky and Tom Wilson
NASA Johnson Space Center, Houston, TX 77058

Great progress has been made in the analysis of interplanetary dust particles (IDPs) over the past few years, as reckoned by short presentations made at meetings of the Meteoritical Society, Lunar and Planetary Science Conferences, and electron microscopy conclaves. However, dust workers have strived for a more focused showcase from which to present recent IDP results with opportunity for consolidation of past work and for the forging of new research collaborations. The recent availability of larger IDPs from the Large Area Collectors and consequent particle analysis consortia have made the necessity of a dedicated workshop even more acute. To satisfy such a need, this first workshop dedicated to the analysis of IDPs was organized by Don Brownlee (University of Washington), John Bradley (MVA Associates), George Flynn (SUNY Plattsburgh), Alfred Nier (University of Minnesota), Frans Rietmeijer (University of New Mexico), and Michael Zolensky (NASA, Johnson Space Center). From the start the principal goal of the workshop was to provide a forum for free and relatively uninterrupted discussion. To establish the maximum degree of participant interplay and productive discussion, the workshop was designed around a few review talks. Each of these talks was intended to pull together past results in a specific branch of IDP research, and suggests future potentially fruitful directions. Following each of these presentations, workshop participants were free to discuss any aspects of the specific subject, and introduce and discuss their own results and ideas. Contributed presentations were made in the form of posters, although these results were folded into the discussions at appropriate times.

For each discussion, one workshop participant served as a summarizer. These summaries, and recordings of the talks and discussions, were used to facilitate production of the present workshop proceedings volume. The invited summary presentations were as follows.

An Overview of the Origin and Role of Dust in the Early Solar System *Don Brownlee*

Modern Sources of Dust in the Solar System *Stan Dermott*

Remote Sensing of Comets and Comparisons to IDPs *Martha Hanner*

What Does the Fine-Scale Petrography of IDPs Reveal about Grain Formation and Evolution in the Early Solar System? *John Bradley*

Solar System Exposure Histories of IDPs *Al Nier*

Changes in IDP Mineralogy and Composition by Terrestrial Factors *George Flynn*

How Are IDPs Related to Larger Extraterrestrial Samples? *Wolfgang Klöck*

The Composition of Primitive Solar System Grains: Major to Trace Elements *Steve Sutton*

Role of Volatile Elements in the Early Solar Nebula *Lindsay Keller*

History of Elements in the Early Solar System Based on Isotope Studies of Dust *Robert Walker*

Classification of Interplanetary Dust Particles *Frans Rietmeijer*

Spacecraft Data from IDPs *David McKay and Herb Zook*

Collection and Curation of IDPs in the Stratosphere and Below *Michel Maurette and Mike Zolensky*

Future Opportunities to Return Primitive Materials Directly from Space *Bill Tanner and Bill Kinard*

These summary presentations were grouped in the following order: (1) Introduction, (2) Observation and modeling of dust in the solar system, (3) Mineralogy and petrography of IDPs, (4) Processing of IDPs in the solar system and terrestrial atmosphere, (5) Comparisons of IDPs to meteorites and micrometeorites, (6) Composition of IDPs (including isotopes), (7) Classification schemes of IDPs, and (8) Collection of IDPs.

Brief overviews of the major points raised at the workshop are arranged at the beginnings of the major sections of this volume, so that they may serve as an introduction to the important points raised at the workshop and discussed in the papers here.

We would like to thank the Director and staff of the Lunar and Planetary Institute for the generous financial and technical support which made the initial workshop possible. The production of the present volume was made possible by the support of NASA.

NASA/LPI Workshop on the Analysis of Interplanetary Dust

ORGANIZING COMMITTEE:

Michael E. Zolensky	NASA, Johnson Space Center
Don Brownlee	University of Washington
John Bradley	MVA Associates
George Flynn	SUNY, Plattsburgh
Alfred Nier	University of Minnesota
Frans Rietmeijer	University of New Mexico

SPONSORS:

National Aeronautics and Space Administration
Lunar and Planetary Institute

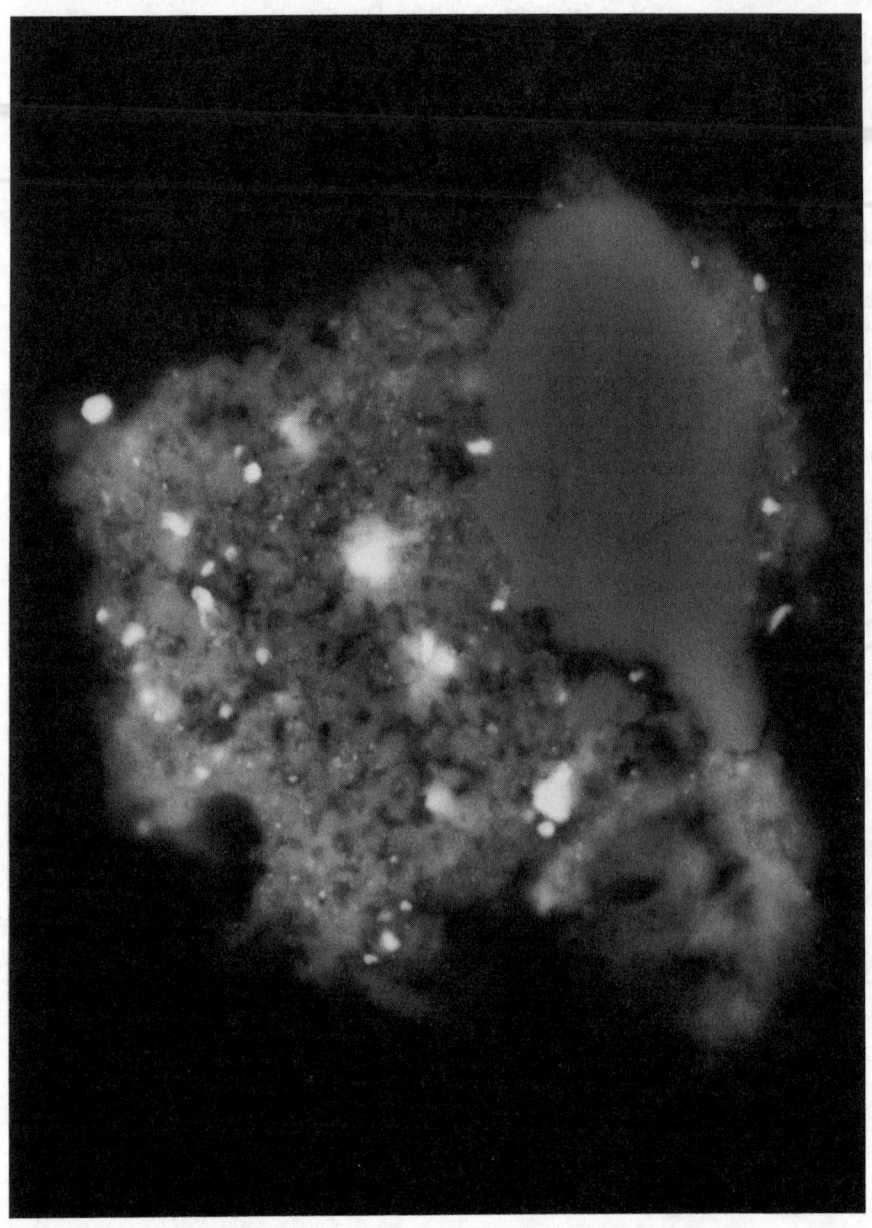

A backscattered electron image of a "polished" surface of anhydrous chondritic IDP L2005 E36. The matrix is a heterogeneous mass of fine-grained ferromagnesian phases, sulfides (white) and amorphous material. The large crystal at upper right is enstatite. The particle measures approximately 15 μm across. (Photo by M. Zolensky)

INTRODUCTION

Zodiacal light photographed from Haleakala. This phenomenon is caused by the scattering of sunlight from dust in interplanetary space. (Photo by P. Hutchison, University of Hawaii, from *The Zodiacal Light and the Interplanetary Medium*, NASA SP-150, Ed. J. Weinberg (1967)).

INTRODUCTION

At the workshop, Don Brownlee presented an overview of the current state of understanding concerning IDPs, and most of the points he raised were addressed at greater length by subsequent presentations and discussions. In particular, Don made the points that IDPs include probably the only samples now available of outer-belt asteroids and comets, including Kuiper belt objects. IDPs could have turned out to be merely additional fragments of H6 chondrites, but (fortunately for us) detailed work over the past two decades has shown them to be compositionally and mineralogically distinct from meteorites. These important distinctions include greater porosity, aggregate structure, higher volatile content, and unique mineralogy. Great progress has been made recently in the development of microanalytical techniques, permitting measurements to be made of IDP bulk compositions (including trace elements by four different and complementary techniques, noble gasses, and some organic compounds), infrared spectra (both transmittance and reflectance), isotopes and physical properties. Important problems remain, including the establishment of a useful classification scheme, resolving differences in terminology (e.g. tar balls vs. granular units, etc.), and elucidation of the relationships between different types of IDPs (chondritic vs. refractory vs. basaltic; hydrous vs. anhydrous). Finally, future collection technologies were discussed, including replacement of silicone oil in stratospheric collection, and the role of dust collection in space. Many of these important points are presented by Brownlee in the following paper.

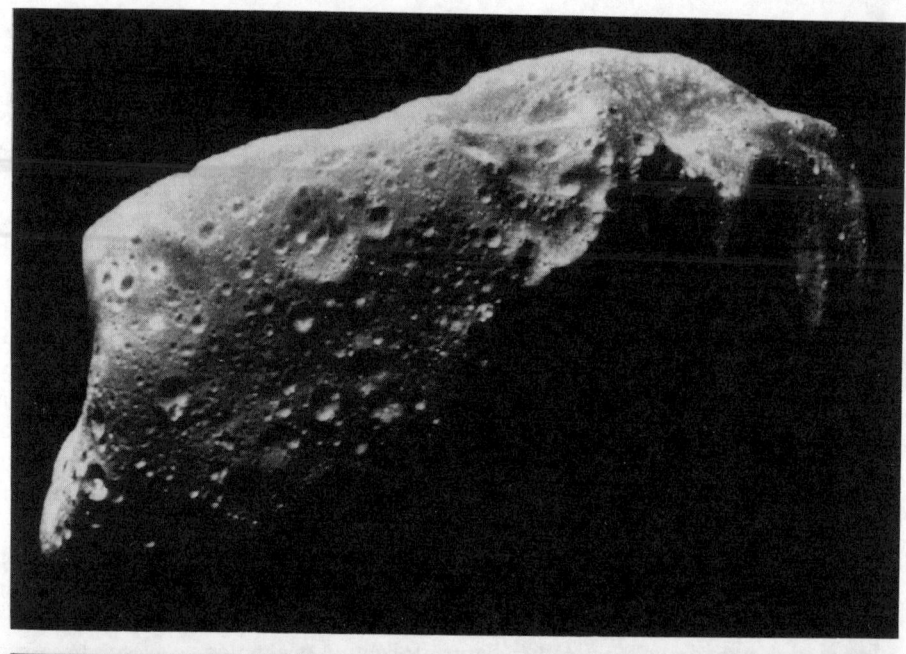
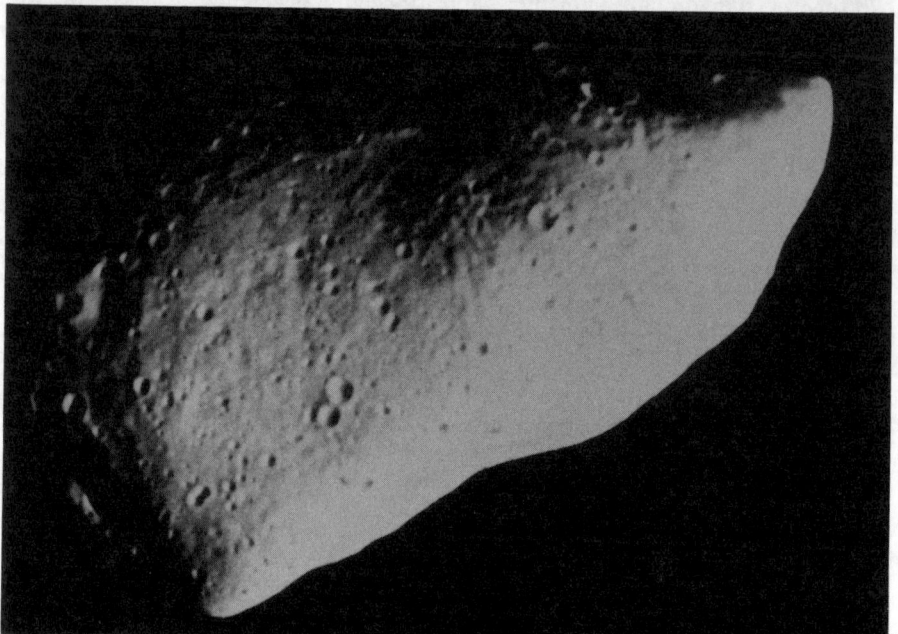

Images of the main belt asteroids Ida (top) and Gaspra (bottom). Asteroids are believed to be the principal sources of IDPs. Ida and Gaspra have lengths of 52 and 19 km, respectively. (Images taken by the Galileo spacecraft, and courtesy of the Lunar and Planetary Institute and JPL.)

THE ORIGIN AND ROLE OF DUST IN THE EARLY SOLAR SYSTEM

Don Brownlee
Dept. of Astronomy, University of Washington, Seattle, WA 98195, USA

Dust is believed to have been the fundamental building material of the planetesimals that accumulated to form minor planets, solid planets and the cores of giant planets. In this model the first generation particulates were pre-solar interstellar grains. At the end of the 10 K free-fall collapse phase that preceded the solar nebula[1], these grains contained nearly all of the condensable atoms now present in the solar system. The grains formed in circumstellar environments around other stars and were partially modified both in the interstellar medium and the local environments that preceded the Solar System.[2] In the nebula, pre-solar grains were modified, and in many cases, destroyed by a variety of processes such as heating, sublimation, sputtering, collisions, irradiation, accretion, adsorption, condensation and chondrule formation. Solids of purely nebular origin formed from the alteration products derived from pre-solar grains. Following accretion into larger bodies both pre-solar and nebular solids were further altered by parent-body processes such as compaction, heating, shock and aqueous alteration. The nebular and parent-body effects had a broad range of severity related to the location and time of their occurrence. It is likely that some primordial pre-solar grains were preserved without major modification while in others they were totally reworked and transformed to purely nebular materials. Inside parent-bodies, alteration processes were in some cases so extreme that nearly all properties of the progenitor solids were erased.

It is likely that nearly all of the major modification processes that influenced solid materials in the nebula had strong radial gradients. Collected samples of interplanetary dust provide a unique source of sample-derived information on these processes. The search for effects due to these processes may provide new insight into some of the properties of different IDP types. Collected IDPs are materials that accreted over a range of radial distance extending from the inner edge of the asteroid belt to 100's of AU in the Kuiper belt. This diverse range of origin is in strong contrast to conventional meteorites whose origins appear to be limited to the asteroid belt[3] with a strong preference for derivation from near 2.5 AU where objects in circular orbit have a 3:1 resonance with Jupiter.[4] Small IDPs are a broader sampling of the Solar System's population of minor planets because it is easier for such small objects to reach Earth-crossing orbits and survive atmospheric entry without fragmentation. Dust orbits rapidly decay due to light pressure effects and nearly all particles generated by comets and asteroids can evolve towards Earth-crossing orbits. In contrast, the orbits of rock-size objects evolve only by gravitational perturbation. The atmospheric entry advantage for dust is that it decelerates at altitudes above 90 km where the ram pressure is so low that even materials as weak as typical cometary meteors can survive without fragmentation. Kilogram rocks penetrate deep into the atmosphere at hypervelocity where they typically fragment and only moderately strong objects survive to become conventional meteorites. The maximum atmospheric ram pressure on nanogram dust is more than three orders of magnitude smaller that which must be survived by conventional meteorites.

The two general source regions of nanogram IDPs collected in the atmosphere are the asteroid belt and the short period comets.[5] The collected asteroid particles are likely to be rather representative samples of material that accreted into planetesimals in the 2.2 to 3.3 AU region of the solar system. After orbital decay, by Poynting-Robertson drag, these particles arrive at the Earth on flat, circular orbits that are favorable conditions for gravitational concentration and atmospheric survival.[5] The comet samples that survive atmospheric entry are biased towards particles from short period comets that impact the Earth at only moderate velocity. The majority of these short period comets are believed to be derived from the Kuiper belt, a flattened distribution of cometary bodies that extends outwards from just beyond the outer planets.[6] Atmospheric entry calculations show that nanogram comet particles entering at velocities up to about 25 km/s can survive entry without melting.[7] These two source regions provide interesting boundary samples for nebular materials and processes. The mean distance to main belt asteroids is about 2.7 AU while the inner radius of the Kuiper belt is about 35 AU, a distance limited by perturbation from the outer planets. The materials in collected IDPs thus sample regions separated by more than an order of magnitude of radial distance with essentially a clear dichotomy of origins. If materials from the inner and outer regions of the solar nebula have drastic differences this should be clearly detectable within the IDP collections.

There are many reasons why typical IDPs from asteroids might differ from those from typical comets. In the solar nebula, the midplane density, pressure, temperature and irradiation effects must have decreased greatly between these two source regions. Because the orbital velocity drops by a factor of five over this distance, shock and other collisional effects should have been much greatly reduced beyond 100 AU. In the asteroid region of the nebula, temperatures were apparently high enough to destroy the majority of pre-solar solids while beyond 100 AU the temperature probably did not exceed 100 K.[8] Estimates based on composition indicate that comets accreted at temperatures of 60 K or below.[9] At cometary distances it is likely that a large fraction of pre-solar grains survived intact although little is really known about grain destruction mechanisms in this region. The abundance of surviving pre-solar grains in some meteorites is as high as 1800 ppm[10] and one would expect that the survival in cometary materials would be much higher. Another key process that may have strongly varied between the inner and outer solar system is chondrule formation. However chondrules formed, the process was highly efficient in melting more that 75% of the solids in broad regions of the asteroid belt region. It is likely that the transient heat sources that formed these bodies was electromagnetic and was confined within the inner regions of the nebula.

The asteroid belt marks the transition zone between the terrestrial and Jovian planets and it was a critical region in the nebula where many potentially complex processes may have occurred. The wide range of meteorite properties are evidence that this was a complex region. The fractionation of metal, refractories, volatiles and range in oxidation state are remarkably variable for meteorites that appear to have formed within a small region of the nebula. Spectral reflectance studies show that the asteroids are compositionally zoned with the more distant asteroids dominated by the P and D reflectance types (that have no meteorite analogs) while the main and inner belt are dominated by C and S classes.[11] There is also a radial variation in the hydrated silicate content with the more distant asteroids showing little or no evidence for their presence.[12&13] The "dryness" of these asteroids indicates that they were not heated sufficiently to melt ice and produce the extensive aqueous alteration seen in many carbonaceous chondrites.[14] The radial composition variations among the asteroids may in part be due to heating effects that were most prominent near the center of the solar system. These heating effects may have been due to inductive heating[15] which would have a strong radial variation or they may have been caused by heating by ^{26}Al and ^{60}Fe decay.[16] This would have more strongly effected inner asteroids because they could accrete before these short-lived isotopes decayed.[16] Many of the most distant minor planets (the Centaurs) have steep red reflectances suggestive of high organic content.[17]

Another process that may have played a major and complex role in the asteroid region but not in the outer solar system is large variation in the O/H ratio that affects the formation and oxidation state of condensing solids. Vaporization of ice and silicates that have become highly concentrated at the nebula midplane can result in highly oxidizing conditions consistent with the observed Fe^{2+} content of meteoritic olivine.[18] Special complications may have occurred in the asteroid region because this is the approximate location of the "snow line" within which it was too hot for ice to condense. It is possible that this ice condensation zone acted like a giant cold finger that extracted and highly concentrated water from the inner Solar System.[19]

Many of the Solar System effects on solid materials should have been more pronounced or even restricted to the <5 AU region where the asteroids formed. Although little is really known about comets it is reasonable to expect that they are much less processed than asteroids due to their small size and isolation from the most plausible alteration processes. It is a major challenge to IDP research to distinguish comet and asteroid samples reliably and use their properties to provide fundamental information on the solid building materials of the planets. Strong cases have been made that at least some of the hydrated IDPs are asteroidal[20&21] although it is by no means assured that this is the origin of all of them. It is possible that all of the IDPs with extensive aqueous alteration are derived from inner solar system parents while the anhydrous particles are derived from both cometary and asteroidal sources but at present it not possible to prove this. Because of similar components in different IDP types it has been suggested that cometary materials and some asteroidal materials may have strong similarities in spite of differences in their sites of accretion.[22] If this is the case then some of the asteroids may be nearly identical to cometary matter. This could occur if conditions become benign and relatively simple beyond some critical distance that may have been within the bounds of the asteroid region. It will be fascinating to see how IDP research in the coming years can better define the differences and possible links between material that formed comets and asteroids.

REFERENCES

1. Hayashi C., Nakazawa K. and Nakagawa Y. (1985) in *Protostars and Planets II*, 1100-1153, eds. D.C. Black and M.S. Mathews.
2. Mathis J.S. (1993) *Rep. Prog. Phys.* **56**, 605-652.
3. Anders E (1978) In *Asteroids: An Exploration Assessment.* 57-75, eds. D. Morrison and W.C. Wells, NASA CP-2053.
4. Wetherill G.W. and Chapman C.R. (1988) in *Meteorites and the Early Solar System* 35-67, ed J. F. Kerridge, U. Arizona Press.
5. Flynn G.J. (1989) *Icarus* **77**, 287-310.
6. Quinn T., Tremaine S. and Duncan M (1990) *Astrophysical J.* **355**, 667-679.
7. Love S.G. and Brownlee D.E. (1991) *Icarus* **89**, 26-43.
8. Boss A.P. (1993) *Astrophysical Journal* **417**, 351-671.
9. Mumma M.J., Weissman P.R. and Stern S.A. (1992) in *Protostars and Planets III*, 1177, ed E. H. Levy, U. Arizona Press.
10. Anders E. and Zinner E. (1993) *Meteoritics* **28**, 490-514.
11. Gaffey M., Burbine T.H. and Binzel R.P. (1993) *Meteoritics* **28**, 161-188.
12. Jones T.D., Lebofsky L.A., Lewis J.S. and Marley M.S. (1990) *Icarus* **88**, 172-92.
13. Vilas F. and Gaffey M.J. (1989) *Science* **246**, 790-792.
14. Zolensky M. and McSween H.Y. Jr. (1988) in *Meteorites and the Early Solar System*, 114-143, ed J. F. Kerridge, U. Arizona Press.
15. Sonnett C.P. and Reynolds R. (1979) in *Asteroids*, 822-848, ed T. Gehrels, U. Arizona Press.
16. Grimm R.E. and McSween H.Y.Jr. (1993) *Science* **259**, 653-529.
17. Binzel R.P. (1992) *Icarus* **99**, 238-240.
18. Wood J.A. and Hashimoto A. (1993) *Geochimica et Cosmochimica Acta.* **57**, 2377-2388.
19. Stevenson D.A. and Lunine J.I. (1988) *Icarus* **75**, 146-155.
20. Bradley J.P. and Brownlee D.E. (1991) *Science* **251**, 549-552.
21. Keller L.P., Thomas K.L. and McKay, D.S. (1992) *Geochimica et Cosmochimica Acta.* **56**, 1409-1412.
22. Zolensky M. and Barrett R. (1993) *Microbeam Analysis* **2**, 191-197.

REMOTE SENSING AND DYNAMICS
OF INTERPLANETARY DUST

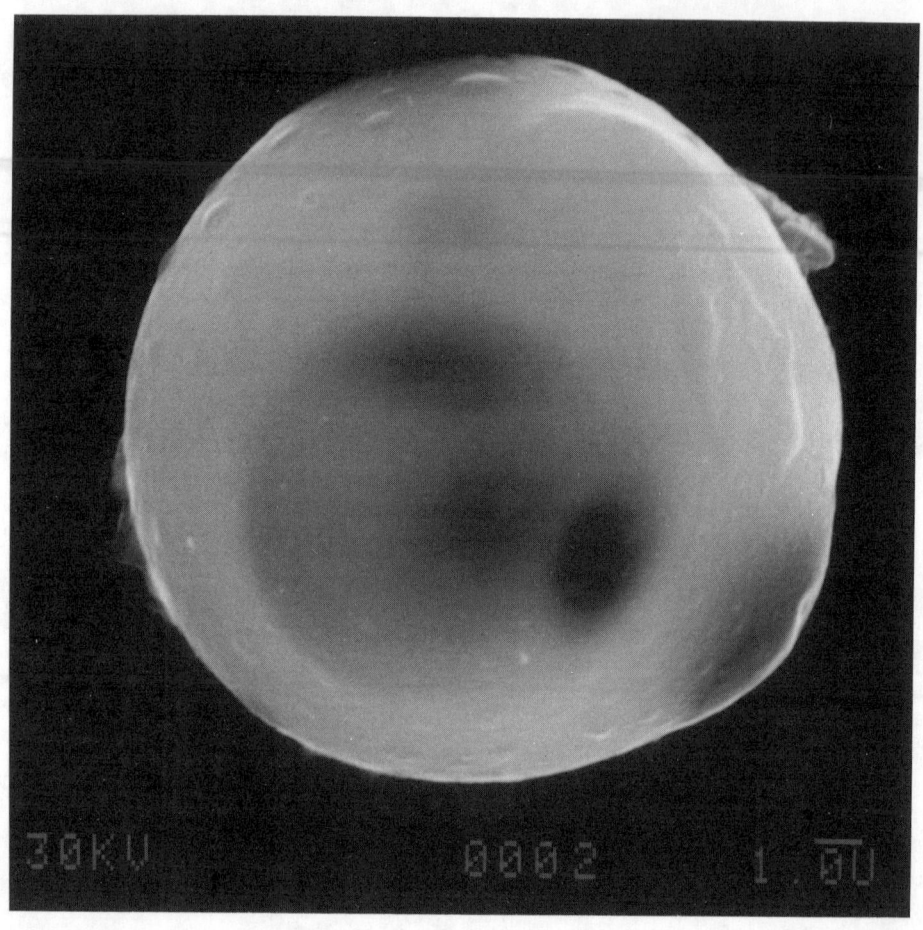

A secondary electron image of a stratospheric particle rich in calcium, aluminum and silicon, generally called a "CAS sphere". David McKay has suggested that these particles could have been derived from impacts into the anorthositic lunar highlands (personal communication, 1984). See the paper by Tom Wilson in this volume. The scale bar measures 1 μm. (Photo by M. Zolensky)

REMOTE SENSING AND DYNAMICS OF INTERPLANETARY DUST

Stan Dermott presented the results of analysis of IRAS and COBE data at the workshop, which indicates that main-belt asteroids are the source of approximately 40% of the dust providing zodiacal light. The remainder must be provided by near-Earth asteroids, comets and interstellar sources. In particular, it would seem to be useful to make detailed dynamical calculations of the evolution of grains from near-Earth asteroids. Apparently, dust grains derived from some asteroid families are distinguishable from one another, in inclination space. Differences in orbital elements for asteroidal grains from different families also cause different encounter velocities at Earth, and hence differential atmospheric entry-heating levels which will vary temporally. Dermott and Liou summarize their results in a paper here. Al Jackson and Herb Zook also presented results (at the workshop) of calculations of which indicate that cometary and asteroidal particles can be distinguished, if their velocities and trajectories are measured in space. This is a critical requirement for proposed dust collection efforts in low-Earth orbit. Herb Zook presented data from the Ulysses spacecraft cosmic dust experiment indicating the presence, in the Solar System, of streams of interstellar dust focused by interaction with Jupiter.

Martha Hanner reviewed the available spectroscopic information on comets, of which there is all too little. The main point of this presentation (repeated in this volume) was that the comets observed to date appear to differ significantly from one another in composition and, presumably, mineralogy. While it is possible that these differences are due to the differential aging of comets, it is also likely that there is considerable inherent intercomet heterogeneity. Recent measurements of the reflectance spectra of chondritic IDPs by John Bradley reveal some spectral features found in some comets, including features possibly due to non-crystalline phases. Definite ties between specific comets and IDPs have yet to be demonstrated by spectroscopic work, although this is clearly a promising line of research. Still unknown is the relationship between cometary dust and interstellar grains, and whether hydrated materials are found on comets. This latter point is currently the subject of considerable contention. Finally, Tom Wilson has contributed a paper to this volume suggesting novel sources for some IDPs.

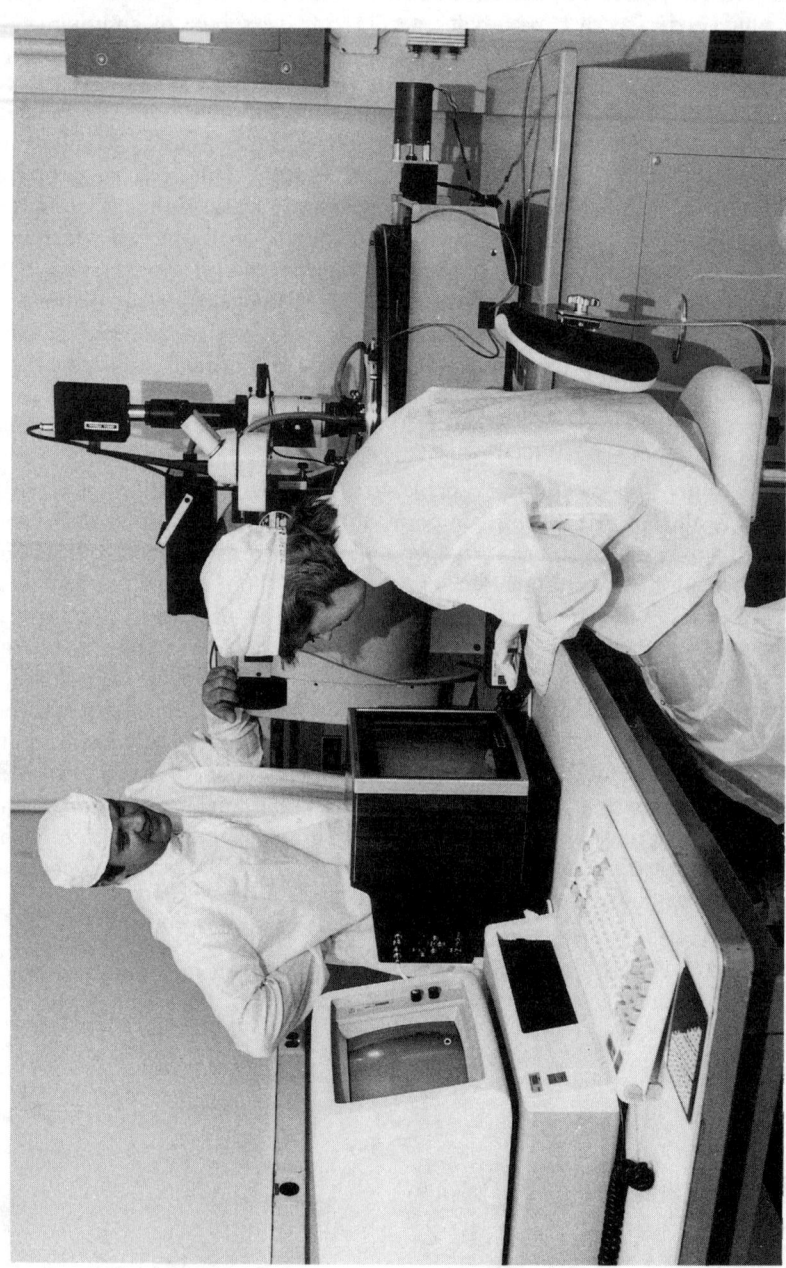

View of the class clean room in NASA's Facility for the Optical Inspection of Large Surfaces (FOILS Lab) at the Johnson Space Center, with scientists Jack Warren and Robbie Marlow. In this laboratory spacecraft parts returned from orbit are characterized with respect to the damage they have received from interplanetary dust, micrometeoroids and spacecraft debris particulates. A stable scanning table (background) translates spacecraft surfaces while scientists locate and characterize particulate impact features using a stereomicroscope interfaced to digitizing hardware. (NASA photo S85-36302)

DETECTION OF ASTEROIDAL DUST PARTICLES FROM KNOWN FAMILIES IN NEAR-EARTH ORBITS

S. F. Dermott and J. C. Liou
University of Florida, Gainesville, FL 32611, USA

INTRODUCTION

Eos, Koronis, and Themis are the most prominent families in the asteroid belt. They are quite distinguishable in their ($e \cos \varpi$, $e \sin \varpi$) and ($I \cos \Omega$, $I \sin \Omega$) phase spaces. Naturally the dust particles produced from these families will have quite distinguishable initial orbital elements. However, when the dust particles spiral in towards the sun, due to Poynting-Robertson light drag, planetary perturbations and passages through resonances may cause their orbital elements to change to such an extent that they are no longer distinguishable when they reach the Earth. Secondly, forced inclinations and eccentricities are imposed on their orbits with the result that the osculating orbits in the vicinity of Earth are quite different from those in the asteroid belt. The questions we would like to answer in this paper are: When dust particles arrive at the near-Earth orbit, (1) is there still any difference between the orbits of dust from different families, and (2) can we still recognize dust from different families by knowing their orbital elements?

METHOD

We have studied the orbital evolution of dust particles with two different sizes (diameters equal to 4 μm and 9 μm) originating from the Eos, Koronis, and Themis asteroidal families. We first wind the solar system back in time numerically in such a way that dust particles being released from their origin at that epoch will reach the Earth in 1983. We calculate the initial forced elements for each family (by using the orbital elements of 7 planets - Mercury and Pluto are not included in our calculation) and combine these with the proper elements to determine the starting orbital elements using the "particles in a circle" method (Dermott *et al.*, 1992). We then numerically integrate 249 particles in each family with RADAU (E. Everhart, 1985) on an IBM ES/9000. All the planetary perturbations, radiation pressure, Poynting-Robertson light drag, and corpuscular solar wind effects are included in the calculation. We record the orbital elements of each particle while they are moving toward the sun and analyze the data.

RESULTS

4 μm Particles

We analyze the dust particles as they approach and pass the Earth. In Figure 1 we plot the positions of dust particles in ($I \cos \Omega$, $I \sin \Omega$) phase space during their passages through the Earth-crossing region. It is obvious that the Eos particles are quite distinguishable from the Themis and Koronis particles in inclination space. The proper inclinations of the dust particles are unchanged by orbital evolution[1] even though their forced inclinations and nodes vary quite considerably. On the other hand, all three families are no longer distinguishable in eccentricity space. This is primarily due to the fact that the radiation pressure and Poynting-Robertson drag change the eccentricities of dust particles, both from the very beginning when they were released, and throughout their orbital evolution. Passage through mean motion resonances with Jupiter or the Earth and trapping in the near-Earth resonances are not significant for particles with this size (the drag rate, $\dot{a}$, is too high such that they do not gain enough angular momentum from the resonance region to counterbalance the loss due to orbital decay.)

The off-center distributions of Themis and Koronis particles in inclination space leads to another interesting result: The seasonal variation of particles being collected by an Earth-orbiting detector. In Figure 2, we show the variation in relative particle numbers which intersect the Earth at their ascending nodes during different months in 1983. These are Themis particles. The difference between the maxima and minima is larger than a factor of three. The seasonal variation for Eos particles, which has a larger proper inclination, is not so large.

9 μm Particles

Particles with approximate diameters of 9 μm are the main contributor to the zodiacal cloud observed by the Infrared Astronomical Satellite (IRAS) at 25 μm wavelength.[2] These particles move with approximately the velocity of 4 μm particles; thus the number of particles being trapped in the near-Earth resonance increases somewhat. However, the overall trapping in resonance is still not significant.

In the eccentricity space, dust particles from all three families mix up in the area between zero eccentricity and 0.1 eccentricity. In the inclination space, Eos particles are still quite distinguishable from the Koronis and Themis particles (see Figure 3). The off-center distribution for Koronis and Themis particles again results in the seasonal variation similar to that in Figure 2.

DISCUSSION

From the distribution of dust particles in the inclination space we can clearly see that some families are distinguishable from the others. Our study covers particles with diameters equal to 4 μm and 9 μm. This is equivalent to a factor of eight in particle masses. Thus our conclusions here are valid for a wide range of particles.

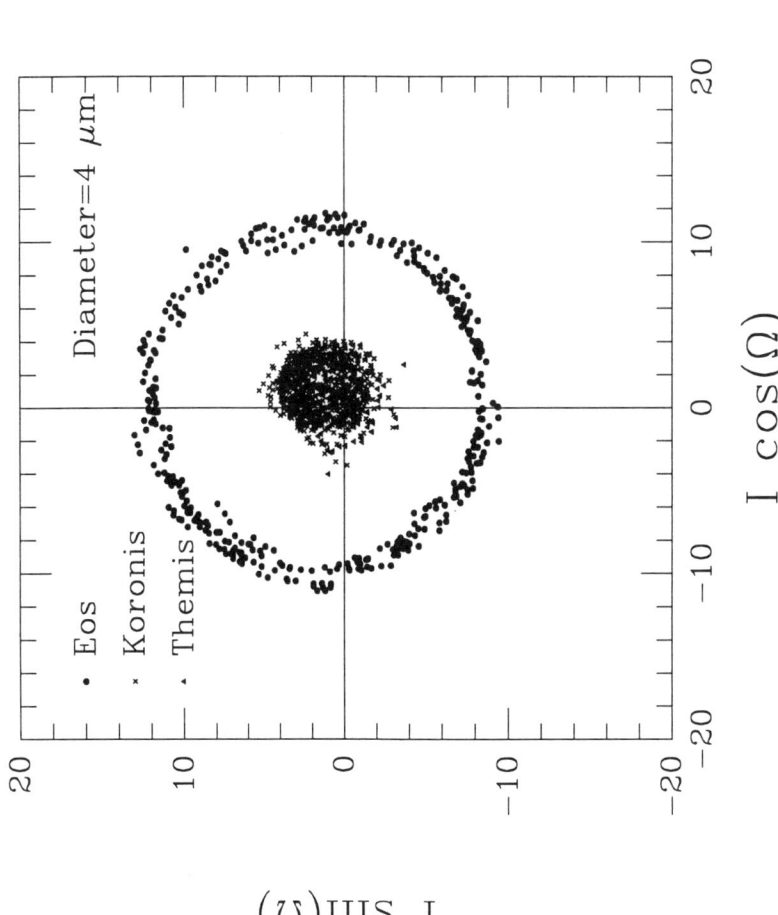

Figure 1. Positions in ($I \cos \Omega$, $I \sin \Omega$) space of 4 μm diameter particles from all three families in the Earth-crossing region.

16 Detection of Asteroidal Dust Particles

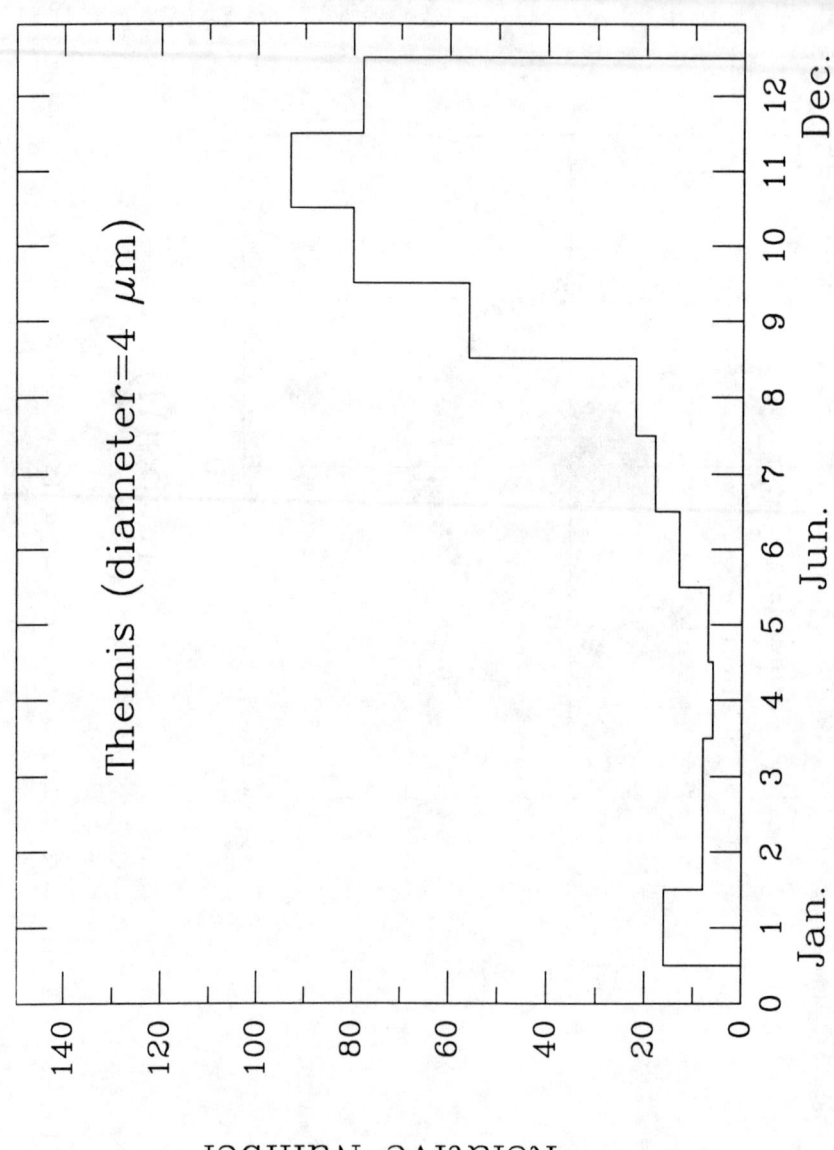

Figure 2. A histogram showing the variation in relative number of particles as a function of different months of the year. These are 4 μm diameter Themis particles.

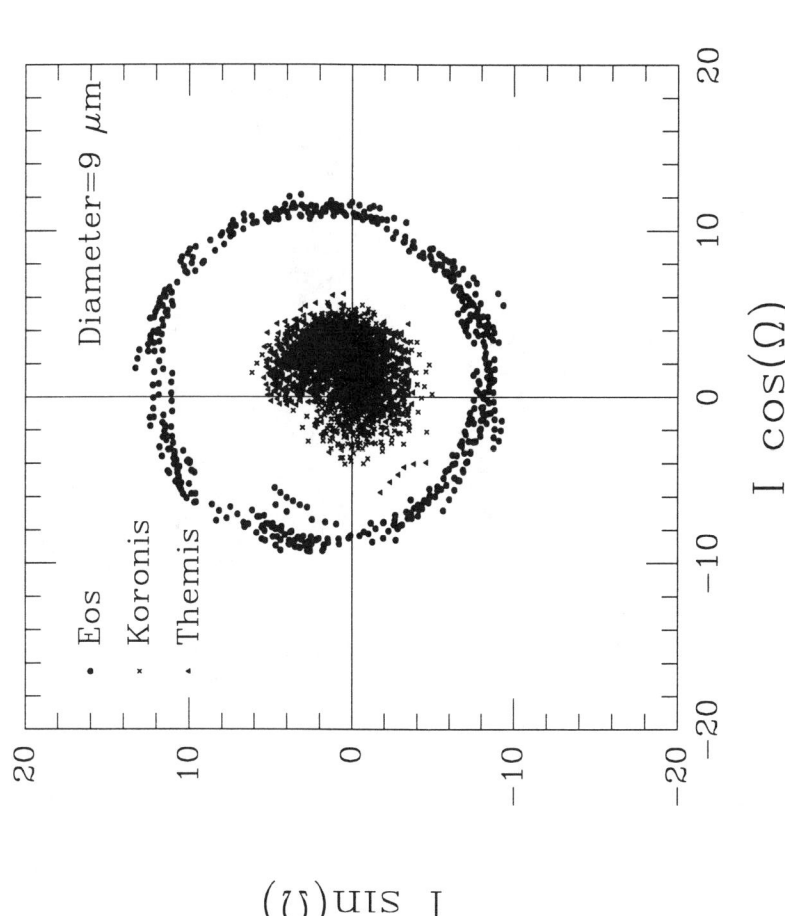

Figure 3. Positions in ($I\cos\Omega$, $I\sin\Omega$) space of 9 μm diameter particles from all three families in the Earth-crossing region.

18 Detection of Asteroidal Dust Particles

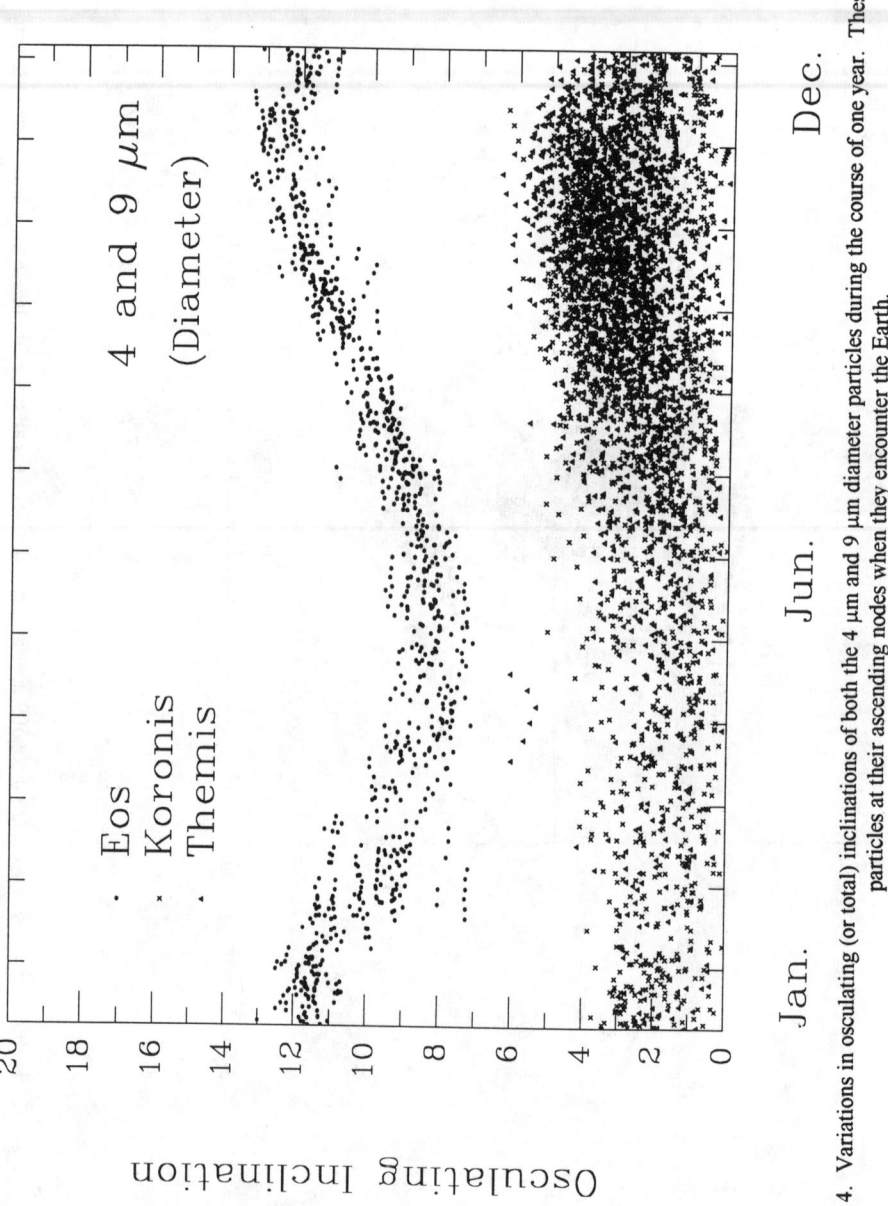

Figure 4. Variations in osculating (or total) inclinations of both the 4 μm and 9 μm diameter particles during the course of one year. These are the particles at their ascending nodes when they encounter the Earth.

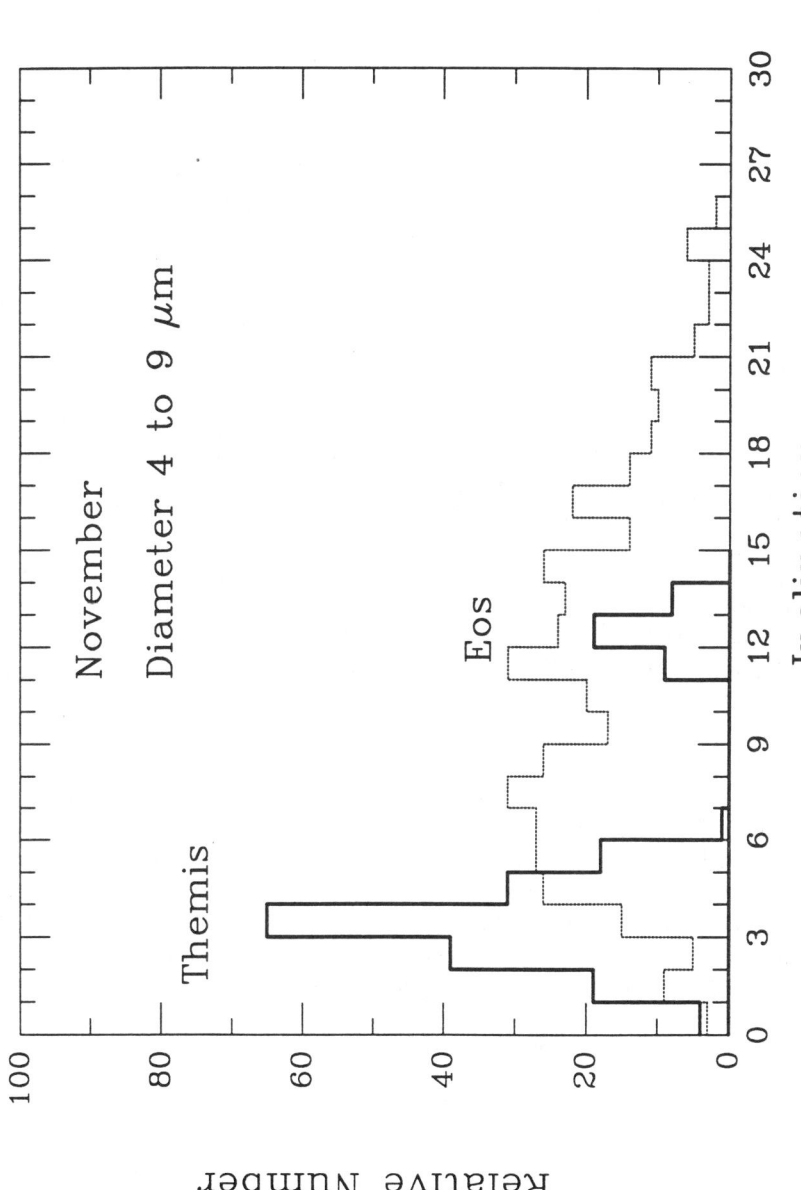

Figure 5. Relative number of particles versus possible inclination range from three different sources being collected at near-Earth vicinity in November. These are 4 μm and 9 μm diameter particles from Themis family (bold line), Eos family (bold line), and all-asteroid Eos and Themis families (dotted line).

When particles are released from their origins and are spiraling towards the sun, their forced inclinations and nodes vary due to planetary perturbation while their proper inclinations remain unchanged. The total inclinations of particles are the vectorial sum of the forced and proper inclinations. In Figure 4 we show the variation in total inclinations as a function of time for 4μm and 9 μm particles from three families. From the diagrams we can see that even though the inclinations of particles vary considerably (for Eos particles, the inclinations vary from 7° to 12°; for Koronis and Themis particles, they vary from 0° to 6°), Eos particles are well separated from Koronis and Themis particles. This is our first conclusion.

Because of the existence of forced inclination and node, the centers of the circle in Figure 1 and Figure 3 are not at the origin. This off-center distribution in ($I \cos \Omega$, $I \sin \Omega$) space produces an important observational consequence, the seasonal variation of the number of dust particles intersecting the Earth. For example, from Figure 2 we can conclude that the best time to collect Themis particles is around April and October.

If Eos, Themis, and Koronis are the only sources which produce asteroidal dust, then the question of identifying the origin of asteroidal dust is very simple. Unfortunately, all the main belt asteroids produce dust. So the question now is: Can we still identify the origin of a dust particle when we consider all asteroids as being possible sources? The answer is: Yes, to a certain extent. In Figure 5 we show the relative number versus possible range in inclinations of particles which intersect the Earth in November. The particles are from three different sources: (1) Eos (bold line), (2) Themis (bold line), and (3) all asteroid background except Eos and Themis (dotted line). It is quite clear that in November the low inclination particles are dominated by Themis particles, and half of the medium inclination (around 12°) particles are of Eos origin. This provides us with a very important clue to identify the origin of dust particles. For example, when we collect dust particles with inclinations around 12° in November, half of them must be from Eos based upon our dynamical study. Then, we can analyze the chemical composition of all the particles. We will find that half of those particles have similar compositions. These must be Eos particles. And their composition is the composition of the Eos family. At very low inclinations (around 3°), most of the particles being collected in November will have similar composition. These are Themis particles. So our final conclusion here is the following: By collecting a sample of particles, we can identify particles form asteroid families. From their chemical composition, we can learn the composition of asteroid families without going to the asteroid belt.

REFERENCES

1. Dermott, S.F., Gomes, R.S., Durda, D.D., Gustafon, A.S., Jayaraman, S., Xu, Y.L., and Nicholson, P.D. (1992). Dynamics of the Zodiacal Cloud. In *Chaos, Resonance, and Collective Dynamical Phenomena in the Solar System* (S. Ferraz-Mello, Ed.), 333-347, Kluwer Academic Publishers, Dordrecht.

2. Dermott, S.F., Durda, D.D., Gustafon, A.S., Jayaraman, S., Liou, J.C., Xu, Y.L., (1993). The Origin of the IRAS Dustbands. In *Meteoroids and Their Parent Bodies* (J. Stohl and I.P. Williams, Eds.), 357-366.
3. Everhart, E. (1985). An Efficient Integrator that Uses Gauss-Radau Spacings. In *Dynamics of Comets: Their Origin and Evolution* (A. Carusi and G.B. Valsecchi, Eds.), 185-202. Reidel, Boston.

22 Detection of Asteroidal Dust Particles

Images of comets, believed to be a major source of chemically primitive interplanetary dust particles. (Top): Image of the nucleus of Comet Halley (measuring 16 x 8 km), obtained by the Giotto spacecraft and the Halley Multicolor Camera Team. (Bottom): Two views of Comet Kohoutek obtained by Skylab astronauts (NASA photo MSFC-74-SL-7200-401A).

REMOTE SENSING OF COMETARY DUST AND COMPARISONS TO IDPS

M.S. Hanner
Jet Propulsion Laboratory, California Institute of Technology
Pasadena CA 91109, USA

INTRODUCTION

Comets provide a link to the composition of the primitive solar nebula. Except for cosmic ray processing of the outer few meters, they have remained relatively unaltered since their formation in the outer, colder parts of the solar nebula.

IN-SITU SAMPLING

We have in-situ measurements of the dust composition of only one comet, from the impact ionization time-of-flight mass spectrometer on the Halley probes.[1&2] The composition of the major rock-forming elements is chondritic within a factor of two, but the comet dust is enriched in the elements H, C, N, O compared to carbonaceous chondrites, implying that the dust is more volatile- rich and more "primitive".[3] Within the so-called CHON particles, there is a wide spread in the abundance ratios, a result that is not consistent with the proposal that there is a refractory organic component consisting mainly of polyoxymethylenes. Among silicate particles, the Fe/(Fe+Mg) ratio ranges from 0 to 1, with Mg-rich silicates predominating. The wide distribution of Fe/(Fe+Mg) does not agree with the narrow range measured in most carbonaceous chondrites, but does resemble anhydrous IDPs.[4&5]

INFRARED SPECTROSCOPY

Infrared spectroscopy is the best means of remotely studying the composition of cometary solids. The spectra can help to establish links to interstellar grains and to identify classes of interplanetary dust particles which might have originated from comets. Transitions within various organic molecules produce features in the 3 μm region, while the 10 μm spectral region contains information about the mineral content of the grains.

Spectral Features from Organic Molecules

A broad emission feature at 3.36 μm was discovered in Comet Halley (Figure 1).[6-10] It has subsequently been detected in every bright comet observed in the 3 μm region.[11-16] The feature is generally assigned to a C-H stretch vibration, but it is not obvious from the shape of the feature whether it originates in the gas or in small grains. The detailed structure varies among comets, with secondary peaks at 3.28 μm, 3.41 μm, and 3.52 μm having varying strengths relative to the main 3.36 μm peak; the 3.28 µm peak is particularly visible in the long period comet Levy 1990 XX.[15] The 3.28 µm peak resembles one of the set of unidentified infrared bands seen in H II regions, planetary nebulae, and a few young stellar objects, but there is no analogous 3.36 µm emission feature in any astronomical source.[17]

Because the feature is seen in both new and evolved comets, it can not be the result of cosmic ray irradiation of ices in the outer few meters of Oort Cloud comets. Brooke et al.[14] demonstrated that the strength of the 3.36 µm band emission correlates better with the gas production rate than with the dust continuum, implying a gas phase carrier. The 3.52 µm feature is consistent with the ν_3 band of methanol (CH_3OH).[18] But methanol also has bands at 3.3 - 3.4 µm. Reuter[19] has computed the contribution of methanol to the main 3.36 µm feature. Depending on how one normalizes the 3.52 µm feature, the accompanying contribution to the 3.36 µm feature ranges from 10 - 50 %. Thus, the spectral shape and the heliocentric distance dependence of the residual "unidentified" 3.36 µm feature may be somewhat different from the total observed flux and we are back to the question whether the carrier is in the gas or solid phase.

Silicates

A broad emission feature near 10 µm is observed in intermediate bandpass filter photometry of most dynamically new and long-period comets.[20&21] The strength of the feature is quite variable, being strongest in "dust rich" comets with strong scattered light continuum. The 10 µm emission feature is most likely due to the Si-O stretching mode vibration in small silicate particles. Elemental abundances indicative of silicates were common in particles detected by the dust mass spectrometer during the Halley flybys. Spectra at 8-13 µm with high signal/noise now exist for eight comets. Four are dynamically new comets, that is, thought to be coming in from the Oort cloud for the first time, two are long-period comets, and two (Halley and Brorsen-Metcalf) have periods ~70 years. These 8 spectra are analyzed in more detail in Hanner, Lynch & Russell.[22]

The spectrum of Comet Halley is shown in Figure 2, from Campins & Ryan.[23] There is a broad maximum at 9.8 µm and a narrower peak at 11.25 µm. A Halley spectrum by Bregman et al.[24] showed similar structure. The 11.25 µm peak agrees with that seen in olivine IDPs[25], as illustrated in Figure 2 and in laboratory olivine samples.[26&27] Thus, crystalline olivine is the probable source of the cometary peak. Other possible explanations for the peak at 11.25 µm, such as SiC or an organic component can be ruled out from the width of the peak, abundance arguments, or for lack of corresponding features, such as the 7.7 and 8.6 µm features present when an 11.3 µm feature is associated with the set of unidentified interstellar emission bands. Two long period comets, Bradfield 1987 XXIX and Levy 1990 XX have similar double-peaked spectra.[28-30] Levy's spectrum is shown in Figure 3.

Koike et al.[27] have obtained infrared transmission spectra of olivine samples with differing Mg/Fe abundance. They find that the peak lies at 11.3 µm for $Mg/(Mg+Fe) = 0.9$ and shifts toward 11.5 µm as the Mg abundance decreases. The 11.25 µm position in the comets implies a high Mg/Fe abundance, a conclusion consistent with the dust analyses from the Halley probes.[3&4]

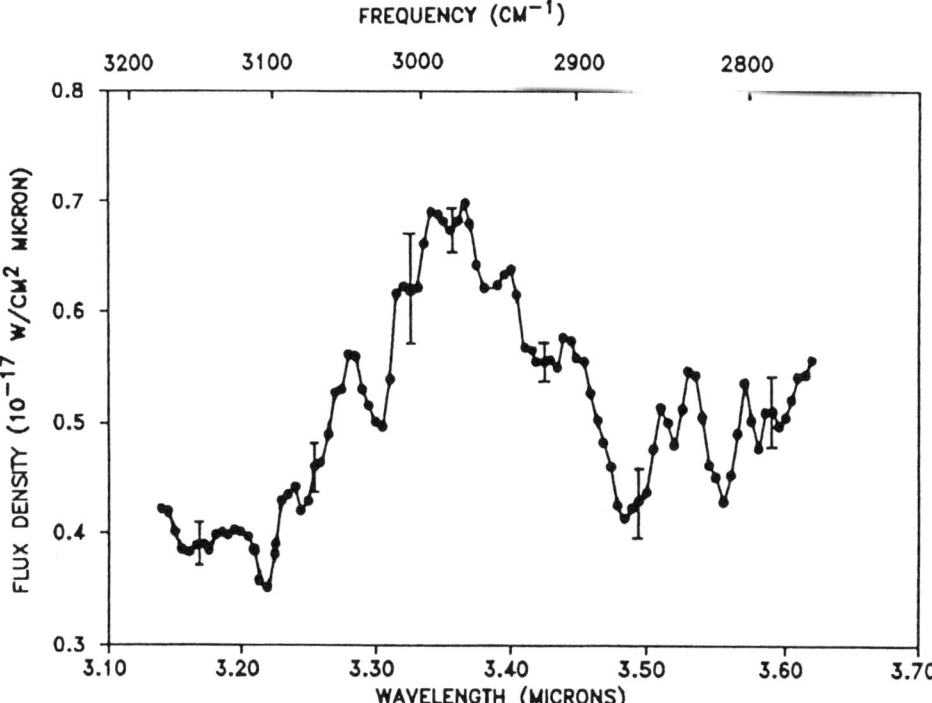

Figure 1. 3 μm spectrum of Comet P/Halley on 25 April 1986 at r = 1.54 AU.[9]

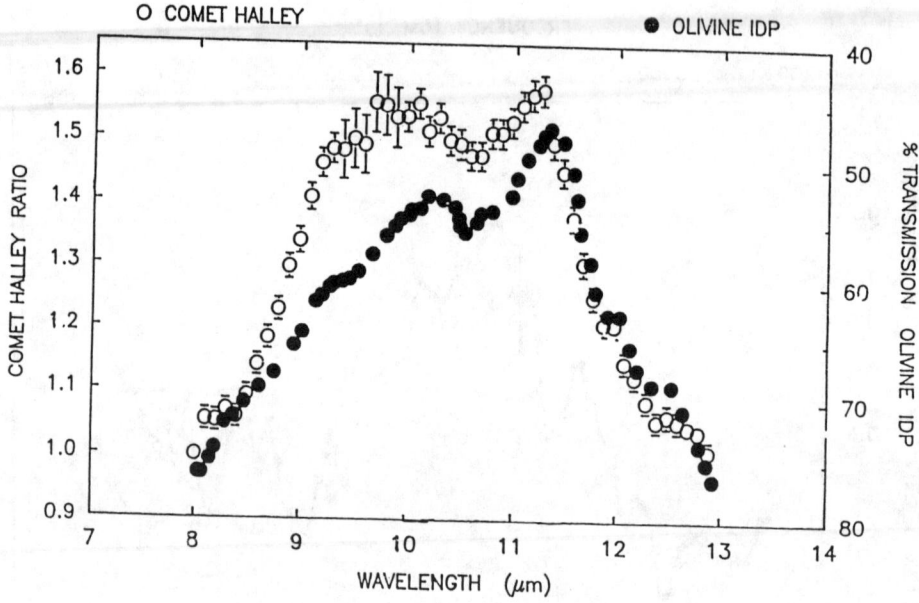

Figure 2. 8-13 μm spectrum of Comet P/Halley at r = 0.79 AU [23] compared with olivine type IDP.[25]

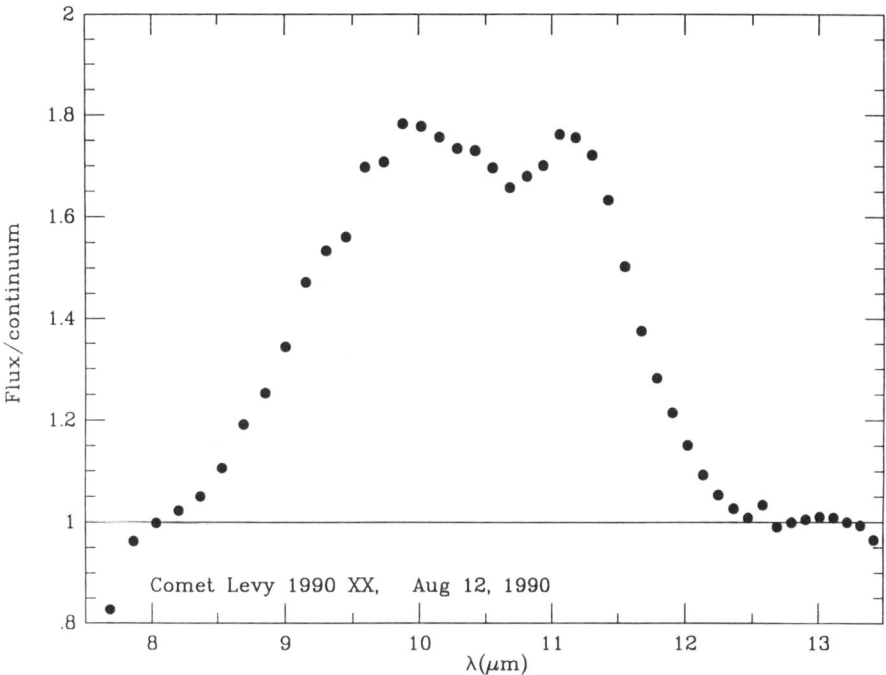

Figure 3. 8-13 µm spectrum of Comet Levy 1990 XX at r = 1.54 AU perihelion.[40] The total flux has been divided by 270 K blackbody continuum.

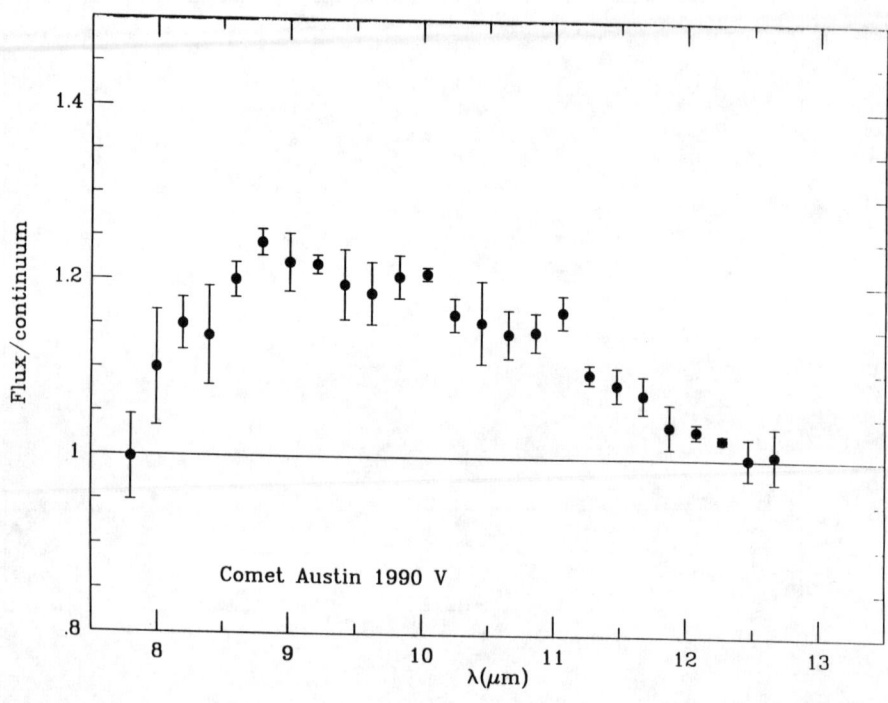

Figure 4. 8-13 μm spectrum of Comet Austin 1990 V at r = 0.78 AU perihelion.[31] The total flux has been divided by 300 K blackbody continuum.

However, another component must give rise to the broad maximum near 9.8 µm. There are several possibilities[31]:

1. Mix of crystalline minerals: Both olivine and pyroxene are high-temperature condensates. Pyroxenes display more variety than olivine in their spectral shapes and the peaks are generally at $\lambda < 11$ µm. Bregman et al.[24] were able to fit their Halley spectrum with a mix of predominately olivine and pyroxene-rich IDP spectra, although these classes of IDPs do not necessarily originate from comets.

2. Glassy silicates: Interstellar silicate grains are thought to be amorphous because of their broad, structureless 10 µm feature. From the emission spectra measured by Stephens & Russell[26], amorphous olivine can explain the 9.8 µm cometary peak, while a mixture of amorphous olivine and amorphous enstatite can account for the rise between 8 and 9 µm. This explanation is attractive because it requires the least alteration of interstellar silicates before their incorporation into comets.

3. Hydrated silicates: Type II carbonaceous chondrites have about equal proportions of hydrated silicates and olivine. Their spectra actually look rather similar to the comet spectra.[32] Nelson, Nuth, and Donn[33] suggested that amorphous silicate grains on the surface of a comet nucleus could absorb one or more monolayers of water molecules from the outflowing gas and could be converted to hydrated silicates if exposed to temperatures of 300 K or above for a few weeks. Yet, the silicate feature was strongest in Halley at times of strong jet activity when the silicate dust appeared to be emanating from deep vents in the nucleus, rather than the surface. While the Mg/Fe/Si distribution in most carbonaceous chondrites is much narrower than that in the Halley dust[4], at least one [Kaba] shows a broad distribution.[34] Some otherwise anhydrous pyroxene- or olivine-rich IDPs do contain minor amounts of hydrated silicates.[35&36] Rietmeijer[37] pointed out that layer silicates exist in dry Antarctic valleys, where chemical alteration of silicate-ice mixtures is aided by the presence of thin water layers between the solid grains and ice at $T \leq 195$ K. He suggested that a similar process could have taken place in IDP parent bodies, including comet nuclei.

Of the 4 new comets for which good quality spectra are available, each has a unique spectrum that differs from that seen in Comets Halley, Bradfield, and Levy. Kohoutek[38] has a strong emission feature similar to that in Halley, except that the 11.3 µm peak is lacking. Wilson, 1987 VII[28] showed a broader emission feature with an unidentified peak at 12.2 µm. Okazaki-Levy-Rudenko 1989 XIX[38] had a weak feature with a maximum between 10.5-11.5 µm, while Austin 1990 V[31] showed a weak feature with a possible small peak at 11.1 µm (Figure 4). Comets OLR and Austin were both dust-poor comets, based on the strength of their scattered and thermal continuum radiation; that is, the total dust cross-section was low, but not necessarily the total dust mass, if the dust is concentrated in larger particles.

P/Brorsen-Metcalf showed the danger of generalizing. With a 70.6 year period and perihelion at 0.48 AU, P/Brorsen-Metcalf has an orbit similar to that of P/Halley and one might have expected their spectra to be similar. Yet, Brorsen-Metcalf showed no emission feature at all in six days of observing.[40] Most short-period comets do not display a silicate feature. Because silicate particles larger than a few microns in size do not produce a feature, P/Brorsen-Metcalf and other short-period comets are most likely lacking small grains, rather than being deficient in silicate material; other indicators of small grains, such as a dust tail and strong 3 μm emission are also missing.

Clearly, then, there is not a single kind of "cometary" silicate dust. Our sample of comets is too small to generalize about the characteristics of new versus evolved comets. Spectra of one or more "dusty" new comets will be needed before we can conclude whether small crystalline olivine grains are ever present in new comets.

RELATIONSHIP TO IDPS

No doubt, we already have examples of cometary dust among the IDPs. But which ones are they? No single class of IDPs exhibits infrared spectra resembling any of the cometary spectra. Even a spectral match would not prove a generic relationship. Olivine IDPs show evidence of atmospheric heating, implying fairly high entry velocities, while hydrated IDPs experience minor heating, implying entry velocities ~12 km/sec, consistent with prograde circular orbits.[41] Pyroxene IDPs appear to be intermediate. This might suggest an asteroidal origin for the hydrated particles and a cometary origin for the olivine and pyroxene classes. There may be a strong selection effect against cometary dust particles which have fragile structure or contain material that is volatile at temperatures of a few hundred degrees.

Bradley et al.[42] have identified two very porous aggregates whose spectra are qualitatively similar to the spectra of Halley, Bradfield, and Levy. These IDPs contain abundant glass, along with tiny (<0.1 μm) crystals, mainly enstatite. A few larger (0.1 -1.0 μm) olivine crystals give rise to a 11.25 μm signature. Bradley et al.[42] emphasized that such particles are rare, although they seem to be related to the pyroxene-rich class.

Thus, the link between comet dust and IDPs is ambiguous. If cometary silicates are predominantly glassy, in order to produce the 8-9 μm rise and the broad 9.8 μm maximum, why are glass-rich IDPs relatively rare? Do only the sturdiest cometary IDPs survive? If only the "primitive" pyroxene IDPs are cometary, why do the olivine IDPs have the highest entry velocities? If both olivine and pyroxene IDPs are from comets, why are the two types so distinct? Does this imply that some comets are pyroxene-rich while others are olivine-rich? If so, we should see significant differences among comets in the relative strength of the 11.25 μm peak. We will be searching for such differences as we examine a larger sample of new and evolved comets.

ACKNOWLEDGMENT

This research was carried out at the Jet Propulsion Laboratory, California Institute of Technology, under contract with NASA.

REFERENCES

1. Kissel, J. et al. 1986. *Nature* **321**, 280, 336.
2. Fomenkova, M. and Chang, S. 1994. Carbon in Comet Halley dust particles. This volume.
3. Jessberger, E.K., Christoforidis, A. & Kissel, J. 1988. *Nature* **332**, 691.
4. Lawler, M.E., Brownlee, D.E., Temple, S. & Wheelock, M.M. 1989. *Icarus* **80**, 225.
5. Zolensky, M.E. and Barrett, R.A. 1994. Olivine and pyroxene compositions of chondritic Interplanetary Dust Particles. This volume.
6. Combes, M. et al. 1986. *Nature* **321**, 266.
7. Wickramasinghe, D. & Allen, D. 1986. *Nature* **323**, 44.
8. Knacke, R.F., Brooke, T.Y. & Joyce, R.R. 1986. *Astrophys. J.* **310**, L49.
9. Baas, F., Geballe, T.R., & Walther, D.M. 1986. *Astrophys. J.* **311**, L97.
10. Danks, A.C., Encrenaz, T., Bouchet, P., Le Bertre, T. & Chalabaev, A. 1987. *Astron. Astrophys.* **184**, 329.
11. Allen, D. & Wickramasinghe, D. 1987. *Nature* **329**, 615.
12. Brooke, T.Y., Knacke, R.F., Owen, T.C. & Tokunaga, A.T. 1989. *Astrophys. J.* **336**, 971.
13. Brooke, T.Y., Tokunaga, A.T., Knacke, R.F., Owen, T.C., Mumma, M.J., Reuter, D., & Storrs, A.D. 1990. *Icarus* **83**, 434.
14. Brooke, T.Y., Tokunaga, A.T. & Knacke, R.F. 1991. *Astron. J.* **101**, 268.
15. Davies, J.K., Green, S.F. & Geballe, T.R. 1991. *MNRAS* **251**, 148.
16. Green, S.F., Davies, J.K., Geballe, T.R., Brooke, T.Y., & Tokunaga, A.T. 1992. *Asteroids, Comets, Meteors 1991*, p 211.
17. Tokunaga, A. T. & Brooke, T. Y. 1990. *Icarus* **86**, 208.
18. Hoban, S., Mumma, M., Reuter, D.C., DiSanti, M. 1991. *Icarus*, **93**, 122.
19. Reuter, D.C. 1992. *Astrophys. J.* **386**, 330.
20. Ney, E.P. 1982. In *Comets*, ed. L.L. Wilkening (Tucson: University of Arizona Press), p. 323.
21. Gehrz, R.D. & Ney, E.P. 1992. *Icarus* **100**, 162.
22. Hanner, M. S., Lynch, D. K. & Russell, R. W. 1994. *Astrophys. J.* (April 10), in press.
23. Campins, H. & Ryan, E.V. 1989. *Astrophys. J.* **341**, 1059.
24. Bregman, J., Campins, H., Witteborn, F.C., Wooden, D.H., Rank, D.M., Allamandolla, L.J., Cohen, M. & Tielens, A.G.G.M. 1987. *Astron. Astrophys.* **187**, 616.
25. Sandford, S.A. & Walker, R.M. 1985. *Astrophys. J.* **291**, 838.
26. Stephens, J.R. & Russell, R.W. 1979. *Astrophys. J.* **228**, 780.
27. Koike, C., Shibai, H. & Tuchiyama, A. 1993. *MNRAS* **264**, 654.

28. Lynch, D.K., Russell, R.W., Witteborn, F.C., Bregman, J., Rank, D. & Cohen, M. 1989. *Icarus* **82**, 379.
29. Hanner, M.S., Newburn, R.L., Gehrz, R.D., Harrison, T., Ney, E.P. & Hayward, T.L. 1990. *Astrophys. J.* **348**, 312.
30. Lynch, D.K., Russell, R.W., Hackwell, J.A., Hanner, M.S. & Hammel, H.B. 1992. *Icarus* **100**, 197.
31. Hanner, M. S., Russell, R. W., Lynch, D. K. & Brooke, T. Y. 1993. *Icarus* **101**, 64.
32. Zaikowski, A., Knacke, R.F. & Porco, C.C. 1975. *Astrophys. Space Sci.* **35**, 97.
33. Nelson, R., Nuth, J.A. & Donn, B. 1987. *Proc. 17th Lunar Plan. Sci. Conf. JGR* **92**, p E657.
34. Scott, E. R. D., Barber, D. J., Alexander, C. M., Hutchinson, R. & Peck, J. A., in *Meteorites and the early Solar System*, 1988. ed. J. F. Kerrridge & M. S. Matthews, Univ. Ariz. Press, p. 718.
35. Rietmeijer, F. J. M. 1991. *Earth & Plan. Sci. Lett.* **102**, 148.
36. Zolensky, M. & Barrett, R. 1993. *Microbeam Analysis* **2**, 191.
37. Rietmeijer, F. J. M. 1985. *Nature* **313**, 293.
38. Merrill, K. M. 1974. *Icarus* **23**, 566.
39. Russell, R. W. & Lynch, D. K. 1990. In *Workshop on Recent Comets*, ed. Huebner, Rahe, Wehinger, Konno, Albuquerque, p. 92.
40. Lynch, D.K., Hanner, M.S. & Russell, R.W. 1992. *Icarus*, **97**, 269.
41. Sanford, S. A. & Bradley, J. P. 1989. *Icarus* **82**, 146.
42. Bradley, J. P., Humecki, H. J. & Germani, M. S. 1992. *Astrophys. J.* **394**, 643.

VOLCANISM AND LEVITATION: UNUSUAL ORIGINS OF INTERPLANETARY DUST

Thomas L. Wilson
National Aeronautics and Space Administration, Johnson Space Center,
Houston, Texas 77058, USA

ABSTRACT

The transport of charged dust and dusty plasma into planetary orbit from volcanic eruption and levitation is discussed as an additional origin for interplanetary dust particle's (IDP's). Although complicating the issue of the origin of IDP's, it would appear that volcanic dust should have very specific signatures which delineate this process from the other collisional sources such as meteoroids, and debris spread from comets while disclosing much about the extent of volcanism in the early solar system. Lunar dust as levitated "blow-off" dust should be detectable downwind at the Earth's bowshock and possibly in the Earth's stratosphere depending upon the plasma circulation model assumed.

INTRODUCTION

Upon entering the plasma and radiation environment of space, small dust grains can become charged. This has the consequence that their motion is determined by electromagnetic as well as gravitational forces. Similarly, any energetic event such as volcanism which can create a plasma-like condition would also subject charged dust to both of these forces. In the case of planets and satellites having no appreciable atmosphere, the result of these "dusty plasmas" is that they can migrate and be transported by magnetic and electric fields into places where heavier, neutral debris cannot. They can be accelerated into planetary orbit by electrostatic surface potentials as blow-off dust, or be swept away by moving magnetic fields as pick-up dust.

A prevelant notion regarding ejecta from volcanism is that there is not enough energy in a volcanic eruption to transport planetary material into orbit. This circumstance is unlike the situation for impact ejecta produced by collisional bombardment from interplanetary meteroids[1] where enough kinetic energy exists to knock surface material beyond its escape velocity. As a consequence, some scientists have taken the view that volcanism is an unlikely source. Rather, the source mechanisms for interplanetary dust particles (IDP's) are comets, asteroids, and meteroids because these all have the energy to place or maintain dust beyond its escape velocity and out of planetary gravitational potential wells.

Obviously, such dust is a relic whose signature tells a great deal about the extent of volcanism in the early history of the solar system. A brief discussion of this subject will be presented here, and it will be demonstrated that volcanism can be a measurable source of dusty plasmas which are subsequently swept out into space or levitated into orbit. Linked to thermal evolution arguments, Earth-stratospheric dust measurements might reveal much about the history of volcanism in the solar system if dusty plasma can migrate up the Earth's plasma sheet and be precipitated there.

DUSTY PLASMAS: HISTORICAL PERSPECTIVE

The mechanisms involved are certainly not new. One is the Lorentz force (see "Magnetic Fields", below), and is akin to pick-up ion transport[2] from planetary atmospheres by the solar wind in the absence of magnetospheres. The second is levitation (see "Levitation", below), which can actually transport dust into space without the kinetic aid of volcanism.

That space plasmas can transport dust is the subject of dusty plasmas.[3&4] This subject involves dust-plasma interaction[5], which has been analyzed extensively in the context of Jovian dust[6-8] and general interplanetary dust[9], and even goes back to ion transport off the Moon[10-12] using the gyro-radius and guiding-center approximation of adiabatic invariants theory in space physics. Electrostatic levitation has been discussed for Mercury[13], Jovian dust[14], cometary blow-off dust[15], and the related electrostatic potentials have even been measured for the Moon.[16-19]

Although several aspects of the fundamental physics of space plasma theory have only been indirectly supported by observation (such as the "frozen in flux" approximation below), experiments on Earth have clearly demonstrated the existence of dusty plasma transport in laboratory glow-discharge plasmas.[20]

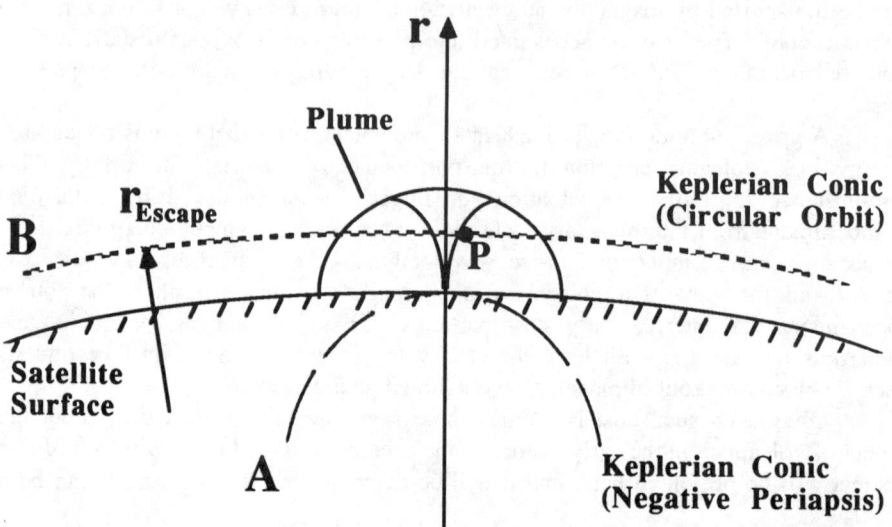

Figure 1. An actual volcanic plume photographed by Voyager 1 while passing Io.

DYNAMICS OF VOLCANIC PLUMES

In order to portray the energetics of volcanic dust for an actual satellite in a planetary plasma environment, we will use Io. Several discussions of simple ballistic models have been published[21&22], based upon actual photographs by Voyager 1 of the volcanic plume events [references 21, 22 - Plates 5-7].

Figure 1 illustrates such a ballistic plume where a particle of dust P, initially sitting on the satellite surface, is vented or kicked along a ballistic trajectory within the gravitational well of Io. To escape, it must have zero total energy E_T. The dust particle P is entrapped as long as it has negative energy ($E_T < 0$), and therefore a negative periapsis (such that its orbital ellipse falls within Io itself.).

The blow-off and pick-up mechanisms which follow (see the following two sections below) simply boost E_T until the elliptical conic A circularizes as dashed orbit B by bringing the periapsis above the planet's surface on the opposite side of the planet. Then $E_T = 0$, giving the escape velocity $v_{surface} = 2.55$ km/s for Io. Approximately 10^{10} tons of volcanic material are erupted each year on Io[23], indicating that even if a fraction of a percent were charged dust, a respectable source of small dust grains could result.

MAGNETIC FIELDS: CONSEQUENCES FOR CHARGED DUST

Whenever dust becomes charged, interplanetary electromagnetic forces will alter the dynamics of such charged grains of matter in unexpected ways. Instead of following gravitational trajectories subject to Kepler's laws modified by Poynting-Robertson drag[24&25], charged dust can spiral about magnetic field lines much like electrons and protons. It then behaves like the charged ion constituents (pick-up ions) of a planetary atmosphere or exosphere interacting directly with the solar wind[8-10] as in Figure 2.

The motion of a charged particle in an electric field **E**, a magnetic field, and no gravitational field, is subject to the Lorentz force **F** [27]

$$F = q(E + \frac{1}{c} v \times B) \quad (1)$$

where q is the total charge and c is the speed of light. Thinking of the velocity **v** as the fraction **V** = v/c in (1) is equivalent to adopting units where c = 1. (a)

(a) The unique contribution of Lorentz in (1) was the second term **VxB** which he discovered when studying invariance in the theory of relativity using the so-called Lorentz transformations. It is a force which changes the direction of the charged particle q so as to orbit a magnetic field line **B** in a circle. This term is a motionally induced "electric" field because it is dimensionally equivalent to **E** in (1), and for a closed system in equilibrium it performs no net work on q (such as raising q out a gravitational potential well). Ironically, the frozen-in-flux assumption results in work being performed on q.

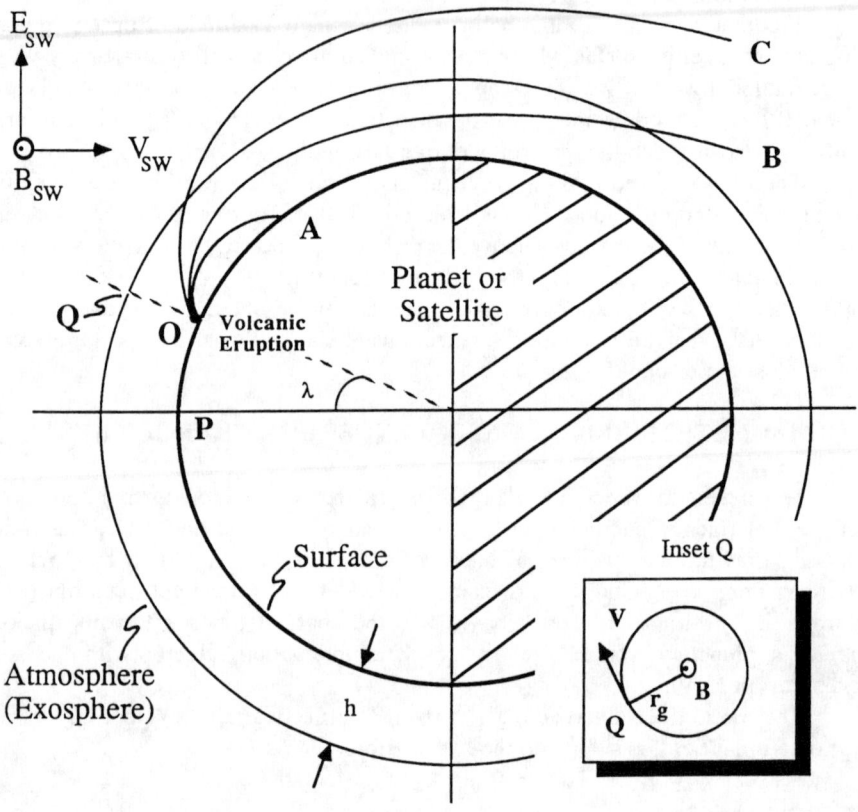

Figure 2. Illustration of the orbits of positively charged dust originating at O as volcanic ejecta from a planetary surface in the solar wind, for some longitude λ about the sub-solar-wind point P. The surface could be that of the Moon or Mercury, for example. In the case of a Jovian satellite (such as Io) within Jupiter's magnetosphere, the source of the motionally induced electric field $\mathbf{E} = -\mathbf{V} \times \mathbf{B}$ is the plasma torus[26] and not the solar wind. The inset is for a cyclotron trajectory at Q with height h above the surface, velocity vector $\mathbf{V}$, and r_g (4).

This classical picture of the electrodynamics of q was changed, however, with the advent of space plasma physics[28] where large-scale magnetic fields are in fact being formed and altered by magnetohydrodynamic effects. An interplanetary space environment modifies the simple closed-system concept of Lorentz in (1) into an open system subject to external forces and currents of moving plasma and moving magnetic field lines **B**. An important example in the solar system is the solar wind and the Sun's interplanetary magnetic field.[29]

As plasma flows from the Sun a condition referred to as "frozen in flux"[28] occurs whereby the magnetic field lines are tied to the plasma and move with it, creating the interplanetary magnetic field. At the same time, an apparent electric field arises due to the **VxB** term in (1) which is sometimes described as an interplanetary electric field.[8-10] This is understood by performing a simple Lorentz transformation (to first order in V=v/c) from a planetary rest frame (e.g., the satellite in Figure 2) to the coordinate system flowing with the plasma gas. The result modifies the electric field **E** as[28]

$$\mathbf{E} = \mathbf{E'} - \mathbf{V} \times \mathbf{B} \quad . \quad (2)$$

Alternatively, a space plasma is believed to be a condition of high, even infinite conductivity σ, and the current from Ohm's law **J** = σ**E'** = σ(**E** + **VxB**) must remain finite. σ can approach infinity only if (**E** + **VxB**) goes to zero. The consequence is an interplanetary electric field

$$\mathbf{E}_{SW} = - \mathbf{V}_{SW} \times \mathbf{B}_{SW} \quad (3)$$

seen in the rest frame of a planetary surface illustrated in Figure 2.

A helpful visualization of the trajectories A, B, and C in Figure 2 is to neglect the gravitational forces (as did Lorentz) and imagine only $\mathbf{E}_{SW}$ and $\mathbf{B}_{SW}$. Charged dust will orbit the magnetic field line **B** as depicted in the inset with a circular cyclotron frequency

$$\omega = \frac{qB}{mc} = \frac{V_{sw}}{r_g} \quad (4a,b)$$

and have the cycloidal trajectory

$$x = -(\frac{V_{sw}}{\omega}) \sin \omega t + V_{sw} t \quad (4c)$$

$$y = (\frac{V_{sw}}{\omega})(1 - \cos \omega t) \quad (4d)$$

where c=1 and the x-y coordinates are local horizontal ones at O in Figure 2. The charge-to-mass ratio is q/m, and the energy gained is

$$E_P = qE_{SW} y \quad (4e)$$

which has the maximum value

$$E_P = qE_{SW}V_{SW}/\omega = qE_{SW}r_g = V_{SW}^2 mc \quad .$$

When the ratio B/m is comparable to that for molecular ion species (large B offsetting large m for dust), the charged dust grains can be imagined to orbit **B** with gyration frequency ω and radius of gyration r_g perpendicular to **V**. Using (4), one draws a circular trajectory to determine where the charged dust grain will go (Figure 2, inset). The presence of E_{SW} causes the circular orbit or cycloid (4c,d) to drift slowly in the direction of $E_{SW} \times B_{SW}$ with a velocity $V_{Drift} = E_{SW} \times B_{SW}/B^2$. It is as simple as drawing circles of radius r_g which osculate with **V** at Q, and then allowing them to drift downwind in the direction of $E_{SW} \times B_{SW}$. Examples are illustrated as the sequence A,B, C for bound, grazing, and unbound trajectories - noting that r_g is quite large for dust. All the charged dust grain has to do is get picked up by the solar wind and it is gone or swept away.

Again, the visualization for Eq. (4) is the extreme case where the planetary gravitational field is turned off - the opposite of the conventional assumption where the magnetic and electric fields are turned off. The true physics of the matter is that both effects are present. The equation of motion in inertial coordinates fixed at the planet's center is

$$\ddot{r} = -\frac{GMr}{r^3} + \frac{q}{m}(E + \frac{\dot{r}}{c} \times B) \quad . \qquad (5)$$

which results from a total "F = ma" force which is the sum of the gravitational Newtonian force (first term), and the Lorentz force (second term) in Eq. (1). In Eq. (5), G is Newton's gravitation constant (G = 6.672×10^{-8} cm^3 g^{-1} s^{-2}), M is the planet's mass, **r** is the charged dust grain's position vector, and $\ddot{r}$ its acceleration vector ($\ddot{r} = a$ in "F = ma"). The first integral of motion in (5) is the inertial velocity $\dot{r} = v$. When $\dot{r}$ reaches an escape velocity due to the Lorentz force terms, or when its position r exposes it to certain orientations of the electric field E_{SW} in Figure 2, the charged dust grain is swept out of and away from the gravitational well by the interplanetary electric field which is basically performing work (as an external energy source) to get it off of the planet's surface. This happens when

$$E_T = E_P + E_{Grav} = E_{Vent} + qE_{SW}y - GMm/r > 0 \quad .$$

LEVITATION: THE EFFECT OF ELECTROSTATIC SURFACE POTENTIALS

There is yet a third mechanism which enters equation (5). It is levitation, which is the ejection of small dust grains from a satellite's surface when very high electrostatic charging has become possible (Figure 3). The electrostatic potential Φ

has actually been measured for the Moon[16-19] to be as high as $|\Phi| \sim 200$ V in Figure 3(a) as it crosses the Earth's magnetospheric tail.

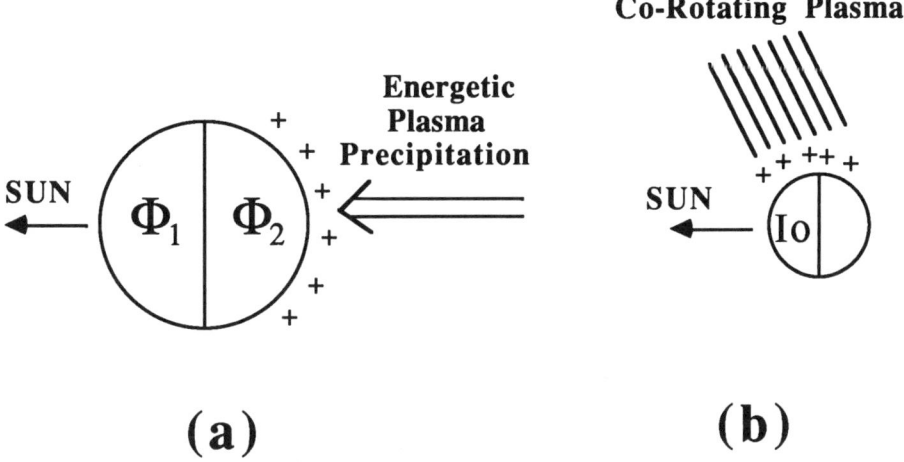

Figure 3. Examples of electrostatic surface charging due to the precipitation and sputtering of energetic plasma. (a) could be the Moon orbiting through the Earth's plasma sheet, or Mercury[13] experiencing its own plasma sheet. (b) illustrates the case of Io interacting with the ambient plasma co-rotating with the Jovian magnetosphere at 57 km/s.[30]

We will assume a uniform hemispherical charge distribution, creating an electrostatic levitation potential $\Phi = k_0 Q/r$, where Q is the total surface charge and k_0 is the Coulomb electrostatic constant. At the planetary surface is $\Phi_0 = k_0 Q/r_0$, whereby we have

$$\Phi(\hat{r}) = \Phi_0 / \hat{r} \qquad (6)$$

and $\hat{r} = r/r_0$ is in units of the planetary radius r_0. Eq. (6) is just an idealized case, realizing that the real field $\Phi(\hat{r})$ must be measured to solve the problem. However, when Φ behaves like (6) it mimics Newtonian gravitation identically since the planetary surface's electric field $\mathbf{E_0}$ is

$$|\mathbf{E_0}(r)| = -\nabla\Phi = \Phi_0/r^2 \qquad (7a,b)$$

$$\Rightarrow \quad \mathbf{E_0} = \Phi_0 r/r^3 \qquad (7c)$$

Substitution of (7c) into (5) with $\mathbf{E} = \mathbf{E_0} + \mathbf{E^*}$ yields the general equation

$$\ddot{r} = -\frac{\mu r}{r^3} + \frac{Z}{m}(E^* + \frac{\dot{r}}{c} \times B) \qquad (8a)$$

$$\mu = (GM - \frac{Z}{m}\Phi_0) \qquad (8b)$$

where $Z = Z_{Dust}q$ represents the effective total charge of the dust grain. E^* is any other ambient electric field such as E_{SW}. Whenever

$$\Phi_0 = GMm/Z \qquad (9)$$

gravitation is cancelled out, and hence the term levitation. In general, the electrostatic levitation potential Φ in (6) is poloidal and must be measured in order to solve (8a,b).

LEVITATION AND PLASMA CIRCULATION MODELS

An example of the unusual behavior of dusty plasmas can be found in the Earth-Moon system. It has been claimed[31] that there exists an horizon glow above the sunrise terminator created by lunar dust. The distinguishing feature of this phenomenon, if it is due to dust, is that it would be an ongoing dust production mechanism much like volcanism. It would not be limited to activity in the early solar system, which might have long since left little measurable trace. Subject to the lunar levitation potentials[16-19] some fraction of any such active lunar dust would necessarily be swept away dynamically as pick-up and blow-off dust. The consequence would be that the resulting dusty plasma must drift downstream until encountering the Earth's magnetospheric bowshock, becoming a part of the plasma model circulation pattern for the Earth, an example of which is illustrated in Figure 4.

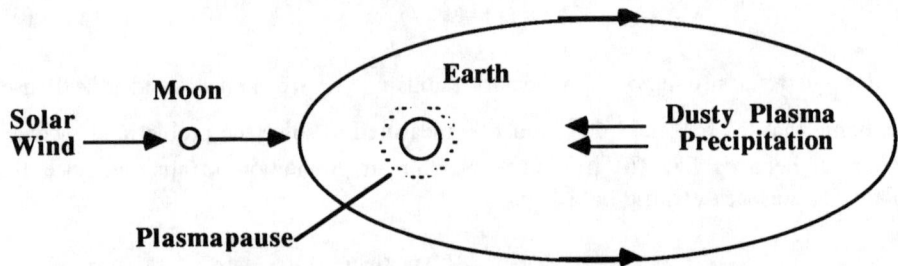

Figure 4. The teardrop plasma circulation model[32], illustrating the unusual manner through which lunar levitation dust could precipitate into the near-Earth environment.

Unexpectedly, the resultant dusty plasma would precipitate into the near-Earth environment upstream via the Earth's neutral sheet.

LEVITATION SCALE HEIGHT OF CHARGED DUST

Levitated dust is actually an atmospheric species which can be either gravitationally bound, blown off by electrostatic repulsion, or swept away as a massive form of pick-up ion. The density n(r) of any conventional gravitational species decreases with radial distance r as

$$n(r) \approx n_0 \, e^{-r/H} \qquad (10a)$$

for an isothermal atmosphere in hydrostatic equilibrium, where n_0 is the surface density of the species and H is its scale height

$$H = \frac{kT}{mg} \qquad (10b)$$

with k being Boltzman's constant, T the temperature, m the mass of the dust grains, and g the gravitational acceleration, modified by the levitation potential Φ as in (8b), of the planetary object in Figure 2. The dust of a particular grain size can be thought of as having such a scale height H, behaving much like molecular constituents of the atmospheric exosphere.

The effect of the electrostatic levitation potential is to modify g in (10b) by increasing the scale height, as

$$\hat{g} = \frac{GM}{r_0^2} - \frac{Z}{m}\frac{\Phi_0}{r_0} \qquad (10c)$$

giving the levitation scale height in a planetary gravitational field to be

$$\hat{H} = \frac{kT}{m\hat{g}} \qquad (10d)$$

THE JOVIAN DUST

For planetary objects in the solar system which form a bow shock in the solar wind[33], either due to a magnetosphere (such as Earth, Jupiter, Saturn, Uranus, Neptune, and Mercury) or an electrically conductive ionosphere (Venus), exospheric volcanic dust is shielded from the interplanetary electric and magnetic fields. Nevertheless, one of the most interesting cases of planetary dust is Jovian dust.[6] The situation here is that the large, tidally excited satellite Io generates a plasma torus[26,30,33] consisting of sulfur dioxide, hydrogen sulfide, and other gases vented by volcanic eruption and discovered by the Voyager spacecraft. Such venting would be the source of kinetic energy for transporting dusty plasma.

The same physical arguments in Eqs. (1)-(5) and Figure 2 still apply, except that the electric field **E** and magnetic field **B** are associated with the plasma torus and

Jupiter's dipole field. The motionally induced electric field VxB in (1) and (5) is generated by Jupiter's local magnetic field, and there is a co-rotational electric field E as in (2) and (3) for use in (5) and (8), given by

$$E = -VxB \tag{11a}$$

$$= (rx\Omega)xB/c \tag{11b}$$

where $|\Omega| = 1.76 \times 10^{-4}$ s^{-1} is Jupiter's rotation rate and $|B|$ can be represented by a simple, slanted magnetic dipole.[6&33]

CHARGED ASTEROIDAL & COMETARY DUST

Any asteroidal surfaces with dusty regoliths are subject to the same physical mechanisms of Figure 2, even without volcanism or outgassing since the solar wind can strip away charged dust by the pick-up transport mechanism discussed. Comets, though more complicated[34] involving bow shocks and comas, can likewise lose dust in this fashion which is one of the more obvious sources of the debris they spread into interplanetary space.

CONCLUSION

The motionally induced electric fields introduced by Lorentz as necessary consequences of the relativity of charge in motion have been shown to play a pertinent role in the dynamics of charged dust on planetary and interplanetary objects. The notion of electrostatic pick-up dust, akin to pick-up ion transport in interplanetary plasmas, has been introduced in order to characterize this process for the transport of volcanic dust into outer space from a planetary surface. The general equation of motion involved is (8a,b) and not (5). Much of the unexpected behavior of charged dust grains can now be understood because (8a,b) cannot be solved without an *in situ* measurement of the electrostatic levitation potential Φ. The commonplace assumption of (5) will always give an incorrect result when $\Phi \neq 0$.

ACKNOWLEDGEMENTS

The author would like to express his gratitude to H. Newsom, J.M. Rietmeijer, and H. Zook for making several helpful comments upon reading the manuscript. He would also like to thank D. McKay and M. Zolensky for sharing the unpublished result that they had found a few lunar-like spheres or glass nodules of micron size in NASA's stratospheric dust collection (conceivably ablated anorthositic grains from the lunar highlands). D. Lindstrom first pointed their results out to the author.

REFERENCES

1. Banaszkiewicz, M., and Ip, W.-H. *Icarus* **90**, 237 (1991).
2. Kecskeméty, K., and Cravens, T.E. *Geophys. Res. Lett.* **20**, 543 (1993).
3. Xu, W., d'Angelo, N., and Merlino, R.L. *J. Geophys. Res.* **98**, 7843 (1993).
4. Rao, N.N. *Planet. Space Sci.* **41**, 21 (1993).
5. Rosenberg, M., and Mendis, D.A. *J. Geophys. Res.* **97**, 14,773 (1993).
6. Grün, E., Morfill, G., and Johnson, T.V. *Geophys. Res. Lett.* **7**, 305 (1980).
7. Grün, E., Morfill, G., and Johnson, T.V. *Planet. Space Sci.* **28**, 1087, 1101, 1111, 1115 (1980).
8. Grün, E., Morfill, G., and Horanyi, M. *Nature* **363**, 144 (1993).
9. Morfill, G.E., and Grün, E. *Planet. Space Sci.* **27**, 1269, 1283 (1979).
10. Manka, R.H., and Michel, F.C. *Proc. of the Second Lunar Science Conference*, Vol. **2**, 1717 (MIT Press, Cambridge, 1971).
11. Manka, R.H., et al., *Lunar Science-III*, 504 (Lunar Science Institute, Houston, 1972).
12. Manka, R.H., and Michel, F.C. *Science* **169**, 278 (1970).
13. Ip, W.-H., *Geophys. Res. Lett.* **13**, 1133 (1986).
14. Grün, E., Morfill, G.E., and Mendis, D.A., in *Planetary Rings*, Greenberg, R., and Brahic, A., eds. (University of Arizona Press, Tucson, 1984), 275, Sect. II.B and II.C.
15. Mendis, D.A., Hill, J.R., Houpis, H., and Whipple, E.C., *Ap. J.* **249**, 787 (1981).
16. Vondrak, R.R., Freeman, J.W., and Lindeman, R.A., *Proc. 5th Lunar Conf.*, Suppl. 5, Geochimica et Cosmochimica Acta **3**, 2945 (1974).
17. Reasoner, D.L., and Burke, W.J., *J. Geophys. Res.* **77**, 6671 (1972).
18. Freeman, J.W. and Ibrahim, M., *The Moon* **14**, 103 (1975). See also Schneider, H.E., and Freeman, J.W., *The Moon* **14**, 27 (1975).
19. Fenner, M.A., Freeman, J.W., and Hills, H.K., *Proc. 4th Lunar Science Conf.*, Geochimica et Cosmochimica Acta **3**, 2877 (1973).
20. Barnes, M.S., Keller, J.H., Forster, J.C., O'Neill, J.A., and Coutlas, D.K. *Phys. Rev. Lett.* **68**, 313 (1992).
21. Cook, C.F., Shoemaker, E.M., and Smith, B.A., *Nature* **280**, 743 (1979).
22. Smith, B.A., Shoemaker, E.M., Kieffer, S.W., and Cook, A.F., *Nature* **280**, 738 (1979).
23. Johnson T.V., Cook, A.F., Sagan, C., and Soderblom, L.A., *Nature* **280**, 746 (1979).
24. Robertson, H.P. *Mon. Not. Roy. Astron. Soc.* **97**, 423 (1937).
25. Burns, J.A., Lamy, P.L., and Soter, S. *Icarus* **40**, 1 (1979).
26. Dessler, A.J., and Sandel, B.R. *Geophys. Res. Lett.* **19**, 2099 (1992).
27. Lorentz, H.A. *The Theory of Electrons*, 2nd edition, 14 and 198 (Dover reprint of 1915 edition, New York, 1952).
28. See, for example, Parker, E.N. *Ap. J.* **160**, 383 (1970); Cowling, T.G. *Magnetohydrodynamics* (Interscience, New York, 1957); Spitzer, L. *Physics of Fully Ionized Gases* (John Wiley & Sons, New York, 1962).

29. By virtue of the theory of relativity, one can always find a frame of reference in which both electric and magnetic fields are present.
30. Clouthier, P.A., Daniell, R.E., Dessler, A.J., and Hill, T.W., *Ap. Space Sci.* **55**, 93 (1978).
31. Zook, H.A., and McCoy, J., *Geophys. Res. Lett.* **18**, 2117 (1991).
32. Hill, T.W., and Dessler, A.J., *Science* **252**, 410 (1991).
33. Van Allen, J.A., in *The New Solar System*, Beatty, J.K., and Chaikin, A., eds., 3rd edition (Cambridge University Press, New York, 1990), 29.
34. Grewing, M., Praderie, B., and Reinhard, R., eds. *Exploration of Halley's Comet* (Springer-Verlag, Berlin, 1988).

MINERALOGY AND COMPOSITION
OF INTERPLANETARY DUST

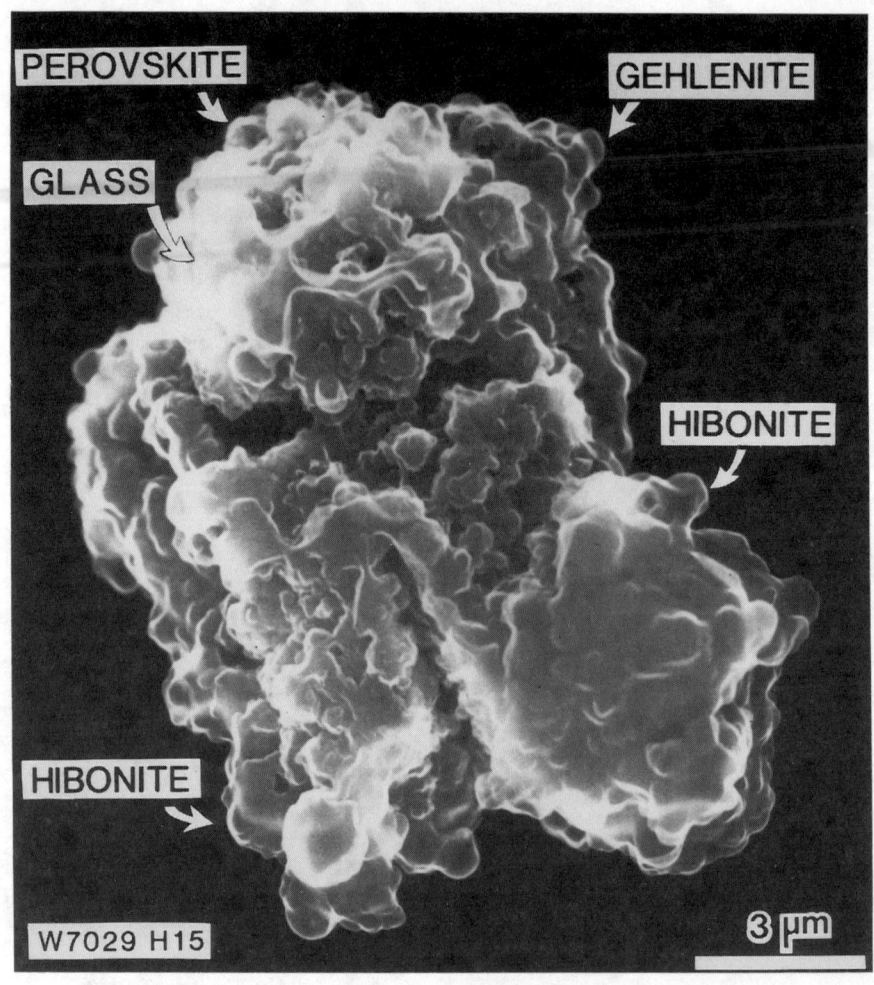

A secondary electron image of a refractory interplanetary dust particle, consisting essentially of submicron-sized grains of perovskite, gehlenite, hibonite and amorphous siliceous material ("glass"). Scale bar measures 3 μm. (From Zolensky, M.E. (1987) *Science* **237**, pp.1466-1468)

MINERALOGY AND COMPOSITION OF INTERPLANETARY DUST

Mineralogy and Petrography of IDPs

John Bradley and Wolfgang Klöck summarize what has been learned concerning the mineralogy of chondritic IDPs. A fascinating aspect of IDP petrography is the frequent occurrence of fine-grained aggregates, which can have bulk chondritic composition at the femtogram scale. These objects are still the subject of nomenclaturial disagreement; Bradley has (in the past) called them *tar balls*, Rietmeijer has used the term *granular units*. Bradley has now identified three flavors of aggregates, which he calls unequilibrated-, equilibrated- and reduced aggregates (UAs, EAs and RAs), depending upon the mineral assemblage. (Frans Rietmeijer suggests an alternate classification scheme in much the same vein, see **Classification**, below.) Are these aggregates the products of nebular accretion, with subsequent processing in the cases of the EAs? Lindsay Keller has suggested that some of these aggregates could be agglutinates from parent body regoliths. These objects clearly will receive much more attention in the future, although it is unfortunate that they are so small, being right at the analytical limit for present instruments.

Bradley also reviews the evidence for hydration on the IDP parent bodies, concluding that the frequent occurrence of hydrated and anhydrous phases within the same grain is likely due to the incipient nature of the alteration event. Mike Zolensky and Ruth Barrett describe work on the compositions of olivines and pyroxenes in chondritic IDPs. The common occurrence of diopside in hydrous IDPs probably indicates parent body metamorphism. Patti Jo Burkett and Mike Zolensky have also been measuring the porosities of IDPs, and find that these commonly vary from 5 to 25%, with higher porosities being rarer. Considerable attention is also centering on the non-crystalline component of IDPs. Refractory IDPs were not discussed at length at the workshop, and they clearly deserve more attention. It is almost certain that the current interest in chondritic IDPs is preventing additional IDP types from being recognized from among the stratospheric collections.

Processing of IDPs in the Solar System and Terrestrial Atmosphere

Al Nier has been measuring noble gas contents of individual IDPs by step heating. He has realized that examination of the temperature - gas release profile for a particle will reveal the peak temperature of atmospheric entry heating. This latter value can be used to infer cometary vs. asteroidal origin for the particle, as the cometary grains enter the atmosphere at considerably higher velocities (on the average). This technique appears to be the only current course for distinguishing cometary from asteroidal particles.

George Flynn reviews here the evidence indicating the degree of atmospheric entry-heating experienced by IDPs. Despite the obfuscatory effect this has on nebular and parent body processes, recognition of the degree of heating can provide unique source information (see above). Documented entry-heating effects include formation of magnetite rims, depletion of volatile elements like zinc, dehydration of phyllosilicates, and the changes in the release temperatures of noble gasses mentioned above. At one time it was hoped that inspection of solar flare track densities and their density variation across a grain would provide useful information on space-exposure duration and entry-heating level, however track densities appear to be too low to permit useful estimates to be made of these values. The explanation of these heating effects is still somewhat controversial. Some researchers believe that magnetite rims could form by sublimation onto the particles from atmospheric E-layer gases (see the paper by Maurette et al. later in this volume for a better discussion of this point). However, most workshop participants felt that this sublimation process would be unlikely, due to the low iron concentration of the E-layer, and the short atmospheric residence time of the particles. Some researchers are unsure of the origin of the bromine and zinc enrichments reported for many chondritic IDPs, making heating/depletion estimations potentially uncertain. All participants concluded that we need to understand the concentration and sources better of zinc, bromine and iron in the upper atmosphere (above the tropopause).

Discussion has also centered on the potential for particulate contamination during collection in silicone oil on the inertial impaction collectors currently used by NASA. Although workers have not found indications of contamination from this material (except for Si, which can apparently be removed to large degree), lingering uncertainties on this subject indicate that care must be taken during the interpretation of compositional analyses. George Flynn presents a discussion of potential particulate contamination sources in the present volume.

Comparisons of IDPs to Meteorites and Micrometeorites

This topic is well covered in the present volume by Wolfgang Klöck's paper. Basically, most IDPs appear to exhibit significant differences from meteorites, although certain IDPs appear to be identical to some carbonaceous chondrite matrix materials. Although these differences do not preclude derivation from the same parent bodies as meteorites, it would require that different lithologies are being sampled. In general, low-strength, poorly-consolidated materials like IDPs would not be expected to survive as meteorite-sized bodies, only as dust. Klöck also compared IDPs to the larger (≥ 100 μm) micrometeorites collected in the oceans (deep sea spheres), and polar ice caps. These latter materials are generally highly contaminated, and severely heated by atmospheric entry, which often precludes useful comparison. However, the most pristine of these larger materials bear considerable resemblance to CI, CR and CM chondrites. There are sufficient differences between these micrometeorites and IDPs to warrant continued study of the former, particularly since they are considerably easier to handle and are available in abundance (thanks largely to the efforts of Michel Maurette).

Composition of IDPs

As discussed by Steve Sutton, numerous complementary techniques are now routinely utilized for the measurement of the bulk compositions of IDPs. Problems linger concerning potential contamination during atmospheric residence and collection, however progress is being made in understanding the actual dimensions of these hazards. Lindsay Keller and Kathie Thomas' papers provide good guides to recent achievements in the measurement of carbon in IDPs; which they find to contain up to 50 wt% carbon, far higher than for bulk chondrites. The highest carbon concentrations appear to come in pyroxene-dominated IDPs.

Robert Walker summarizes ion probe analyses of chondritic IDPs, with a major goal being the location of preserved interstellar material. Isotopic measurements have been made of H, C, Mg, N and Si on fewer than 100 IDPs. No large isotopic anomalies have been found in C, Mg or Si. Approximately half of the analyzed particles exhibited deuterium enrichments (up to 2000‰). Some IDPs have shown ^{15}N enrichments (up to 411‰), usually correlated with a deuterium enrichment. Hydrous and pyroxene-dominated IDPs appear to be isotopically similar, however no anomalies have been located in olivine-dominated IDPs. Basically, no IDPs have yet been located which are entirely interstellar in origin (as we understand them). While this is a disappointing result, the fact is that only a very small total mass of IDPs has yet to be examined; if interstellar materials are present among IDPs in the same concentration as in the Murchison CM chondrite then (statistically) we should not expect to have seen even one entirely interstellar grain yet. However, isotopic anomalies are found in IDPs, and these remain to be explained satisfactorily.

Classification

Major problems remain with IDP classification. The most widely used scheme classifies IDPs based upon the dominant crystalline silicate phase: olivine, pyroxene, saponite and serpentine. Not included in this framework at all are the refractory IDPs. By the current classification scheme an IDP containing a small but not dominant quantity of phyllosilicates is considered *anhydrous*. Not all workers accept this division, and this disagreement could lead to confusion for the reader of the present volume (so beware). For example, Thomas et al. describe here IDP L2006 B23, which they find to be an anhydrous chondritic IDP with the highest documented carbon concentration (~45 wt.%). Elsewhere in the present volume, Gibson and Bustin find that another portion of this same IDP (called L2006 B16) contains a minor quantity of hydrous phases. So, is this IDP hydrous or anhydrous?

Frans Rietmeijer reviewed these schemes at the workshop (and in the present volume), proposing a new classification based on the composition and petrography of the micrometer-sized aggregates present in most chondritic IDPs (see above). Rietmeijer proposes that these objects be called *granular units* and *polyphase units*, in contrast to John Bradley's suggestion. Actually, both classifications have common features. No common classification scheme for IDPs was found acceptable to everyone, although participants agreed that a radically new classification scheme along the lines of those proposed was desirable. However, lively discussion of these classification schemes established that for any new one to be accepted it must (1) be based on measurements which can be made by numerous investigators, (2) not rely on inferred processes, and (3) not include unsavory terms or acronyms (you had to be there). It is clear that further development work will be necessary on classification schemes.

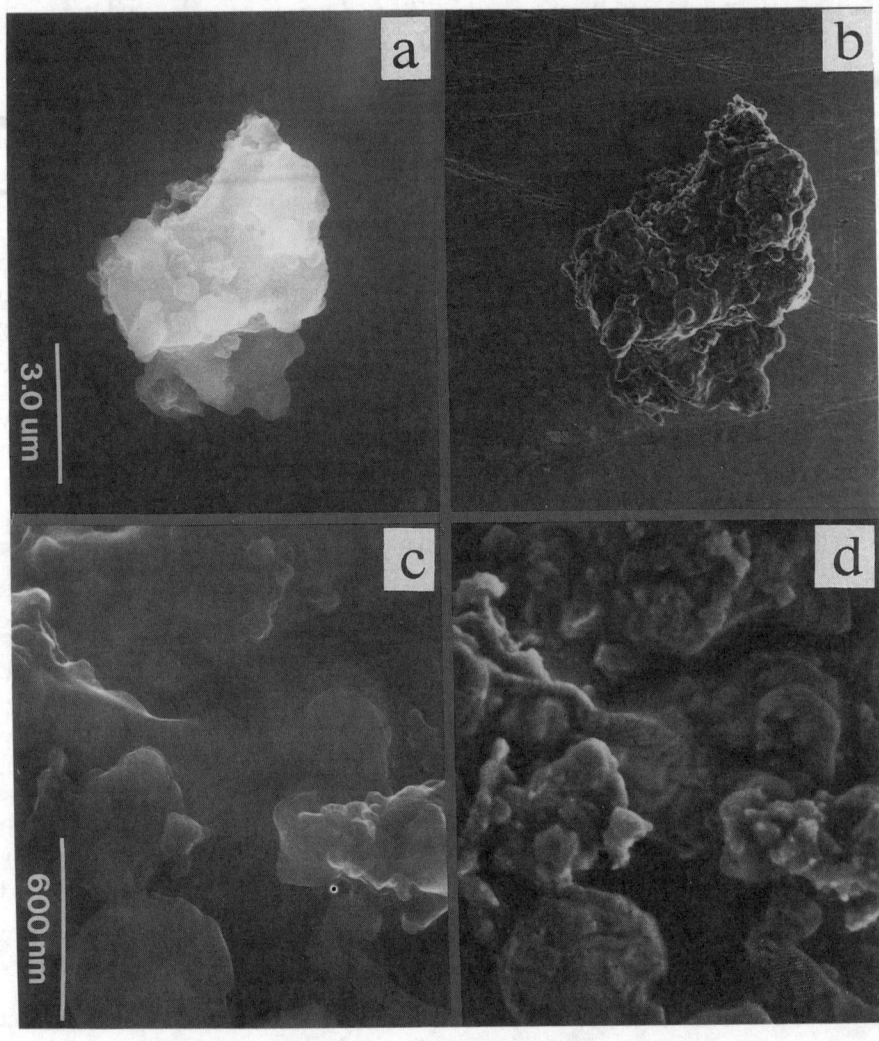

Secondary electron images of a chondritic interplanetary dust particle, showing the additional morphological information available from low-voltage electron microscopy. (a) 30 KeV (conventional voltage) low-magnification image. (b) Same as (a), but recorded at 2 KeV; note additional detail from, presumably, carbonaceous material. (c) 30 KeV high-magnification image. (d) Same as (c), but recorded at 2 KeV. (Figure after D. Blake et al. (1987) *Proceedings of the 45th Annual Meeting of the Electron Microscopy Society of America*. San Francisco Press, pp. 208-209)

MINERALOGICAL AND CHEMICAL RELATIONSHIPS OF INTERPLANETARY DUST PARTICLES, MICROMETEORITES AND METEORITES

W. Klöck
Institut für Planetologie der Universität Münster, Wilhelm-Klemm-Straße 10
D-48149 Munster, FRG

and
F.J. Stadermann
Fachbereich Materialwissenschaft, Technische Hochschule Darmstadt, Hilpertstr. 31,
D-64295 Darmstadt, FRG

ABSTRACT

Interplanetary dust, micrometeorites and meteorites should share some common physical and chemical properties if the small dust particles as well as the larger extraterrestrial samples originate in the asteroid belt. Indeed, interplanetary dust particles (IDPs) and micrometeorites were found showing clear relationships to existing meteorite classes and to specific components of carbonaceous chondrites. Primitive anhydrous IDPs, though mineralogically unlike known meteorites, contain forsterite with a unique chemical signature which are also present in primitive chondrites, and therefore establish a link between IDPs and meteorites. These primitive anhydrous IDPs are most likely derived from objects in the asteroid belt which do not deliver meteorites to Earth. A number of IDPs appear to be samples of strongly heated or melted parent bodies and could establish links to Acapulcoites and the even more reduced silicates in type IAB iron meteorites. Refractory IDPs with isotopic and chemical properties of Calcium-Aluminum-Rich Inclusions (CAIs) are present among stratospheric dust particles. They suggest a relationship of some IDPs with carbonaceous chondrites, although a small number of CAIs are also reported from ordinary chondrites. However, IDPs mineralogically identical to equilibrated ordinary chondrites have not so far been identified. It is therefore most likely that refractory IDPs are samples from similar to carbonaceous chondrite parent bodies.

Among Antarctic micrometeorites is a sizable population of unmelted particles. These unmelted micrometeorites are most likely the products of thermally altered hydrous phases. This population of micrometeorites, originating from aqueously altered parent objects in the asteroid belt, is comparable in abundance to the hydrated class of IDPs among stratospheric dust particles. The majority of hydrated IDPs are however not identical to the known aqueously altered meteorites. One micrometeorite from Antarctica, having its phyllosilicates preserved, establishes a link between unmelted micrometeorites and smectite-type IDPs. A number of aqueously altered parent objects in the asteroid belt could account for the existence of abundant hydrated particles among IDPs and micrometeorites. These hydrous asteroids are however not sampled by the known meteorites.

INTRODUCTION

Interplanetary Dust Particles (IDPs) are, in addition to meteorites and lunar rocks, a third class of extraterrestrial material available for laboratory studies.[1] Larger dust particles (>100 micrometer), which settle too fast to be effectively collected in the stratosphere, are called micrometeorites. Recently, micrometeorites were recovered in large numbers from polar regions of the Earth[2&3] (see also the paper in this volume by Maurette et al.[4]), whereas in the past, they have been extracted from deep sea sediments.[5&6]

One important aspect in understanding the flux of cosmic dust onto the Earth, is the relationship of stratospheric dust to micrometeorites and meteorites. Therefore we will outline here some general properties of stratospheric IDPs and unmelted Antarctic micrometeorites including a few examples of the mineralogical and chemical similarities and differences between IDPs, micrometeorites and meteorites. A detailed review of properties of IDPs, their formation histories and relationships to meteorites and to solar system processes has been published by Mackinnon and Rietmeijer.[7]

MINERALOGICAL PROPERTIES OF INTERPLANETARY DUST PARTICLES

Interplanetary dust particles have been classified according to their IR spectral signatures into olivine-, pyroxene- and layer-silicate classes.[8] In general, this division has been supported by analytical electron microscopy of IDP ultramicrotomed thin sections.[9-11] Olivine- and pyroxene-type IDPs contain abundant mafic silicates and the composition of these mafic minerals in anhydrous IDPs is a potential mineralogical indicator to distinguish primitive and processed dust particles[12-14] (see also the paper by Zolensky and Barrett in this volume[15]).

A subset of the anhydrous class of IDPs consists of particles dominated by very fine-grained material, so-called "microcrystalline aggregates"[16] or "granular units".[17] These aggregates, having approximately chondritic major and minor element abundances (except for carbon), consist of 5-50 nanometer sized grains of Mg-Fe silicates, Fe-Ni sulfides and Fe-Ni metal embedded in a glassy or carbonaceous matrix.[18] Rietmeijer[19] and Bradley[20] also discuss these aggregates elsewhere in this volume

Anhydrous IDPs of that subclass are strongly enriched in bulk carbon contents relative to CI-meteorite abundances.[21-23] High carbon abundances and some additional mineralogical properties suggest that particles of this subclass are among the most primitive solar system material available for detailed laboratory studies.[7,9,10,16,18,21]

A major portion of extraterrestrial particles collected in the stratosphere belong to the layer-silicate class.[7&18] Among these phyllosilicate-rich IDPs, saponite particles dominate over serpentine particles by a factor of at least 4:1.[24] Anhydrous mineral phases, like (Fe,Ni) sulfides, Mg-Fe silicates, Mg-Fe carbonates, chromite and magnetite are uncommon, but certainly present in smectite IDPs. Surprisingly, at present only three hydrated interplanetary dust particles have been found with distinct mineralogical similarities to CI, CR and CM meteorites.[25,26,27] Again, as it is the case with anhydrous IDPs, major element abundances of hydrated IDPs are also approximately chondritic, except for a sizable carbon enrichment of about 4 x CI-chondrite values[24], and Ca depletions relative to the anhydrous IDPs (see reference 15 and references therein). The carbon enrichment indicates that the supposedly asteroidal parent objects (or portions thereof) of hydrated IDPs are chemically and mineralogically unlike the asteroidal sources of the most abundant carbonaceous chondrites.

Common to anhydrous as well as to some hydrated particles is the occurrence of olivines having a striking composition; these forsteritic olivines are characterized by Fe/Mn ratios of ≈1. Olivines with an identical chemical signature also occur in the matrix of carbonaceous chondrites and in matrix of the unequilibrated ordinary chondrite (UOC) Semarkona.[28] Olivines with this unique composition provide a link between primitive meteorites, most likely derived from asteroids, and the sources of cosmic dust particles, which could be comets as well as asteroids.

MINERALOGICAL PROPERTIES OF MICROMETEORITES

Most micrometeorites in the 100 to 1000 micrometer size range are spherules, and their shapes and compositions indicate that they were totally melted during atmospheric entry, in accordance with entry heating calculations.[29-31] A minor portion of these spherules contain relict refractory mineral phases, like forsterite and enstatite, which survived melting, and their minor element signatures suggest a link between deep sea spheres, melted polar micrometeorites and carbonaceous chondrites.[32&33]

Among micrometeorites in the 50-100 micrometer size range, recovered by M. Maurette from Antarctic ice[3], is a sizable population of morphologically irregular and therefore most probably unmelted micrometeorites. Textures, mineral assemblages and bulk chemical compositions of some of these micrometeorites are reminiscent of CI-meteorite matrix material[34], though, for instance, framboidal magnetite also occurs in CV and CR chondrites. Transmission electron microscope (TEM) studies of irregular Antarctic and Greenland micrometeorites[35&36] have, so far, failed to reveal layer-silicates, which probably indicates significant atmospheric entry heating. Recently we were able to unambiguously identify layer-silicates in one irregular, unmelted micrometeorite. The particle was studied in detail by Kurat et al.[34], and they concluded that it is a CI-like micrometeorite. The survival of hydrated phases in that micrometeorite, measuring at least 70 µm in size, clearly argues in favour of an asteroidal origin. According to entry heating calculations[29&31], it appears that this micrometeorite entered Earth's atmosphere at grazing incidence. The majority of the supposedly unmelted micrometeorites appear to be heated to such an extent that any phyllosilicates originally present were transformed into assemblages of fine-grained (grain sizes range from ≈10 nm up to several 100 nm) olivines, pyroxenes and magnetite embedded in a glassy matrix. The population of these polar micrometeorites, heated to various degrees, should, however, preserve even more relict phases and textures, and therefore important information about their source material compared to the melted micrometeorites. Interpretation of their minor and trace element patterns has turned out to be complicated by changes in bulk composition during residence in the Antarctic and Greenland ice.[35,37,38] Terrestrial alteration in the ice causes dissolution of sulfides and loss of siderophile and chalcophile elements. At the same time extraterrestrial particles adsorb lithophile elements from terrestrial sources, such as K and U.[35,37,38]

A few examples are given below illustrating possible relationships of stratospheric dust particles and larger extraterrestrial samples.

MINERALOGICAL AND CHEMICAL ASPECTS OF PRIMITIVE ANHYDROUS INTERPLANETARY DUST PARTICLES

Among the class of anhydrous stratospheric dust particles is a subset of dust particles having abundant fine-grained aggregates, earlier described as "tar-balls"[39] and "granular units" (GUs).[40] An example is given in Figure 1. The unequilibrated nature of these particles is documented by a variable iron content of olivines dispersed throughout the fine-grained aggregates (Fig. 2). These anhydrous IDPs seem to consist of at least three mineralogically distinct components: (i) "Granular units", (ii) coarser-grained Mg-Fe silicates (typically a few hundred nm in size) and (iii) varying amounts of carbonaceous material. These three components are present in variable amounts in these anhydrous IDPs. The mineralogy of primitive chondritic IDPs has been discussed in terms of these basic building blocks by Rietmeijer.[40] In addition, anhydrous IDPs are characterized by high ^{4}He abundances, relatively low ^{4}He release temperatures, chondritic levels of Zn and other volatile elements[41] and the presence of solar flare tracks in Mg-Fe silicates[42] (Fig. 3). Anhydrous particles also have chondritic abundances of volatile elements, like Na, S, K, Mn, Cu, Rb and Cd, as shown by Stadermann.[43] Stadermann reported trace element analyses of IDP NERO (u44-m1-7), which is a typical example of the anhydrous particle class (Figs. 4 and 5), and found CI-meteorite abundances (within a factor of two) for all major, minor and trace elements analyzed. The unfractionated trace element pattern obtained from this particle suggests that it was not heated significantly above 600°C during atmospheric entry.[43]

Other examples of anhydrous IDPs are IDP U2015*B2, (Fig. 6), L2005D35 and W7029*B1 (Fig. 7). Mg-Fe silicates in the first two particles contain solar-flare tracks, limiting their entry heating temperatures to less than 600°C. These particles are fragments of larger aggregates. If the original aggregates were 40 μm in size, like one other cluster-type IDP shown in the literature[44], and did not experience temperatures higher than 600°C during atmospheric entry, it is most likely that the source of these dust particles is in the asteroid belt, though cometary sources with perihelia > 1.5 AU[29&30] cannot be excluded.

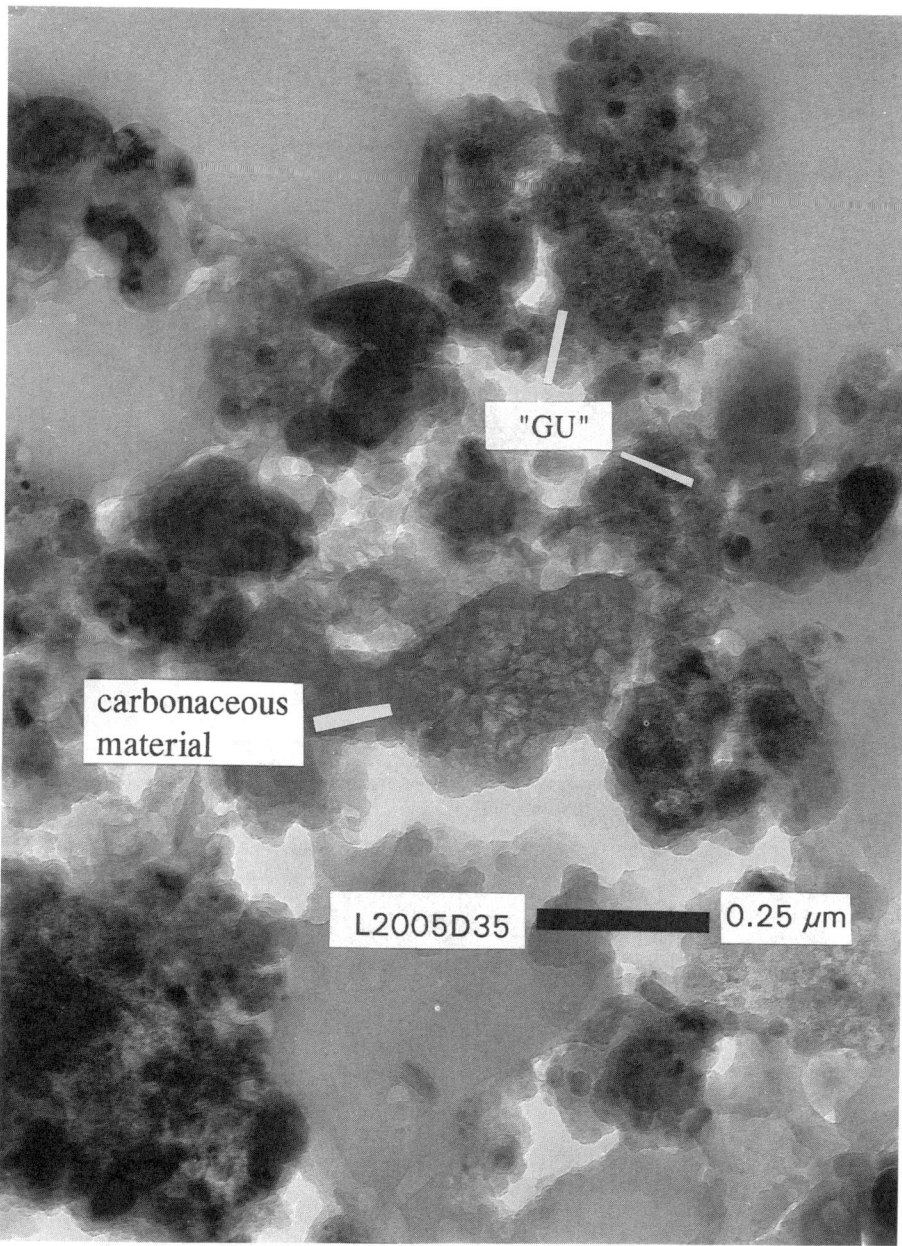

Figure 1. Bright-field TEM photograph of an ultramicrotomed section. IDP L2005D35 belongs to the class of C-rich anhydrous stratospheric dust particles. Discrete clumps of carbonaceous material are mixed with "granular units" and some coarser-grained Mg-Fe silicates and Fe-sulfides. These mineral components have been proposed as basic building blocks of primitive IDPs by Rietmeijer.[40]

56 Mineralogical and Chemical Relationships

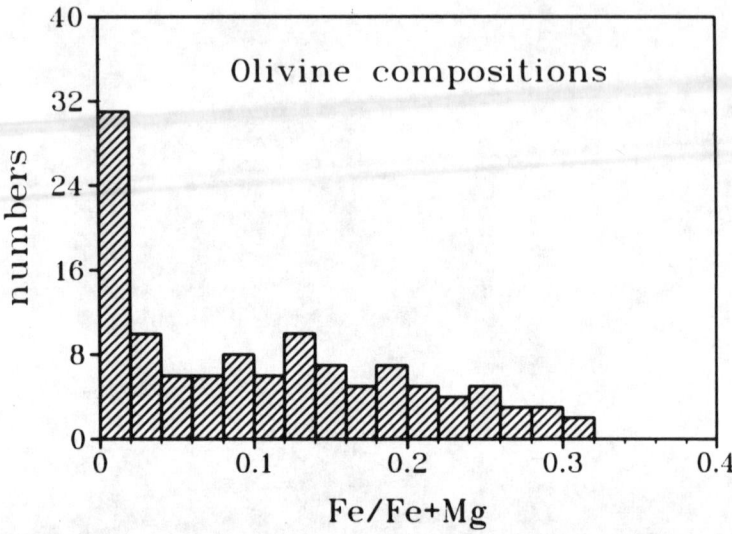

Figure 2. Histogram of olivine compositions of 8 anhydrous IDPs. Olivines analyzed were all in the size range from about 100 nm to 1 μm. These silicate grains are dispersed throughout the fine grained matrix consisting of "granular units". Individual particles display a similar range of olivine compositions supporting the unequilibrated nature of anhydrous particles.

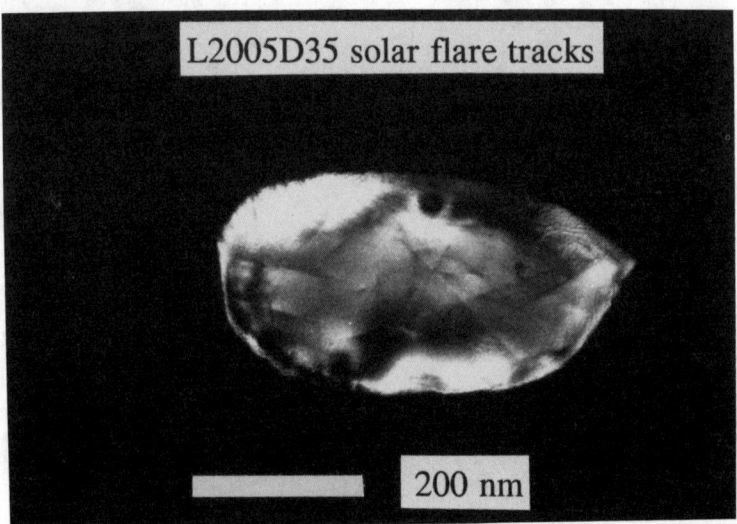

Figure 3. Dark-field TEM-photograph of olivine crystal in particle L2005D35 showing crystal defects due to solar ion bombardment. The survival of solar flare tracks in Mg-Fe silicates in these primitive particles limits their atmospheric entry temperatures to less than 600°C.[45]

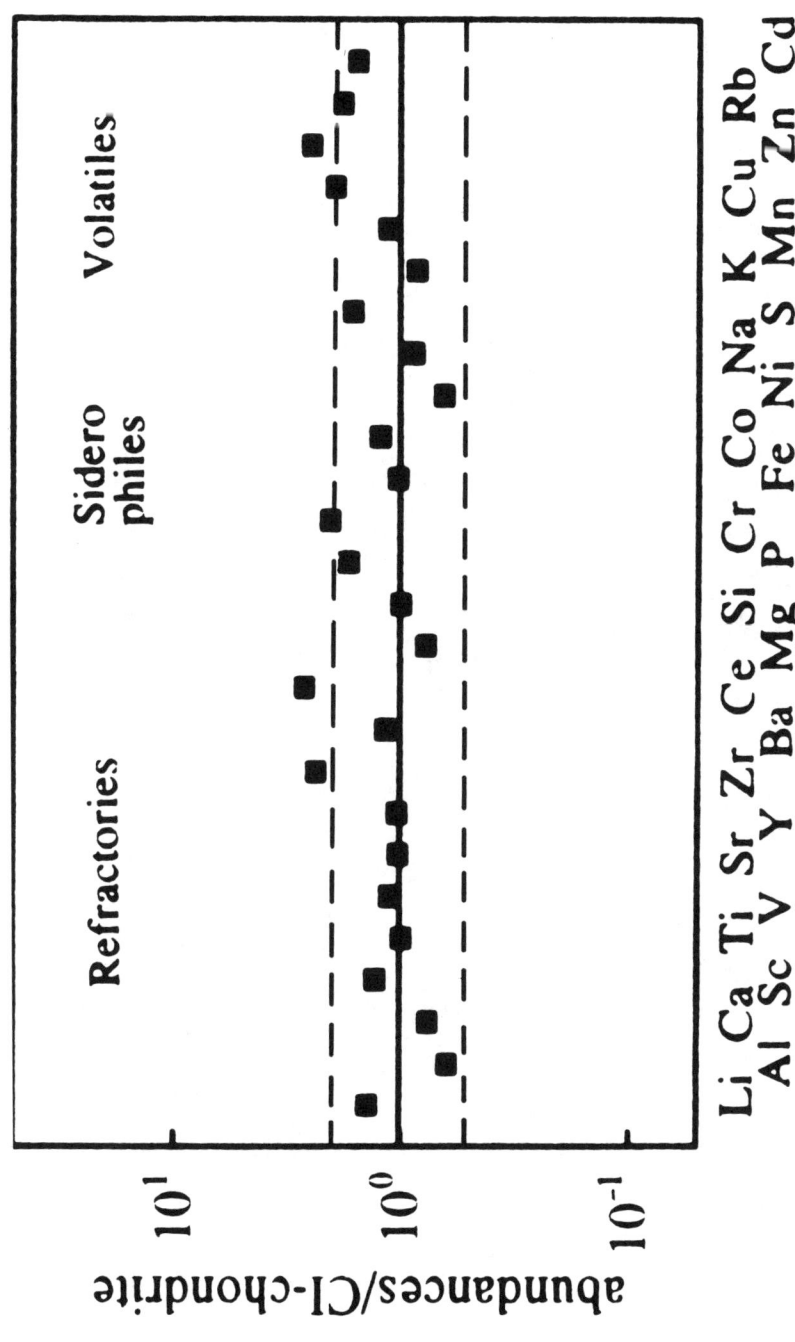

Figure 4. Major and trace elements of interplanetary dust particle "Nero" as determined with the ion microprobe. The data are normalized to CI-chondrite abundances and silicon. Remarkable agreement is found (within a factor of 2) with average CI-chondrite element abundances for refractory, siderophile as well as for volatile elements, considering the nanogram mass of the particle. The unfractionated element pattern clearly supports an extraterrestrial origin of this stratospheric dust particle.

58 Mineralogical and Chemical Relationships

Figure 5. Dark-field TEM photograph of IDP "Nero". Individual crystallites are less than 100 nm in size, except for a few larger Fe-sulfide grains. This particle consists mainly of "granular units".

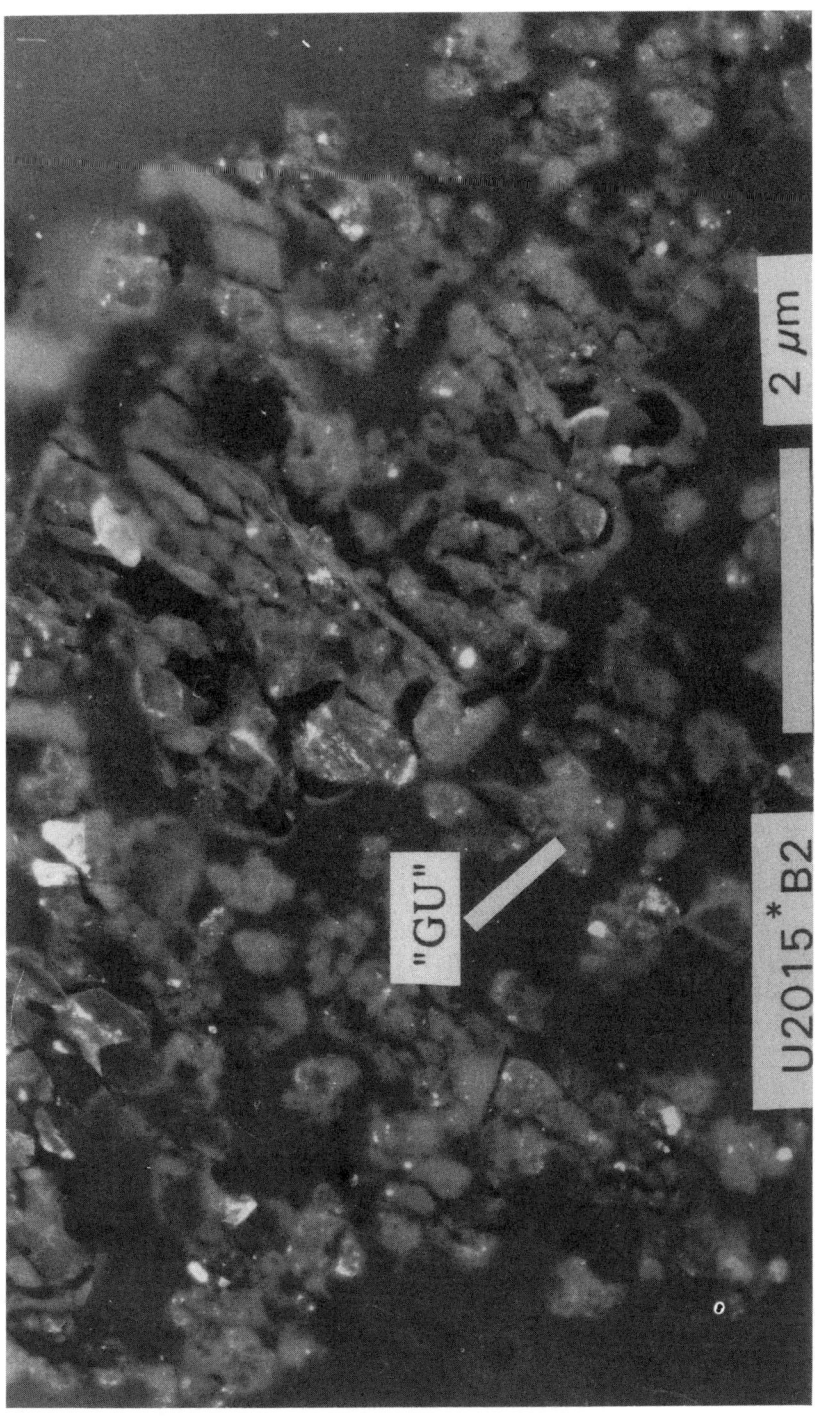

Figure 6. Another particle dominated by "granular units" is IDP U2015*B2. Mineral grains in "granular units" are on the average less than 50 nm in size. Some larger silicates are found in the upper part of the section. This particle is a fragment of a larger aggregate. Solar flare tracks are preserved in Mg-Fe silicates limiting the heating of the dust particle aggregate during atmospheric entry to less than 600°C.

60 Mineralogical and Chemical Relationships

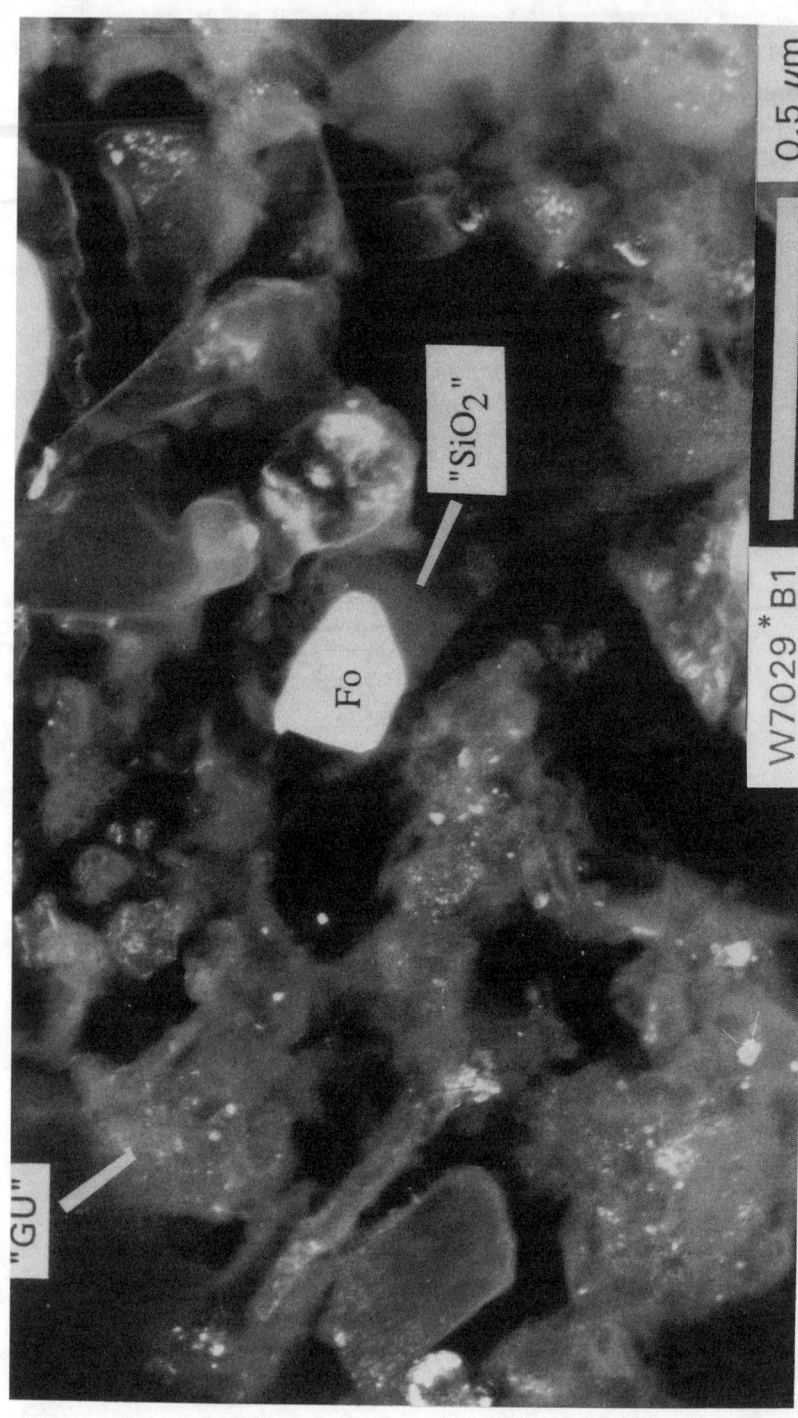

Figure 7. Dark-field TEM photograph of IDP W7029*B1. This particle is another example of an unheated, primitive, anhydrous IDP. Silicate grains are dispersed throughout the fine-grained "GU" matrix. A crystal of forsterite (Fo) is in contact with amorphous silica. This assemblage would have readily reacted to enstatite during prolonged heating; the absence of enstatite is evidence for the primitive nature of the particle. This particular forsterite grain was found to contain more Mn than Fe ($Fo_{98.8}Fa_{0.4}Te_{0.8}$).

MN-RICH OLIVINES IN IDPs AND METEORITES

A further argument in favour of an asteroidal connection for IDPs and micrometeorites is the occurrence of Mn-rich forsterite in dust particles of the above mentioned subset of the anhydrous class, as well as in several carbonaceous chondrites and the unequilibrated ordinary chondrite Semarkona.[28] These Mn-rich forsterite in unequilibrated anhydrous IDPs and primitive meteorites show identical chemical signatures, suggesting that this Mn-enriched forsterite was present at the formation locations of different meteorite parent bodies in the asteroid belt.

Forsterite with this distinct chemical signature is also a minor component of some hydrated interplanetary dust particles. Stratospheric dust particles dominated by serpentine are believed to be genetically linked to the hydrous (CI, CM, CR) carbonaceous chondrites. Mn-rich forsterite in CM2 chondrites displays a prominent iron enrichment from core to rim. This distinct zoning pattern could be used, in addition to the identification of the characteristic hydrated phases, to forge links with some hydrated IDPs to CM2 chondrites. However, Figure 8a-d clearly shows that the few hydrated interplanetary dust particles with Mn-rich forsterite studied so far are *not* genetically linked to CM2 meteorites. The formation of Mn-rich forsterite could be explained by non-equilibrium condensation in the solar nebula[28], but this does not preclude that other processes from being responsible as well. Formation by aqueous alteration however is not likely because Mn-rich forsterite also occurs in anhydrous IDPs which were not affected by aqueous alteration. The iron-enrichment of Mn-rich forsterite (Fig. 8d) in CM meteorites could be due to the reaction of early formed forsterite with an increasingly more oxidizing solar nebula.

COMPOSITIONS OF MICROCRYSTALLINE AGGREGATES

The major building blocks of unequilibrated anhydrous IDPs are very fine-grained components. Bradley[16] used the term "microcrystalline aggregates" and Rietmeijer[15] referred to them as "granular units" ("GUs"). On average, these aggregates are a few hundred nanometer in size[18] (Fig. 9). Individual crystallites in "GUs" are too small (5-50 nm) to analyze individually in TEM thin sections, but some mineralogical information can be obtained from bulk analysis and electron diffraction patterns.[17]

Average bulk major element compositions of "granular units" of a number of primitive anhydrous IDPs point to a depletion of Mg and Fe relative to Si normalized to solar abundances (Fig. 10). Some of these "granular units" are high in Zn, up to about 10 times the chondritic value, most probably due to the presence of tiny Fe-Zn sulfides. Surprisingly, Halley dust particle compositions show similar elemental ratios for Mg and Fe relative to Si.[46&47] Particles analyzed by Giotto and Vega mass spectrometers had masses of about 10^{-17} to 10^{-10} grams (see references 48 and 49, as well as the paper in this volume by Fomenkova and Chang[50]). This mass range corresponds to particles in the size range from less than 100 nm to a few micrometers, which is compatible with sizes of mineral grains and pieces of "granular units" in unequilibrated IDPs.[17] The wide compositional range of Halley

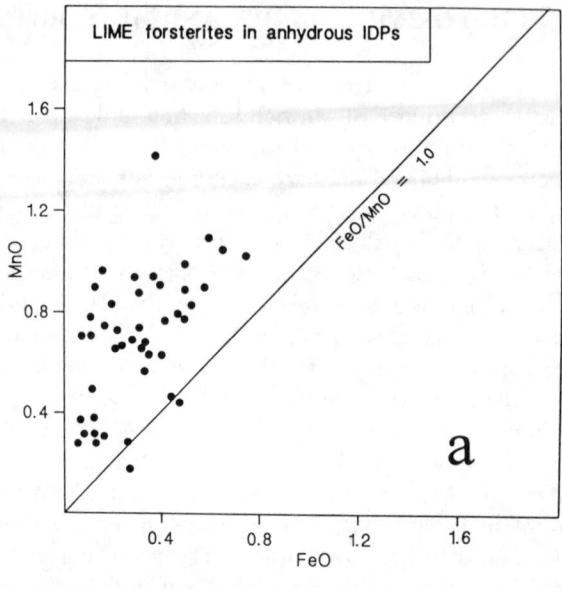

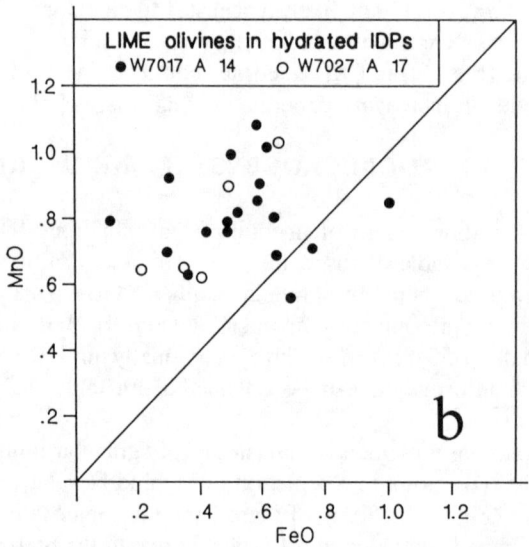

Figure 8. Comparison of iron and manganese contents of sub-micrometer olivines in (a) anhydrous IDPs, (b) hydrated IDPs. Forsterite in IDPs and in Semarkona has a Fe/Mn ratio of about one and shows no evidence of iron enrichment from core to rim. High-Mn forsterite in accretionary dust mantles of EET 83226 shows a prominent zoning in iron content. Core compositions of this forsterite is comparable to IDP and Semarkona forsterite, but olivine-rims are strongly enriched in iron. This iron enrichment is characteristic of CM2 sub-micrometer olivine and is absent in CI and CR chondrites and in Semarkona forsterite.

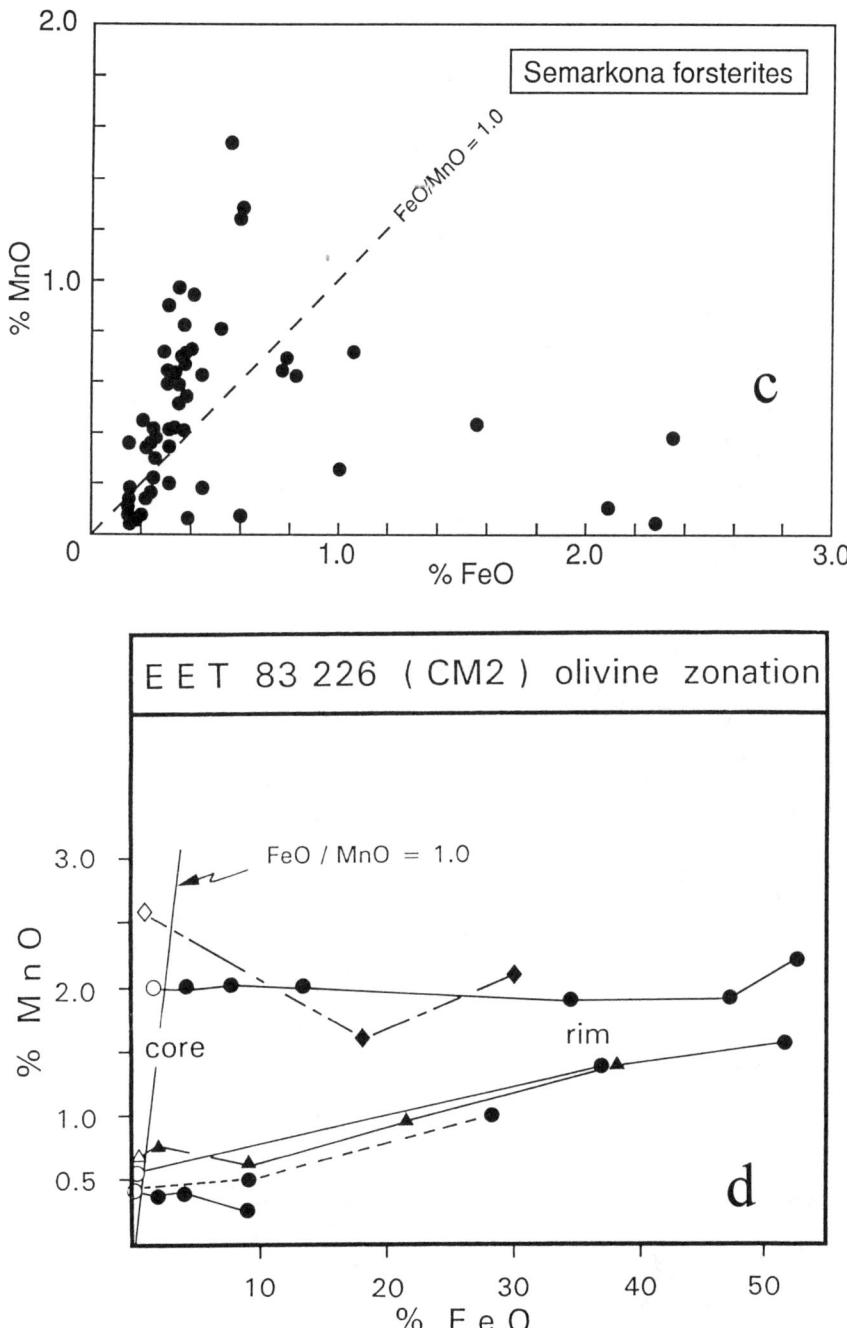

Figure 8. (cont.) Comparison of iron and manganese contents of sub-micrometer olivines (c) in the Semarkona matrix, and (d) in the EET 83226 CM2 meteorite.

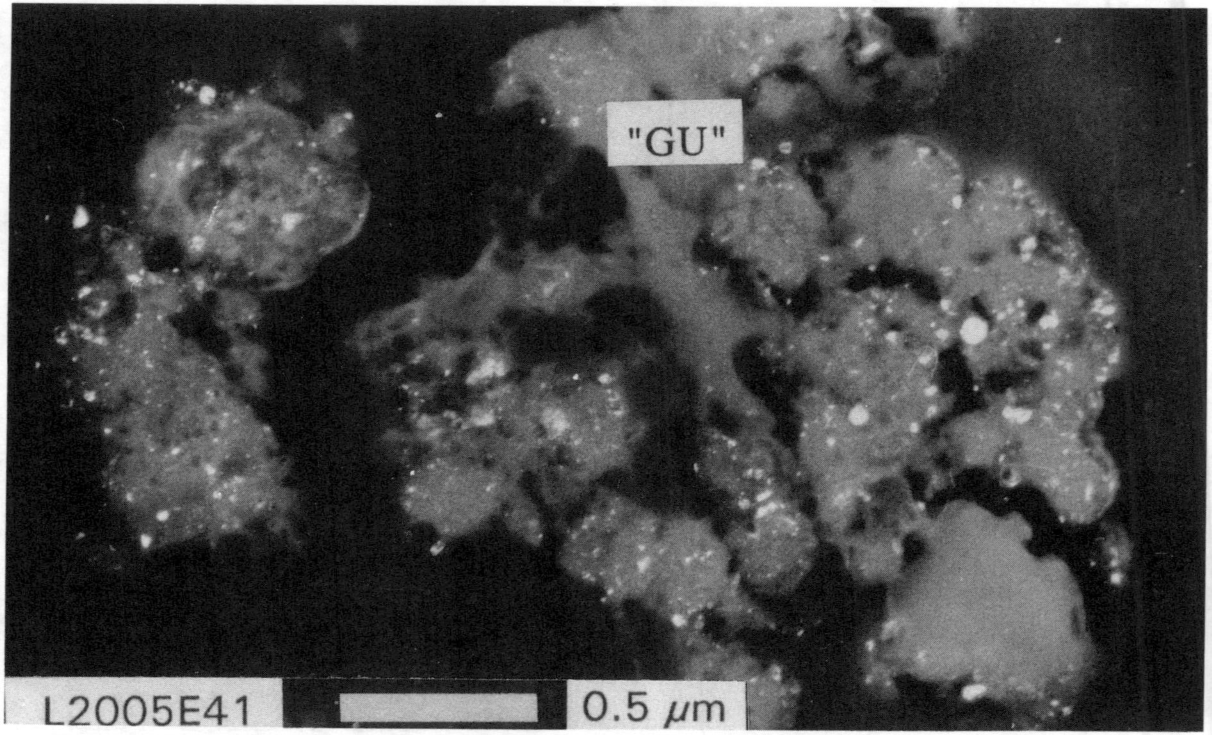

Figure 9. IDP L2005E41 is an example of a carbon-rich primitive anhydrous stratospheric dust particle. It consists entirely of "granular units"; no larger Mg-Fe silicates were found by TEM in ultramicrotomed sections. It is a fragment of a larger aggregate (cluster 28 from collector L2005, in ref. 51, the cluster number is erroneously given there as L2006#28). Two other fragments of this cluster have exhibited bulk carbon abundances of 18 and 20%. The good agreement between these chemical analysis of two diminutive fragments suggests a fine-grained nature for the whole cluster particle. "Granular units" like those in this figure were chosen for bulk analyses.

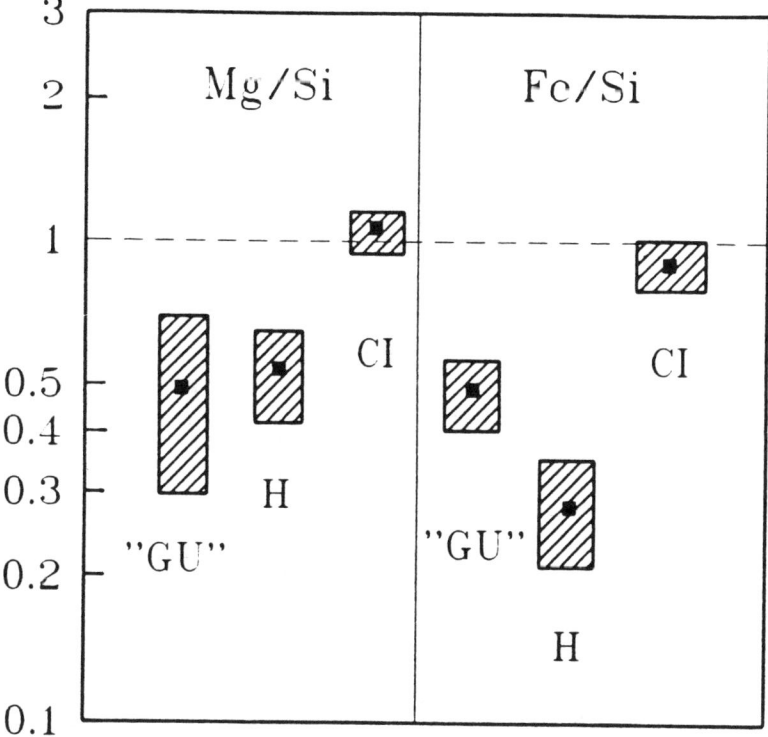

Figure 10. Average Mg/Si and Fe/Si ratios of "granular units" of 8 primitive anhydrous interplanetary dust particles compared to elemental ratios of CI-chondrites and Halley dust particles. Data are normalized to solar abundances.

dust particles is consistent with each dust particle analyzed by the mass spectrometers being an agglomeration of many mineral grains of varying composition and size.[47] However, few single mineral grains (exceptions being iron-sulfides and carbonates[48]) were identified in the Giotto and Vega mass spectra.

Compositions of "granular units" or "microcrystalline aggregates", silicate emission features[16], and other physical properties seem to suggest a cometary relationship for primitive unequilibrated IDPs; the presence of Mn-rich forsterite, however, indicates an asteroidal connection. This inconsistency can presently only be resolved if we accept that sources of primitive IDPs are long-period comets or if we assume that primitive asteroids and comets are compositionally similar.

IDPs WITH LINKS TO PRIMITIVE ACHONDRITES

In addition to porous, fine-grained particles, a number of rather compact and coarse-grained objects were identified among anhydrous IDPs. (This paper is, however, principally concerned with polymineralic aggregates with approximate chondritic bulk compositions.) Rietmeijer and Blanford reported the capture of a single olivine grain on a thermal blanket of the Solar Maximum satellite.[52] Individual mineral grains of olivines and pyroxenes in these dust particles range from about 100 nm up to about 1 µm in size and their textures suggest that these particles suffered intense thermometamorphism, with some being of possible igneous origin (Fig. 11). In most cases mafic silicates are embedded in glass and often adjacent grains have 120° intersections typical of recrystallization. Fe/(Fe+Mg) ratios of coexisting olivines and pyroxenes in these particles are given in Figure 12 and Table 1; olivines and pyroxenes in these IDPs are equilibrated with respect to Fe and Mg.

It is unlikely that flash-heating during atmospheric entry causes complete Fe/Mg equilibration of these micrometer-sized mineral grains. In one equilibrated particle (U2015C16) solar flare tracks in pyroxene were observed (Fig. 13). This particular particle, therefore, did not experience temperatures higher than 600-700°C during atmospheric entry.[45] Olivines and pyroxenes in IDP U2015C16 are micrometer-sized and Fe and especially Mg diffusion at 600-700°C is simply not fast enough to equilibrate micrometer-sized grains within a few seconds. Figure 14 was calculated from the Fe-Mg diffusion data of Buening and Buseck[53] for diffusion at atmospheric conditions; using the relation $t \approx r^2/D$ diffusion distances were obtained. At about 600°C and very oxidizing conditions, Fe-Mg diffusion barely permits equilibration of 100 nm sized olivines within 10 seconds. If conditions are more reducing equilibration times are longer. Therefore, the Fe and Mg equilibration shown by the olivines and pyroxenes in these particles requires metamorphism or melting on their parent objects.

Ranges of iron contents in olivines and pyroxenes are comparable to those observed in silicates in Acapulcoites and in silicate inclusions of IAB iron meteorites (Fig. 15), which experienced intense heating and partial melting causing recrystallization and equilibration of originally chondritic silicates. The grain sizes of mafic minerals in equilibrated IDPs are smaller compared to olivines and

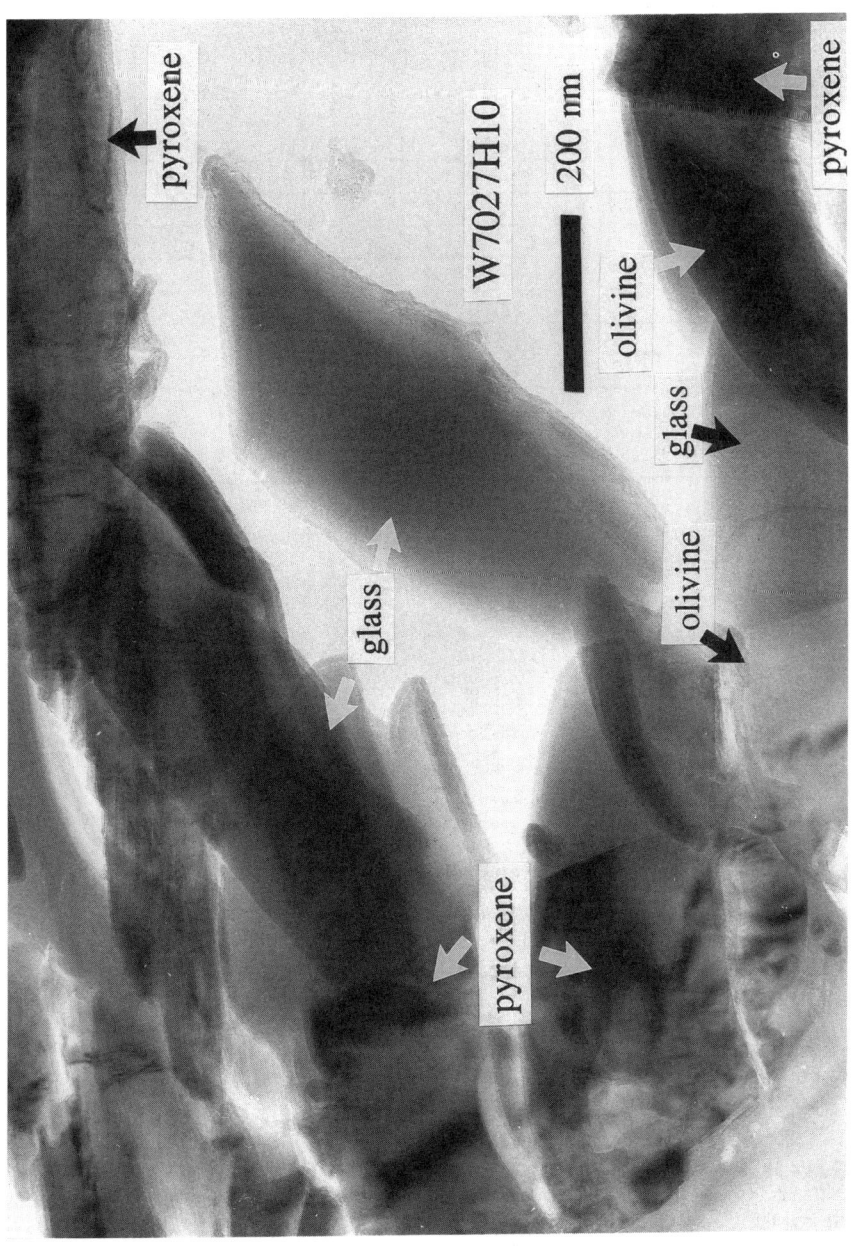

Figure 11. TEM bright-field image of particle W7027H10. This particle consists of olivines and pyroxenes embedded in a glassy matrix. Olivines and pyroxenes are equilibrated with respect to Fe and Mg; grain sizes of Mg-Fe silicates are too large to have permitted equilibration during atmospheric entry.

Table 1
Olivine and Pyroxene Compositions in Equilibrated Anhydrous IDPs

Number in diagram	Particle number	Average Fe/(Fe+Mg) in olivine	Number of analyses	Average Fe/(Fe+Mg) in pyroxene	Number of analyses
1	W7013E17	2.4 ± 0.4	11	2.2 ± 0.2	27
2	W7027H10	4.1 ± 0.9	4	3.8 ± 0.8	15
3	U2015D16	4.3 ± 2.2	6	4.2 ± 1.1	15
4	U2015C16	6.9 ± 0.5	8	5.6 ± 0.6	9
5	W7027E6	12.6 ± 3.2	7	11.2 ± 1.5	2
6	L2005D28	13.2 ± 2.5	20	12.3 ± 2.0	17
7	W7028*C2	15.2 ± 1.5	36	12.8 ± 1.8	18
8	W7013E9	18.1 ± 2.4	18	13.1 ± 2.0	8

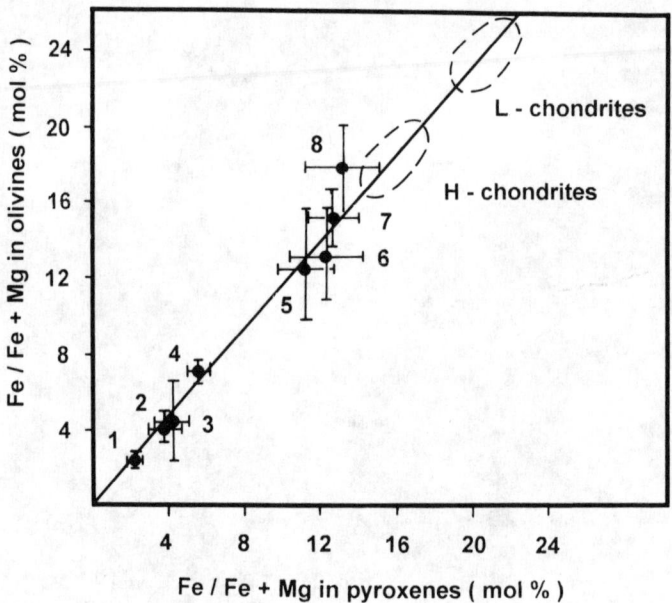

Figure 12. Plot of fayalite (Fe_2SiO_4) content of olivine vs. ferrosilite ($FeSiO_3$) content of low-Ca pyroxene in equilibrated anhydrous interplanetary dust particles. These IDPs are relatively coarse-grained, compact objects. Numbers in the diagram refer to those in Table 1, which contains a listing of fayalite and ferrosilite contents of olivines and pyroxenes of the equilibrated particles. Indicated, for comparison, are the fields of Mg-Fe silicates of equilibrated H-and L-chondrites. The diagonal line represents the experimental equilibrium distribution of Fe and Mg between olivines and orthopyroxenes.[54&55]

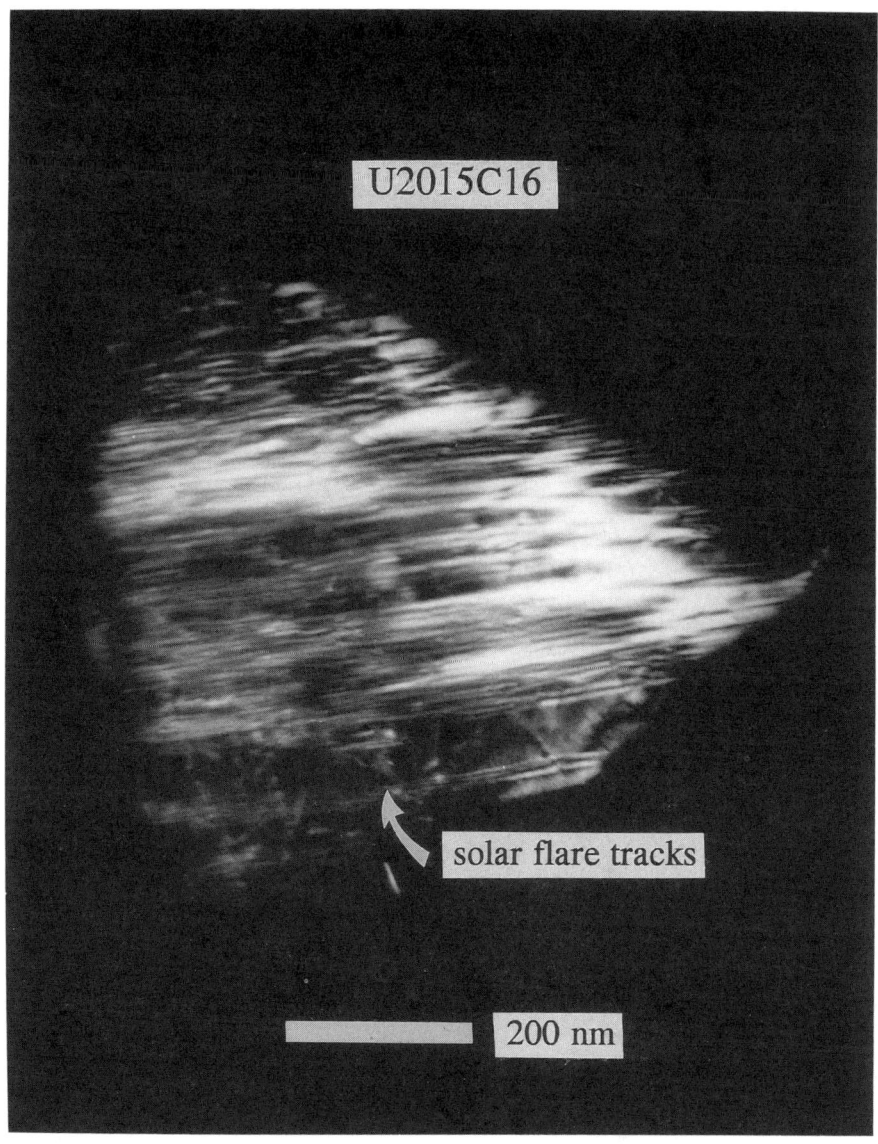

Figure 13. TEM dark-field photograph of pyroxene in IDP U2015C16. This particle is one of the "equilibrated" particles in Figure 14. The presence of solar flare tracks in a pyroxene supports the view that the silicates were not equilibrated during atmospheric entry. Solar flare tracks would have been annealed away if Mg-Fe silicates in this particle were heated and equilibrated during atmospheric entry.

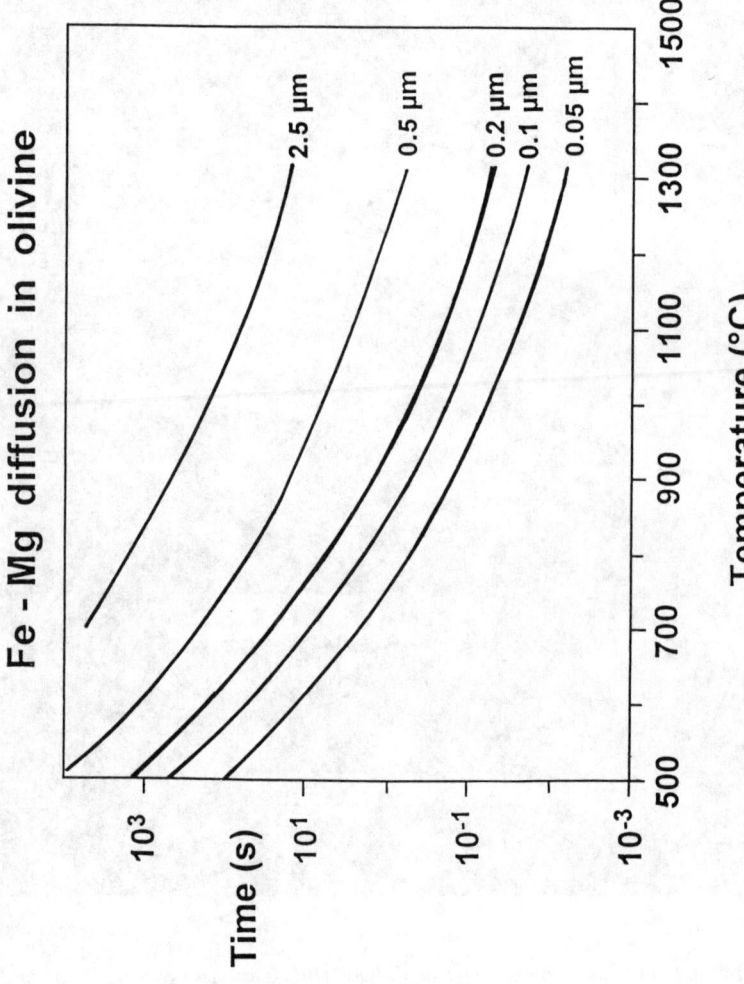

Figure 14. Calculation of diffusion distances in olivine using the diffusion data of Buening and Buseck[53]. The graph shows that Fe and Mg interdiffusion is fast enough to equilibrate olivine grains smaller than 100 nm in diameter within 10 seconds at a temperature of 600°C. The diffusion distances were calculated assuming atmospheric oxygen fugacity. If, e.g. log p O_2 is however in the range of the Ni/NiO buffer the equilibration time increases by almost two orders of magnitude.

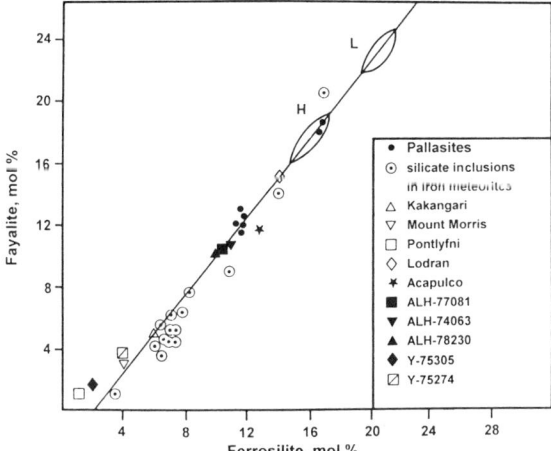

Figure 15. Plot of the fayalite (Fe_2SiO_4) content of olivine vs ferrosilite ($FeSiO_3$) content of low-Ca pyroxene in acapulcoites, pallasites and silicate inclusions of type IAB iron meteorites. Mineral compositions in terms of Fe and Mg of some reduced meteorites fall in the same range as equilibrated silicates in anhydrous stratospheric dust particles. These equilibrated IDPs could be related to the above mentioned meteorite classes. For comparison the compositions of silicates in the equilibrated H-and L-chondrites are also given.

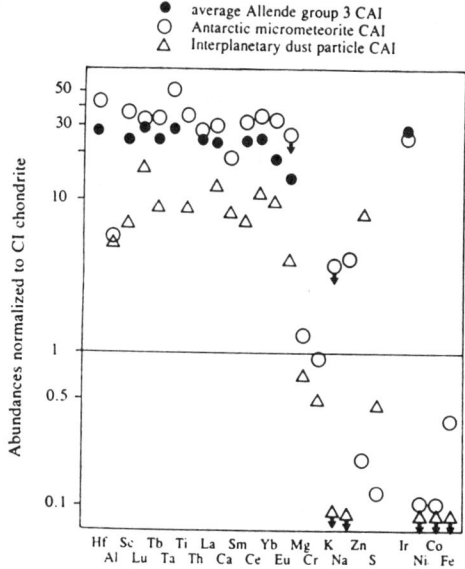

Figure 16. Major and trace elements of CAIs from different groups of extraterrestrial samples. The micrometeorite-CAI (BI 54 B3-31) and the stratospheric dust particle-CAI "Didius" display elemental enrichments and depletions similar to type 3 Allende CAIs. "Didius" was analyzed by F. Stadermann by ion microprobe and its extraterrestrial nature was confirmed by oxygen isotopic measurements.

pyroxenes in chondrules of CM2, CR or C3-chondrites. In addition, olivines and pyroxenes in chondrules of CO3 chondrites are not equilibrated with respect to Fe and Mg.[56] Olivines and pyroxenes in a few CV3 chondrites are almost equilibrated, but variations of Fa and Fs are large compared to those for the equilibrated IDPs.[57]

CAIs AMONG IDPs AND MICROMETEORITES

Most interplanetary dust particles studied are roughly chondritic in composition, but other types have been discovered. Christoffersen and Buseck[58] first reported the presence of refractory minerals in chondritic IDPs. Subsequently, Zolensky[59] reported the identification of entirely refractory IDPs collected in the stratosphere, as verified by the oxygen isotopic work of McKeegan.[60] Stadermann[43] also characterized a suite of non-chondritic stratospheric dust particles by their oxygen isotopic composition and by trace element abundances. Fig. 16 is a comparison of trace element contents of CAIs from three different materials. "Didius" (u47-m1-1a) is a refractory IDP analyzed by F. Stadermann, the Antarctic CAI-like micrometeorite was analyzed by D. Lindstrom.[38] Their refractory and siderophile trace element levels are compared to an average Allende group 3 CAI trace element pattern.[61] Common to all three objects is the enrichment of refractory elements: Didius and the Antarctic CAI-like grain are enriched by factors of 10 and 30, respectively. Non-refractory siderophile elements are strongly depleted in all CAIs compared to CI abundances. The slight negative Eu anomaly in Didius, and possibly also for the Antarctic CAI-like grain, suggests that these particles are related to group 3 CAIs.[61] Figure 17 is a backscattered electron image of the Antarctic micrometeorite CAI BI 54 B3-31, with submicron-sized individual mineral grains. Material from this particle was extracted for TEM studies, but so far only Al-rich clinopyroxenes (Fig. 18) and a paucity of Al-rich glass have been identified.

These examples illustrate that refractory components, primarily known from carbonaceous chondrites, are also present among micrometeorites and among stratospheric dust particles, indicating that at least some IDPs and micrometeorites are directly related to primitive carbonaceous chondrites and their asteroidal parent objects, although the refractory IDPs are generally much finer grained than any CAIs found in meteorites.[59]

MICROMETEORITES WITH RELATIONSHIPS TO IDPs AND PRIMITIVE AQUEOUSLY ALTERED CHONDRITES

About 40% of all stratospheric dust particles consist mainly of hydrated silicates.[62] They experienced an episode (or episodes) of aqueous alteration similar to carbonaceous chondrites and some unequilibrated ordinary chondrites. Up to the present time, however, only very few IDPs have been found showing direct links to carbonaceous chondrites.[25-27] The majority of hydrated IDPs consist of smectite; serpentine dominated particles are rare. This behavior is in contrast to the primitive chondrites, where serpentine is present in minor to major quantities in all but CV and unequilibrated ordinary chondrites (UOCs). The finding of a micrometeorite

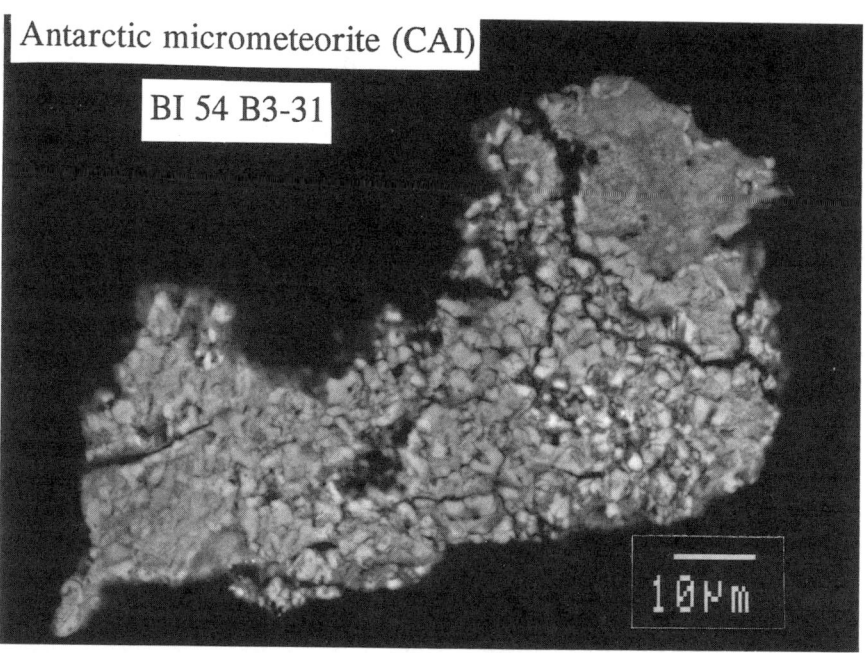

Figure 17. Backscatter electron image of Antarctic micrometeorite BI 54 B3-31. This particle is chemically identical to CAIs and consists mainly of sub-micrometer sized high-Al clinopyroxenes with small amounts of silica-rich glass. This micrometeorite was collected by M. Maurette in Antarctica and analyzed by D. Lindstrom by INAA for trace elements.

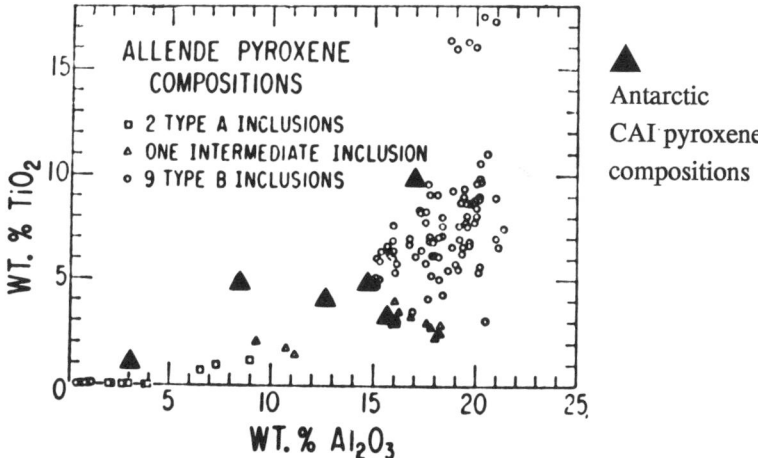

Figure 18. Compositions of Ca-rich pyroxenes of Antarctic CAI particle (BI 54 B3-31) plot in the field of type A and intermediate inclusions. Figure taken from paper by L. Grossman.[63]

from Antarctica, consisting entirely of smectite and framboidal magnetite (Fig.19), supports the existence of a sizable class of extraterrestrial material not present in that form in known meteorites. Textures of several smectite IDPs and the texture of the Antarctic micrometeorite BI 91/3-108 are almost identical to those found in the matrix of the Semarkona UOC. Rounded Fe-sulfides are embedded in a matrix of small smectite crystallites (Figs. 20-22). Compositions of smectites in these IDPs, in the Antarctic micrometeorite, and in the Semarkona matrix cover almost the same range of Fe/Mg ratios (Fig. 23). Smectites in the CV meteorites Kaba and Mokoia are more Mg-rich. Antarctic micrometeorite BI 91/3-108 contains framboidal magnetite as found also in CI, CV and CR-chondrites[64&65]. CI and CR meteorites, however, have abundant serpentine in addition to smectite. Serpentine was not found in the Antarctic micrometeorite. Though CI-like magnetite clusters were so far not reported from the Semarkona matrix, Semarkona is known to contain iron oxides.[66&68]

Therefore, it seems possible that smectite IDPs and the Antarctic micrometeorite 91/3-108 are fragments from the Semarkona parent body. It is, however, equally possible that these extraterrestrial dust particles are pieces of a hydrous asteroidal object which has not yet provided actual meteorites. The latter possibility is supported by the oxygen isotopic composition of IDPs, which is *not* similar to UOCs (see the paper in this volume by Walker[68]). The above mentioned Antarctic micrometeorite BI 91/3-108 is the least heated micrometeorite found so far among the unmelted micrometeorites from Antarctica. Most of the unmelted micrometeorites studied till now were strongly heated during atmospheric entry. During the thermal pulses the originally present layer-silicates were transformed into mineral assemblages of olivines, pyroxenes, magnetite and amorphous materials (Fig. 24). Compositions of these olivines in a number of particles are observed to range from Fa_{20} to Fa_{70} (Fig. 25). In some cases olivines and iron oxides in IDPs (magnetite or maghemite) contain measurable amounts of Ni, indicating possible oxidation of Ni-bearing Fe-sulfides during heating. This process is also known from metamorphosed CK chondrites.[69]

Flash-heating experiments using Orgueil (CI) layer-silicates demonstrated that hydrated phases are converted into assemblages of olivines, pyroxenes, magnetite and glass at temperatures and timescales applicable to the atmospheric entry heating process of micrometeorites.[70] Olivines in the 20-100 nm size range form from Orgueil serpentine/smectite mixtures at 1200°C within 20 seconds (Fig. 26). Fe/Mg ratios of these newly formed olivines are in the same range as Fe/Mg ratios exhibited by unmelted but strongly heated Antarctic micrometeorites (Fig. 27).

Approximately 40-70% of Antarctic unmelted micrometeorites in the size range smaller than 100 μm are fine-grained unmelted particles.[71] Their mineralogical composition indicates that the present mineralogy formed by atmospheric entry heating of originally hydrated phases. The percentage of originally hydrated micrometeorites is comparable to the percentage of hydrated particles among IDPs collected in the stratosphere[72]

Some believe IDPs and micrometeorites to be representative samples of all minor planets in the asteroid belt.[73] Assuming that this is the case, and that particles

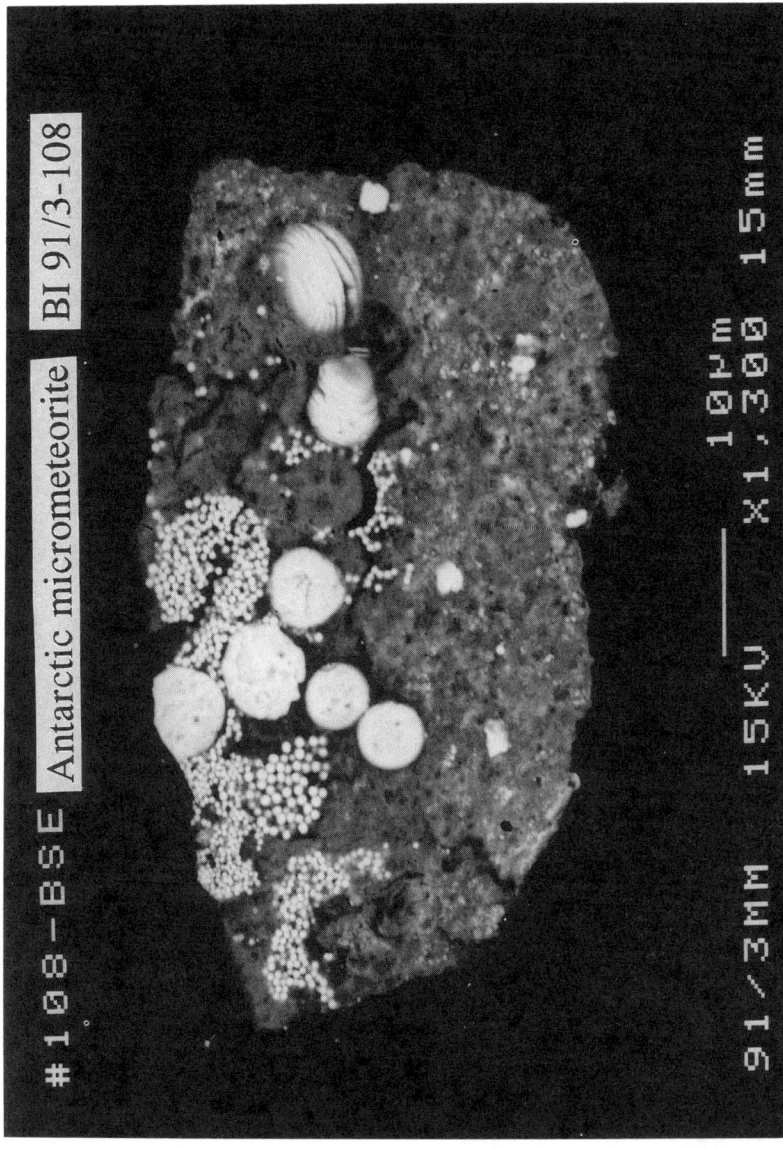

Figure 19. Backscatter electron image of micrometeorite BI 91/3-108. This photograph was taken at the Naturhistorische Museum in Vienna and Prof. Gero Kurat kindly put it at our disposal. The micrometeorite contains framboidal magnetite with morphologies reminiscent of CI-meteorite magnetite (although this is also found in other carbobaceous chondrites), and was therefore described as a CI-like micrometeorite. The gray, fine-grained areas are phyllosilicates.

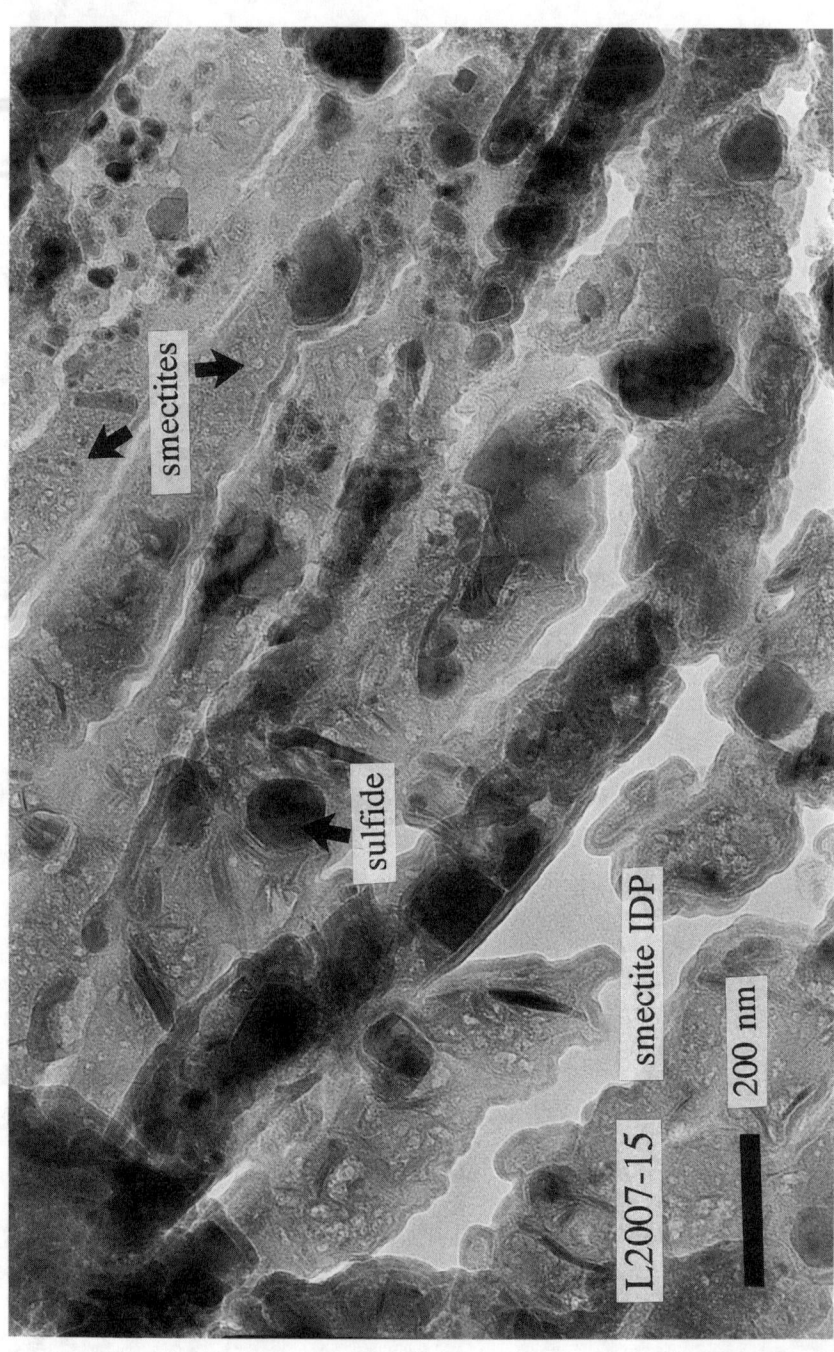

Figure 20. TEM bright-field image of smectite IDP L2007-15. This particle consists mainly of smectites and small Fe-sulfides embedded in the smectite. Anhydrous silicates, rare in this particle, consist mainly of diopside.[15]

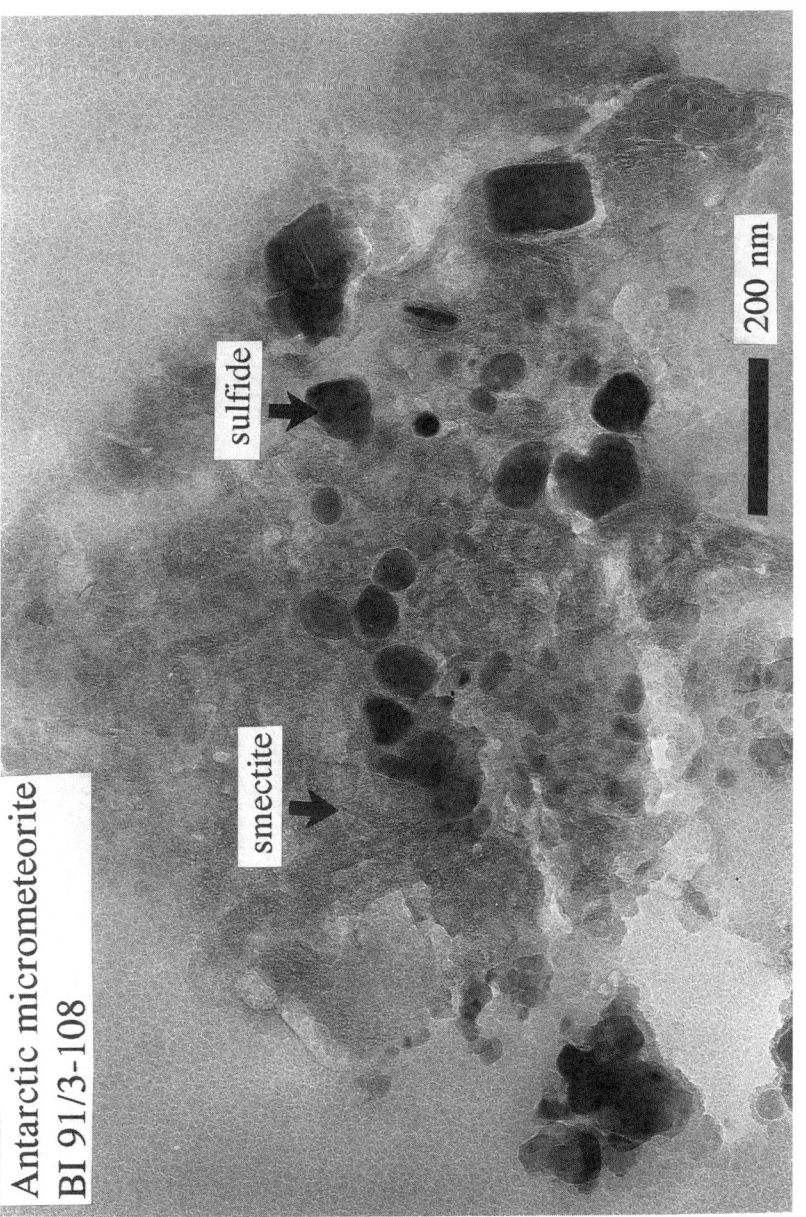

Figure 21. TEM bright-field image of micrometeorite BI 91/3-108. This particle consists entirely of smectites and small Fe-sulfides embedded in the smectite matrix. It is texturally similar to the smectite IDP shown in Figure 20.

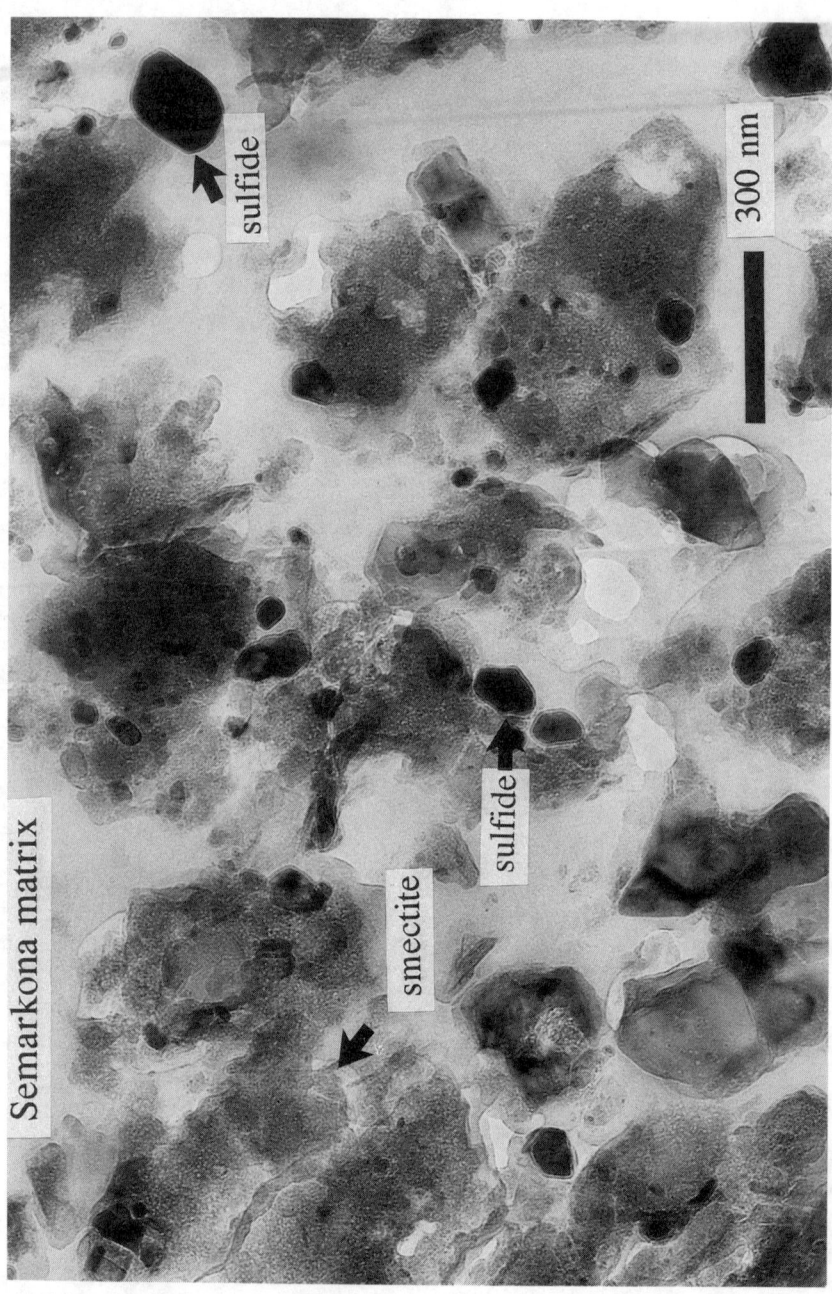

Figure 22. TEM bright-field image of Semarkona matrix material. The hydrous phases are smectites and they contain abundant tiny Fe-sulfides. The textural appearance very much resembles smectite-type IDP L2007-I5 and the smectite-micrometeorite BI 91/3-108.

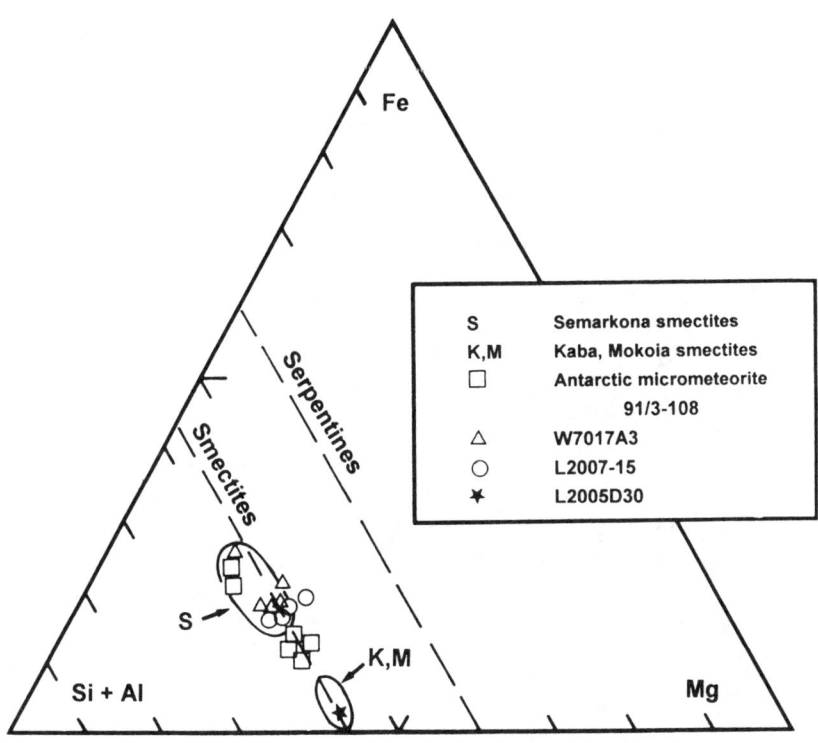

Figure 23. Compositional comparison of layer-silicates in meteoritic materials. Smectites in IDP L2007-15, in the smectite-micrometeorite (BI 91/3-108) and in Semarkona are compositionally very similar. Smectites in CV meteorites Kaba and Mokoia, and in one other IDP (L2005D30) are more Mg-rich. These data support a relationship between smectite IDPs, smectite-micrometeorites and the Semarkona parent object. Data of Kaba and Mokoia smectites are from Keller and Buseck[65] and Tomeoka and Buseck.[74]

Figure 24. TEM dark-field image of a typical Antarctic micrometeorite (BI 54 B1#5). They consist of fine-grained olivines, pyroxenes and magnetite embedded in a glassy matrix. Grain sizes of olivines are somewhat variable and range from less than 50 nm up to several hundred nm.

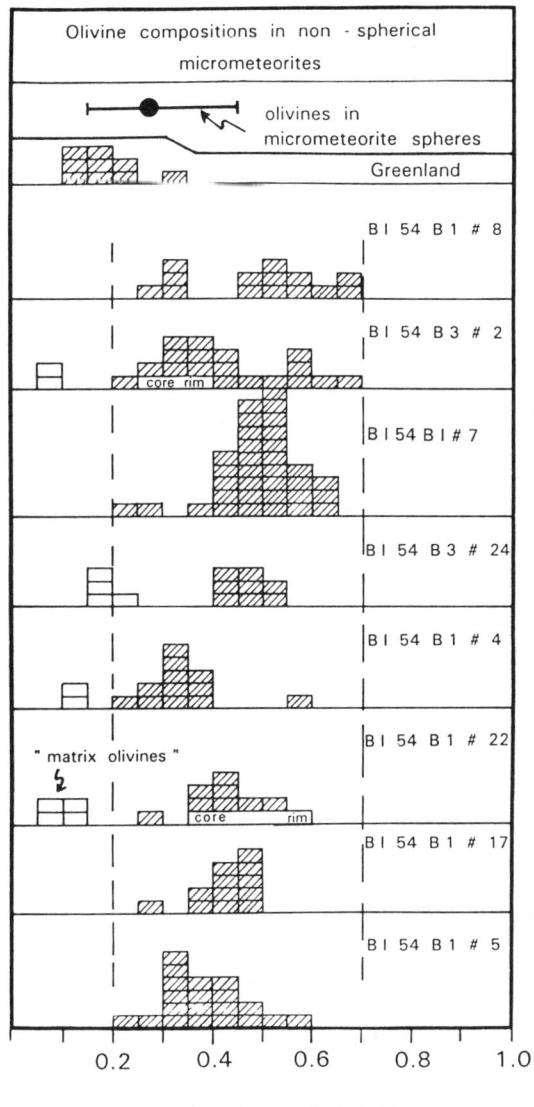

Figure 25. Histogram of the Fa contents of the olivines in 8 fine-grained, unmelted micrometeorites from Antarctica. Compositions of small euhedral olivines are shown as hatched boxes. Olivines embedded in glass (possibly in small melt pockets) and texturally different from the majority of olivines are indicated as unshaded boxes and are labeled matrix olivines. In two cases μm-sized olivines were found and the compositions of their cores and rims are indicated. For comparison, the range of compositions of olivines (excluding relict olivines) in melted micrometeorites are indicated and compositions of olivines in two unmelted Greenland micrometeorites are shown. Except for the "matrix olivines" the range of olivine compositions in the unmelted Antarctic particles is between Fa_{20} and Fa_{67}. The olivines in the Greenland particles are less iron-rich. Olivines in the melted micrometeorites are generally lower in iron than olivines in the unmelted Antarctic micrometeorites.

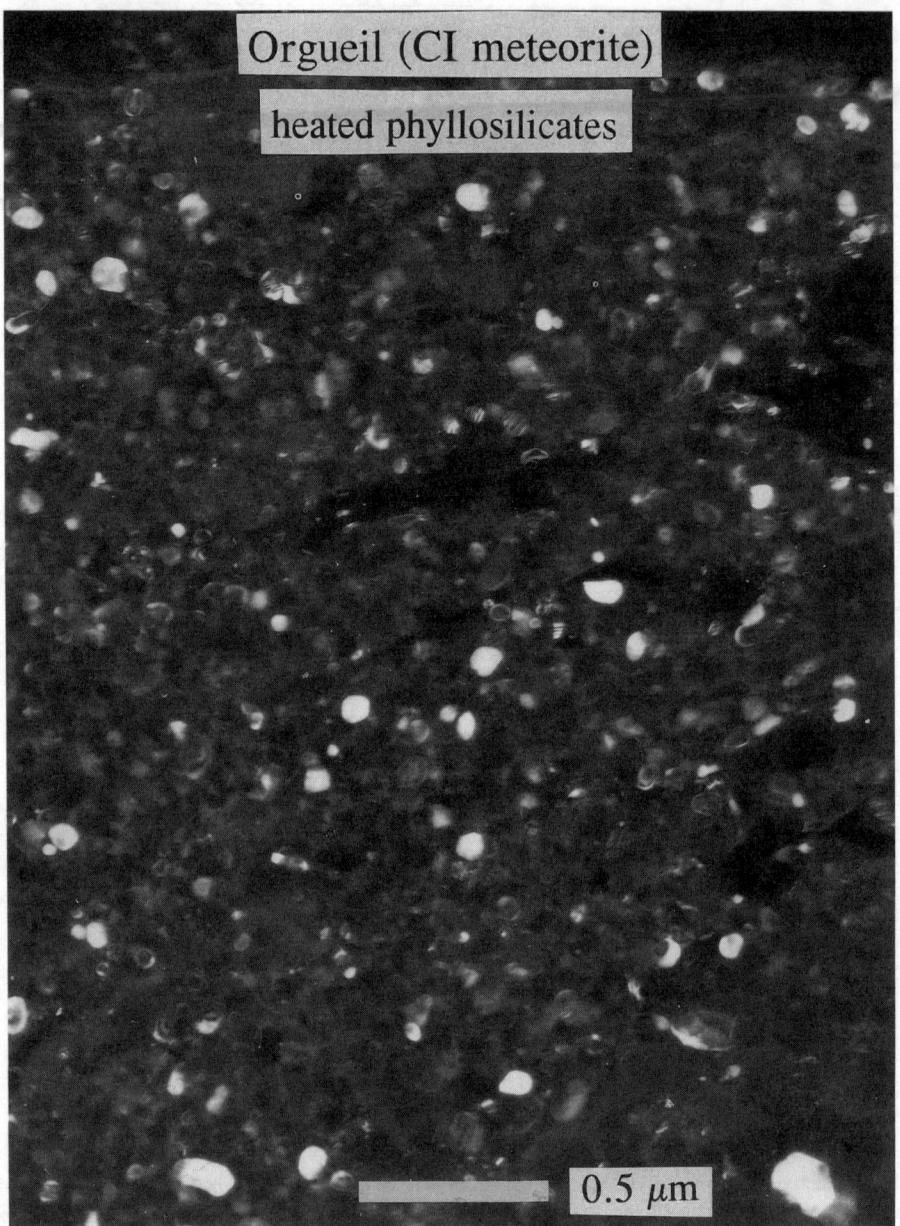

Figure 26. TEM dark-field image of heated Orgueil phyllosilicates. The serpentine/smectite mixture is no longer present because phyllosilicates are transformed within a few seconds to an mineral assemblage of olivines, pyroxenes, magnetite and glass. Experimental conditions were 1200°C and 20 seconds heating time, but we have noted that olivines of somewhat smaller size had already formed at 1000°C within 20 seconds. Grain sizes are on the order of 100 nm, and the texture of the heated sample greatly resembles that of unmelted micrometeorites.

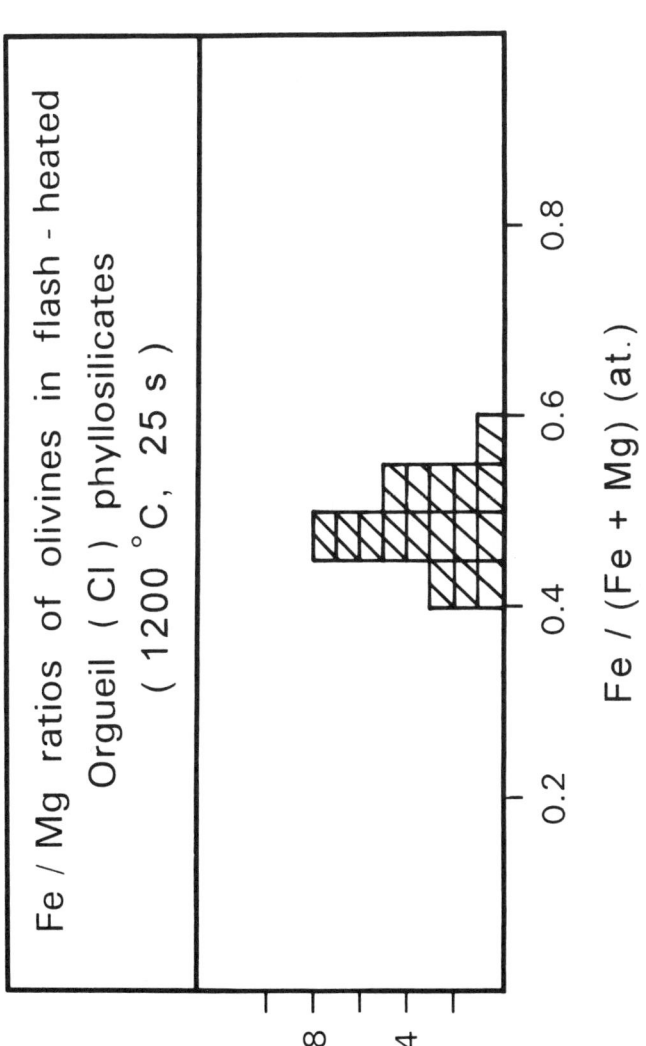

Figure 27. Histogram of Fe/(Fe+Mg) ratios of olivines formed by heating of Orgueil phyllosilicates. The Fa contents of the newly formed olivines are in the same range as Fa contents of olivines in unmelted micrometeorites. The larger spread of Fa contents of heated micrometeorite olivines could be caused by variable composition of the original phyllosilicates.

do not originate mainly from collisions between a small number of asteroids[75], then a major portion of asteroids are mineralogically similar but not identical to the hydrated carbonaceous and unequilibrated ordinary chondrites. These hydrous asteroids are therefore not well-represented among the known meteorite types, and are best sampled by micrometeorites and interplanetary dust particles. The high proportion of hydrous micrometeorite parent objects is compatible with the results of remote sensing studies of asteroids which indicate that low albedo asteroids (carbonaceous-chondrite-like bodies) and high albedo asteroids (stony iron meteorites, ordinary chondrites, achondrites) are present in the asteroid belt in roughly equal proportions.[76&77]

REFERENCES:

1. Brownlee D.E. (1985) *Ann. Rev. Earth Planet. Sci.* **13**,147-173.
2. Maurette M., Hammer C., Brownlee D.E., Reeh N. and Thomsen H.H. (1986) *Science* **233**,869-872.
3. Maurette M., Olinger C., Christophe Michel-Levy M., Kurat G., Pourchet M., Brandstätter F. and Bourot-Denise M. (1991) *Nature* **351**, 44-47.
4. Maurette M., Immel G., Hammer C., Harvey R., Kurat G. and Taylor S. (1994) Collection and curation of IDPs from the Greenland and Antarctic ice sheets. This volume.
5. Millard H.T. and Finkelman R.B. (1970) *J. Geophys. Res.* **75**, 2125-2134.
6. Brownlee D.E. (1981) Extraterrestrial components. In: *"The Sea" Vol.7*, ed. C. Emiliani, Wiley, New York.
7. Mackinnon I.D.R. and Rietmeijer F.J.M. (1987) *Reviews of Geophysics* **25**, 1527-1553.
8. Sandford S.A. and Walker R.M. (1985) *Astrophys. J.* **291**,838-851.
9. Bradley J.P. (1988) *Geochim. Cosmochim. Acta* **52**, 889-900.
10. Germani M.S., Bradley J.P. and Brownlee D.E. (1990) *Earth Planet. Sci. Letters* **101**, 162-179.
11. Christoffersen R. and Buseck P.R. (1986) *Earth and Planetary Science Letters* **78**, 53-66.
12. Klöck W., Thomas K.L., McKay D.S. and Zolensky M.E. (1990) *Lunar Planet. Sci. Conf. XXI*, 637-638.
13. Klöck W. (1992) *Proc. 50th Annual Meeting Electron Microscopy Society of America*, 1718-1719.
14. Zolensky M. and Barrett R. (1993) *Microbeam Analysis* **2**, 191-197.
15. Zolensky M. and Barrett R. (1994) Olivine and pyroxene compositions of Chondritic Interplanetary Dust Particles. This volume.
16. Bradley J.P., Humecki H.J. and Germani M.S. (1992) *The Astrophys. J.* **394**, 643-651.
17. Rietmeijer F.J.M. (1989) *Proc. 19th Lunar Planet. Sci. Conf.*, 513-521.
18. Rietmeijer F.J.M. (1993) *Earth and Planetary Science Letters* **117**, 609-617.
19. Rietmeijer F.J.M. (1994) On the possibility of petrological classification of carbonaceous chondritic micrometeorites. This volume.

20. Bradley J. (1994) Mechanisms of grain formation, post-accretional Alteration, and likely parent body environments of Interplanetary Dust Particles. This volume.
21. Thomas K.L., Blanford G.E., Keller L.P., Klöck W. and McKay D.S. (1993) *Geochim.Cosmochim. Acta* **57**, 1551-1566.
22. Thomas K.L., Keller L.P., Blanford G. and McKay D.S. (1994) Quantitative analyses of carbon in anhydrous and hydrated Interplanetary Dust Particles. This volume.
23. Keller L.P., Thomas K.L. and McKay D.S. (1994) Carbon in primitive Interplanetary Dust Particles. This volume.
24. Keller L.P., Thomas K.L. and McKay D.S. (1993) *Lunar Planet. Sci. Conf.* *XXIV*, 785-786.
25. Bradley J.P. and Brownlee D.E. (1991) *Science* **251**, 549-552.
26. Keller L.P., Thomas K.L. and McKay D.S. (1992) *Geochim. Cosmochim. Acta* **56**, 1409-1412.
27. Rietmeijer F.J.M. (1992) *Lunar Planet. Sci. Conf. XXIII*, 1153-1154.
28. Klöck W., Thomas K.L., McKay D.S. and Palme H. (1989) *Nature* **339**, 126-128.
29. Love S.G. and Brownlee D.E. (1991) *Icarus* **89**, 26-43.
30. Flynn G.J. (1989) *Icarus* **77**, 287-310.
31. Flynn G.J. (1994) Cometary dust: A thermal criterion to identify cometary samples among the Interplanetary Dust collected from the stratosphere. This volume.
32. Steele I.M., Smith J.V. and Brownlee D.E. (1985) *Nature* **313**, 297-299.
33. Beckerling W. and Bischoff A. (1994) Planetary and Space Science (submitted).
34. Kurat G., Presper T. and Brandstätter F. (1992) *Lunar Planet. Sci. Conf. XXIII*, 747-748.
35. Flynn G.J., Sutton S.R. and Klöck W. (1993) *Proc. NIPR Symposium on Antarctic Meteorites No .6*, 304-324.
36. Alexander C.M.O'D., Maurette M., Swan P. and Walker R.M. (1992) *Lunar Planet Sci. Conf. XXIII*, 7-8.
37. Koeberl C., Kurat G., Presper T., Brandstätter F.,and Maurette M. (1992) *Lunar Planet. Sci. Conf. XXIII*,709-710.
38. Lindstrom D.J. and Klöck W. (1992) *Meteoritics* **27**, 250.
39. Bradley J.P. and Brownlee D.E. (1986) *Science* **231**, 1542-1544.
40. Rietmeijer F.J.M. (1992) *Trends in Mineralogy* **1**, 23-41.
41. Flynn G.J., Sutton S.R., Bajt S., Klöck W., Thomas K.L., Keller L.P. (1993) *Meteoritics* **28**, 349-350.
42. Klöck W., Flynn G.J., Sutton S.R. and Nier A.O. (1992) *Meteoritics* **27**, 243-244.
43. Stadermann F.J. (1991) *Lunar Planet. Sci. Conf.* XXII, 1311-1312
44. *LDEF Newsletter Vol. 4*, No.3, June 1993, p. 12.
45. Fraundorf P., Lyons T. and Schubert P. (1982) *Proc. 13th Lunar Planet. Sci. Conf., J. Geophys. Res.* **87**, A409-A412.

46. Jessberger E.K.,Bohsung J.,Chakaveh Sepideh and Traxel K. (1992) *Earth Planet. Sci. Letters* **112**, 91-99.
47. Lawler M.E.,Brownlee D.E.,Temple S. and Wheelock M.M. (1989) *Icarus* **80**, 225-242.
48. Fomenkova M.N., Kerridge J.F., Marti K. and McFadden L.-A. (1992) *Science* **258**, 266-269.
49. Langevin Y., Kissel J., Bertaux J-L. and Chassefiere E. (1987) *Astron. Astrophys.* **187**,761-766.
51. Thomas K.L. et al., (1993) *Meteoritics* **28**, 448-449.
52. Rietmeijer F.J.M. and Blanford G.E. (1988) *J. Geophys. Res.* **93**, B11943-B11948.
53. Buening D.K. and Buseck P.R. (1973) J. Geophys. Res. **78**, 6852-6862.
54. Larimer J.W. (1968) *Geochimica et Cosmochimica Acta* **32**, 1187-1207.
55. Mori T. and Green D.H. (1975) *Earth and Planetary Science Letters* **26**, 277-286.
56. McSween H.Y. Jr. (1977) Geochimica et Cosmochimica Acta **41**, 477-491.
57. McSween H.Y.Jr (1977) Geochimica et Cosmochimica Acta **41**, 1777-1790.
58. Christoffersen R. and Buseck P.R. (1986) *Science* **234**, 590-592.
59. Zolensky M.E. (1987) *Science* **237**,1466-1468.
60. McKeegan K.D. (1987) *Science* **237**, 1468-1470.
61. Kornacki A.S. and Fegley B. Jr. (1986) *Earth Planet.Sci.Letters* **79**, 217-234.
62. Keller L.P. and Thomas K.L. (1991) *Lunar and Planet. Sci. Conf. XXII*, 705-706.
63. Grossman L. (1975) *Geochimica et Cosmochimica Acta* **39**, 433-451.
64. Jedwab J. (1971) *Icarus* **15**, 319-340.
65. Keller L.P. and Buseck P.R. (1990) *Geochimica et Cosmochimica Acta* **54**, 2113-2120.
66. Huss G.R., Keil K., Taylor G.J. (1981) *Geochim. et Cosmochim. Acta* **45**, 33-51.
67. Hutchinson R., Alexander C.M.O., Barber J. (1987) *Geochimica. et Cosmochimica Acta* **51**, 1875-1882.
68. Walker R. (1994) Isotopic constraints on interstellar material in chondritic IDPs. This volume.
69. Geiger T., Bischoff A., Spettel B., Bevan A.W.R. (1992) *Lunar and Planetary Science Conference XXIII*, 401-402.
70. Klöck W. et al., (1994) *Lunar and Planetary Science Conference XXV*.
71. Presper T. *Ph. D. Thesis, Johannes Gutenberg-Universität, Mainz,* 1993.
72. Schramm L.S., Brownlee D.E., Wheelock M.M. (1989) *Meteoritics* **25**, 99-112.
73. Zook H.A. and McKay D.S. (1986) *Lunar and Planetary Science Conf. XVII*, 977-978.
74. Tomeoka K. and Buseck P.R. (1990) *Geochimica et Cosmochimica Acta* **54**, 1745-1754.
75. Flynn G.J. (1993) *Meteoritics* **28**, 349.
76. Gaffey M.J., Bell J.F. and Cruikshank D.P. (1989) in: *Asteroids II*, (eds.: R.P. Binzel, T. Gehrels and M.S. Matthews), The University of Arizona Press, Tucson, p. 98-127.

77. Tholen D.J. and Barucci M.A. (1989) in: *Asteroids II*, (eds.: R.P.Binzel, T. Gehrels and M.S. Matthews), The University of Arizona Press, Tucson, p. 298-315.

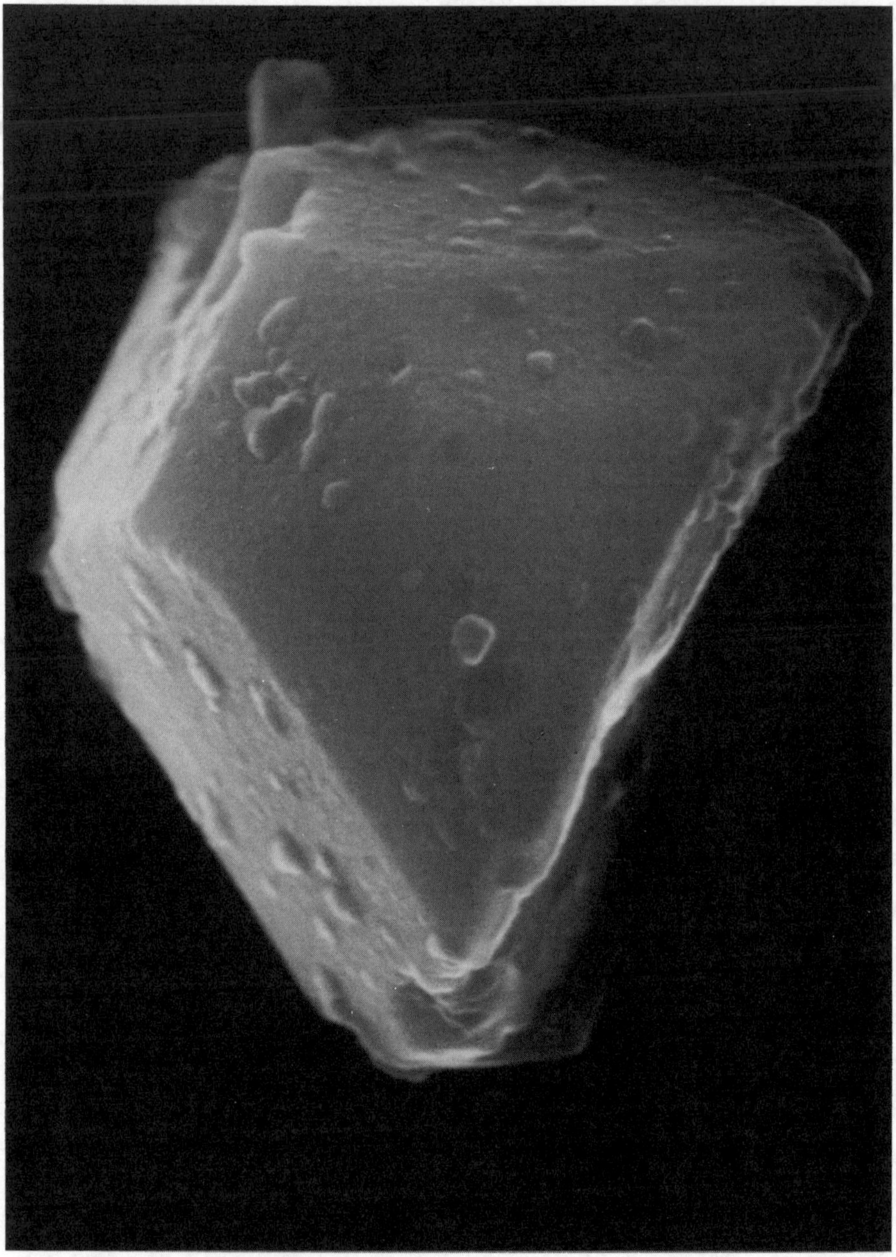

A secondary electron image of a stratospheric particle consisting of a single crystal with the approximate composition FeS (either troilite or pyrrhotite). The crystal measures 6 x 9 μm in size. (NASA photo S82-25727)

MECHANISMS OF GRAIN FORMATION, POST-ACCRETIONAL ALTERATION, AND LIKELY PARENT BODY ENVIRONMENTS OF INTERPLANETARY DUST PARTICLES (IDPS)

John Bradley
MVA Inc.
5500/200 Oakbrook Pkwy, Norcross, GA 30093, USA

ABSTRACT

The fine-scale mineralogy and petrography of the various classes of IDPs provide information about grain-forming reactions, as well as insight into the nature of IDP parent body environments. Many anhydrous, *pyroxene* IDPs are (solar flare) track-rich, have undergone negligible post-accretional alteration, and have preserved mineralogical evidence of primordial gas-phase reactions. Most anhydrous, *olivine* IDPs appear to have undergone post-accretional heating and, since only a few track-rich *olivine* IDPs have been identified, it is possible the heating occurred during atmospheric entry. Infrared (IR) transmission and visible (VIS) reflectance data from track-rich anhydrous IDPs suggest that some of them were derived from outer asteroids or comets. Hydrated, *layer-silicate* IDPs have undergone significant post-accretional aqueous alteration. Three *layer silicate* IDPs have now been linked directly to carbonaceous chondrite petrogenisis and a (main belt) asteroid origin, and it is possible that most of them are from hydrous asteroids. Several *layer silicate* IDPs exhibit (VIS) spectral reflectivities similar to main-belt C asteroids.

INTRODUCTION

One of the major reasons for studying 5-30 µm diameter interplanetary dust particles (IDPs) collected in the stratosphere is the expectation that they include unmelted samples of Solar System bodies (e.g. comets and outer asteroids) that may not be well represented among conventional meteorites.[1&2] Comets and outer asteroids are small bodies located at heliocentric distances where temperatures have probably not exceeded the water-ice solidus.[3] Samples of these bodies could provide evidence of grain formation and aggregation reactions that were important in the early Solar System. This evidence may be particularly well preserved in IDPs because unlike larger (> 50 µm) micrometeorites and conventional meteorites, some IDPs appear to have escaped significant post-accretional (parent body) processing and pulse heating during atmospheric entry. However, IDPs are extremely small objects with nanogram masses. Until comparatively recently, sample handling procedures and methods for analyzing IDPs were not well developed. Within the past few years, there have been major advances in IDP manipulation and analysis techniques. Several important innovations in small particle analysis have been developed around interplanetary dust research, with the result that inorganic (major and trace elements), organic, noble gas, mineralogical, and isotopic measurements of IDPs are now routinely performed by multiple laboratories.[4-11]

The most widely studied IDPs are those with approximately chondritic (solar) bulk compositions.[1] IDPs with non-chondritic compositions have also been recognized.[12] Although the highly porous, fragile structures of some chondritic IDPs are consistent with the expected properties of primitive meteoritic materials (Fig 1a), appearances can be deceiving. Detailed mineralogical examinations have shown that while some IDPs appear to have been collected in essentially pristine condition, others have apparently been altered by frictional heating during atmospheric entry.[7&13] Distinction between pristine and (thermally) altered IDPs is a prerequisite to investigations of indigenous mineralogy and petrography. Solar flare tracks and various mineralogical markers have been used to assess the degree of heating experienced by IDPs[14&15], and recent noble gas measurements suggest that it might be possible to determine peak atmospheric entry heating to within ±100°C.[16] Studies of the least heated IDPs have provided information about grain-formation and particle aggregation processes as well as the nature of the parent bodies from which some of them were derived.

CLASSIFICATION AND MINERALOGY OF CHONDRITIC IDPS

The first classification schemes for chondritic IDPs were based on the physical properties (e.g morphology, texture) and bulk chemistry.[17&18] More recently, particle mineralogy has been incorporated into IDP classification. It has been established, using infrared (IR) and electron microscope studies, that there are at least three mineralogical classes of chondritic IDPs. They are referred to as "*pyroxene*", "*olivine*", and "*layer silicate*" after their most abundant silicate minerals.[19] The *pyroxene* and *olivine* classes are usually porous and contain only anhydrous silicates (Fig. 1a), while the *layer silicate* class are usually compact and smooth and contain hydrous minerals (Fig. 1b). Most IDPs fall within this classification scheme, although IDPs with intermediate mineralogy have been reported. Examples include anhydrous IDPs with similar amounts of pyroxene and olivine[20], porous anhydrous IDPs containing minor amounts of hydrated layer silicates[21], and a smooth *layer silicate* silicate IDP containing a large pyroxene grain (50% of IDP volume).[7] It seems inevitable that existing classification schemes will require reassessment as more IDPs are examined and new data are accumulated.

"*Pyroxene*" IDPs appear to be the most mineralogically primitive class of chondritic IDPs. Because of their distinctive porous, fragile microstructures, they have been the objects of detailed investigations. They contain mineral crystals (e.g. *olivine* IDPs studied to date. Some *olivine* IDPs exhibit melt textures and equilibrated silicate mineralogy, and Fe-sulfides are commonly decorated with magnetite rims [see reference 20]. These observations suggest that this class of chondritic IDPs may contain many strongly heated particles and the absence of solar flare tracks leaves open the possibility that the heating is occurring during atmospheric entry. See also the paper in this volume by Flynn.[22] However, since tracks have been observed in two olivine-rich IDPs[23], the olivine class may include

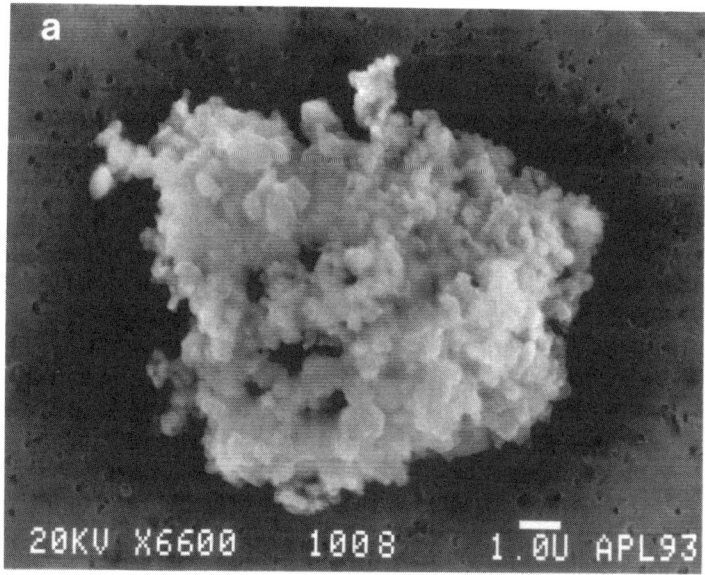

Figure 1. Secondary electron images of chondritic interplanetary dust particles. a) Porous, typically anhydrous IDP (W7035A5) belonging to the *pyroxene* class. (*Olivine* IDPs can be similar in appearance but tend to be more coarse-grained). b) Hydrated *layer silicate* IDP (W7030A15) with characteristically smooth surfaces and compact structure. (see also Figure 5).

IDPs more than one genetic group. Future noble gas measurements may help clarify the thermal histories of *olivine* IDPs [see Nier et al., this volume[24]].

WHAT EVIDENCE LINKS CHONDRITIC IDPS TO COMETS OR ASTEROIDS?

Although it is not possible to state with absolute certainty that some IDPs are derived from comets and others from asteroids, an accumulating body of evidence links certain types of chondritic IDPs to parent bodies similar to those of chondritic meteorites, while other IDPs are more likely to be from more primitive objects like outer asteroids and comets. This evidence comes from (a) observations of the microstructural properties of IDPs, (b) mineralogical and petrographic considerations; (c) comparison of the compositions of IDPs with comet Halley's dust, (d) studies of the optical (infrared and reflectance) properties of IDPs. (Additionally, recent noble gas measurements resulting in distinction between cometary and asteroidal IDPs are described by Nier (this volume[24])) (see also reference 16).

The microstructural evidence concerns the high porosities of some anhydrous IDPs (Fig. 1a). Observations of cometary meteors entering the atmosphere have shown that they are composed predominantly of porous, fragile materials.[25] The Draconid meteors from comet Giacobini-Zinner are an extreme example; they fragment in the atmosphere at ram pressures between 10^3 and 10^4 dynes cm^{-1}. (A 1 cm cube of Draconid material resting on a flat surface would collapse under its own weight). Since cometary IDPs are almost certainly among IDPs collected in the stratosphere, they should be recognizable by their high porosities and fragile microstructures. Highly porous, fragile particles, with densities as low as 0.5 g/cc, are found among IDPs collected in the stratosphere (e.g. Fig. 1a).[26] However, these high porosities are unlikely to be unique to comet dust because there are "comet like" outer asteroids that might also be a source of porous IDPs.[27&28]

Mineralogical and petrographic evidence for a main-belt asteroidal origin for at least some IDPs has been obtained from electron microscopic studies. Although most (chondritic) *layer silicate* IDPs bear a superficial resemblance to the fine-grained matrices of types CI, CM and CR (chondritic) meteorites in that they contain hydrated silicates, there are important mineralogical differences.[7,21,30,31] These differences have made it difficult to conclude that *layer silicate* IDPs are merely small carbonaceous chondrites derived from main belt (Type C) asteroids. Recently, three *layer silicate* IDPs have been linked directly to type CI and CM meteorite petrogenisis. One of these IDPs, RB12A44, contains abundant tochilinite, an ordered mixed-layer mineral found in the fine-grained matrices of type CM meteorites.[31] A second IDP (L2005T12), which is suspected to have been strongly heated during atmospheric entry, was also found to contain a tochilinite-like material.[13] A third IDP, W7013F5, was found to be composed mostly of coherent intergrowths of serpentine and saponite, which have only been previously observed in type CI and CR chondrites.[9&30] These findings establish direct petrographic links to specific

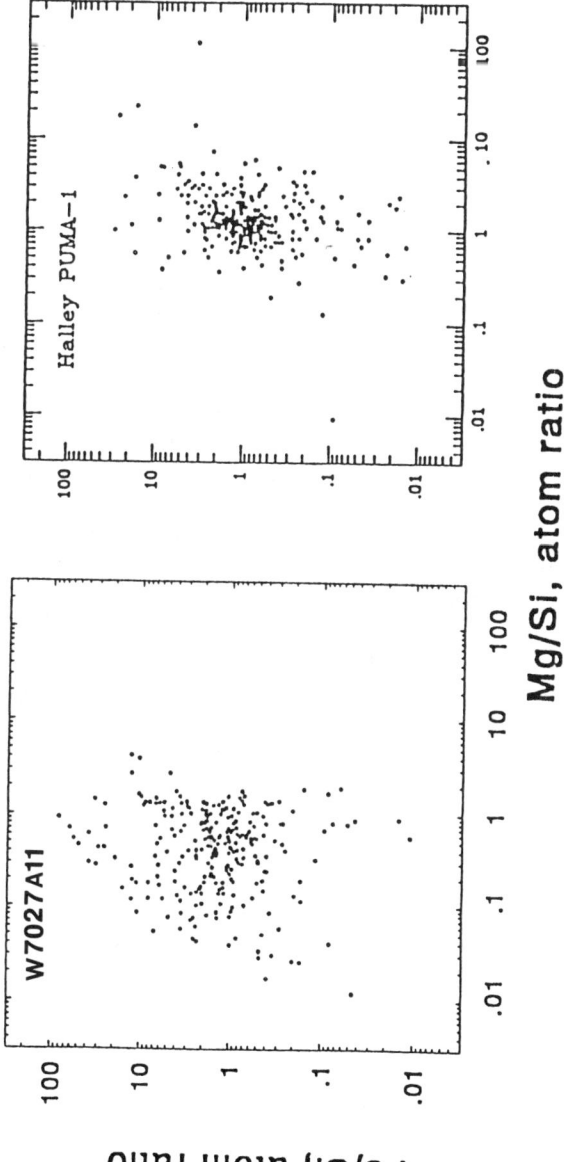

Figure 2. Iron-magnesium-silicon distributions in (glass-rich) anhydrous IDP W7027/A11 and comet Halley's dust. The Halley data, derived by Lawler et al.[32] from 433 selected PUMA spectra, have been corrected to peak at chondritic (solar) abundances.[33]

classes of chondritic meteorites, and establish that at least some *layer silicate* IDPs are derived from (main belt) asteroids.

Comet Halley's dust was analyzed by the PIA and PUMA time-of-flight mass spectrometers on the Giotto and Vega spacecraft. The bulk composition of the dust was found to be approximately chondritic[34], and it has been compared with the fine-grained matrices of chondritic meteorites and (chondritic) IDPs The degree of submicrometer scale heterogeneity in the meteorites and in most IDPs are incompatible with Halley's dust.[35] However, some IDPs are compositionally and mineralogically similar to Halley's dust. For example, Figure 2 shows Mg/Si versus Fe/Si plots for an IDP W7027A11 and Halley's dust. (Fe, Mg, and Si are selected because they major elements in IDPs and Halley's dust, their fine-scale dispersion depends strongly on the chemistry of individual mineral grains, and their measured abundance ratios are least likely to be skewed by uncertainties associated with quantitation of the Halley data [see Lawler et al.[32]). The IDP data are energy-dispersive X-ray analyses collected from submicrometer volumes (of electron transparent) thin sections of W7027A11, and the Halley data are mass spectral analyses of submicrometer dust particles detected by the PUMA 1 instrument. Both W7027A11 and Halley's dust exhibit a diffuse "shotgun" distribution of data points, in contrast to much coarser-grained meteorites and most IDPs whose data points are more tightly clustered. The lack of clustering in Figure 2 reflects extreme compositional and mineralogical heterogeneity on a submicrometer scale, which may be a characteristic of the most primitive (unprocessed) meteoritic materials. However, in-depth interpretation of the Halley PIA and PUMA data is complicated by uncertainties concerning instrument performance and telemetry.[32,34,36,37] See also the paper by Fomenkova and Chang in this volume.[38]

Transmission IR spectroscopy has been employed to classify IDPs and compare their spectral characteristics with those of astrophysical objects.[19] Figure 3 shows transmission infrared (IR) spectra, between 8 and 12 µm, obtained from thin sections of two chondritic IDPs U219C11 and W7027A11. The broad absorption feature, centered around 10 µm, is the Si-O stretching mode for silicates, and the detailed band structure provides information about silicate mineralogy in the IDP. Since, a similar "silicate feature" is observed in astronomical IR spectra of several comets [see Hanner in this volume[39]], IR spectroscopy provides a means of comparing the (silicate) mineralogy of IDPs and comets. Although most chondritic IDPs produce IR spectra that are incompatible with those of comets, there are IDPs whose silicate emission characteristics are similar to those of specific comets [Fig. 3]. It may be significant that these two IDPs are highly porous, fragile particles with mineralogy similar to Halley's dust.[33]

Reflectance spectroscopy has been widely employed to study the surfaces of asteroids and compare their spectral properties with those of meteorites [see Gaffey et al.[40]]. Based on their observed spectral characteristics, the asteroids have been classified into "famillies". Approximately 15 famillies have been catalogued, but the most abundant are the S, C, P, and D asteroids[41] (Fig. 4). S class asteroids dominate the inner main belt, C-class the outer main belt, and P and D asteroids dominate the outer belt. Comparison of reflectance spectra from meteorites and asteroids have

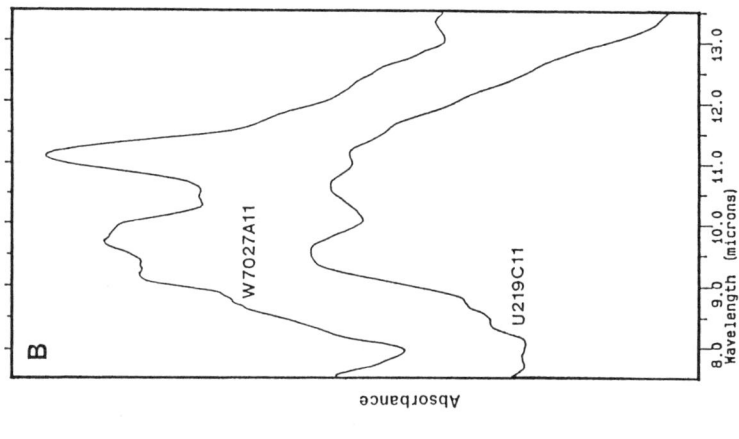
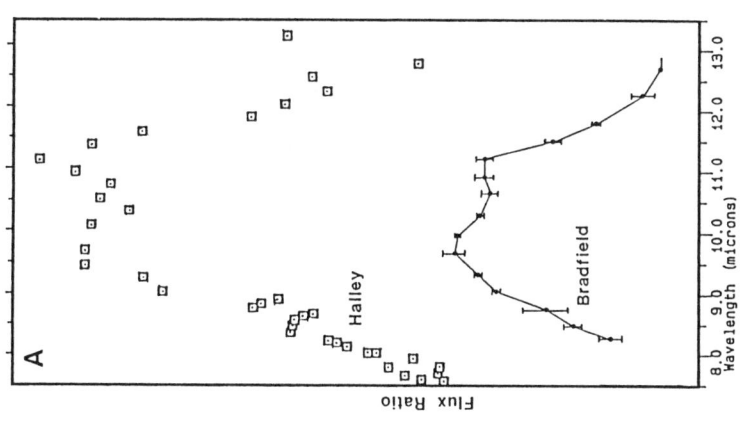

Figure 3. Silicate IR emission from comets and chondritic IDPs. (a) Comets Bradfield and Halley (data from Bregman et al.[42] and Hanner et al.[39]), (b) Glass-rich anhydrous IDPs U219C11 and W7027A11 (see also Fig. 2).

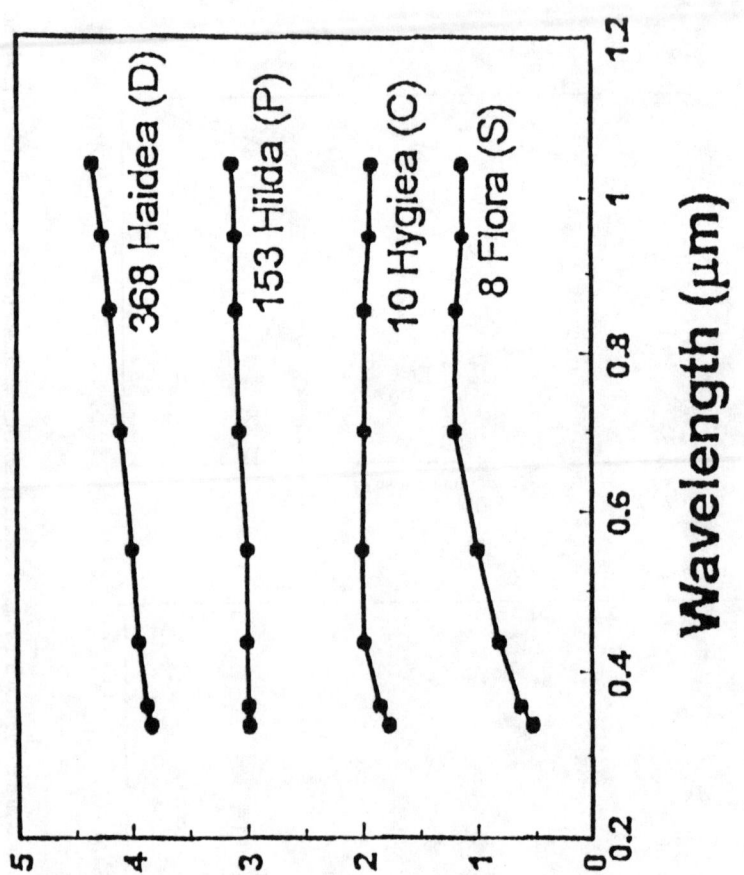

Figure 4. Typical reflectance spectra for S, C, P, and D type asteroids (data from Zellner et al.[43]).

shown that CI and CM meteorites are most likely derived from main belt C-type asteroids.[3,44,45] However, the ordinary chondrites (OC's), which account for >75% of meteorite falls, have no known spectral analogue(s) among the asteroids. Conversely, whole famillies of asteroids, including the outer P and D asteroids, have no spectral analogues among meteorites.

The most striking feature of the asteroid spectral classification scheme is that there is a trend in spectral properties with increasing heliocentric distance (Fig. 4). It has been suggested that this gradation may persist throughout the Solar System out to the dusty component of comets.[46] S and C-class asteroids have generally flat spectra within the visible wavelength range, but outer P and D asteroids exhibit a pronounced rise into the near-IR. This redness is believed to be due to the presence of organic material on the surfaces of the asteroids.[27,46,47] Samples of these spectrally "red" asteroids are of special interest because, like comets, they are likely to contain well preserved materials from the early Solar System.

If IDPs are indeed derived from a range of dust producing Solar System objects, it might be possible to determine their most likely heliocentric distance of origin from their reflectance characteristics. Recently, the first reflectance spectra have been collected, within the visible wavelength range 380-850 nanometers, from individual 3 - 20 µm IDPs. Preliminary results are shown in Figure 5. Several smooth *layer silicate* IDPs exhibit generally flat spectra longwards of 500 nm (e.g. Fig. 5a), similar to CI and CM meteorites and main belt C-asteroids. (The significance of the UV drop-off (Fig. 5a) is yet to be determined.) Some highly porous anhydrous IDPs exhibit a pronounced rise into the near-IR (e.g. Fig. 5b), the first meteoritic objects to display this "reddening" typical of outer P and D asteroids. Although these data are preliminary, they suggest that reflectance spectroscopy may be useful for determining the source regions of the various classes of chondritic IDPs.

EVIDENCE OF (PRIMORDIAL) GRAIN-FORMING REACTIONS IN IDPS.

Anhydrous IDPs have provided mineralogical data about grain-forming reactions that were involved in their formation and accretion. Gas-phase reactions are a recurring theme among these data, and gas-to-solid condensation is the fundamental grain-forming mechanism within the astrophysical environment. The most distinctive and easily recognized crystals in anhydrous "pyroxene" IDPs, enstatite ($MgSiO_3$) whiskers and platelets (Fig. 6), were probably formed by high-temperature (>1000°C) condensation in a relatively low pressure nebular gas.[48] Some whiskers contain screw dislocations resulting from spiral growth. Such whiskers have not been found in any other class of meteoritic materials. In a study of FeNi grains and carbonaceous material, a low temperature carbide (ε-$[FeNi]_3C$) was identified in several IDPs.[49] ε-Fe carbide has been synthesized in the laboratory by Fischer-Tropsch type (FTT) reactions between finely-divided metal grains and carbon-containing gases (e.g. CO and CH_4). The presence of ε-carbide in IDPs suggests that some of the carbonaceous material was emplaced by low-temperature (~200°C) catalytic reactions. A study of the crystal chemistry of Mg-rich olivines and

98 Mechanisms of Grain Formation

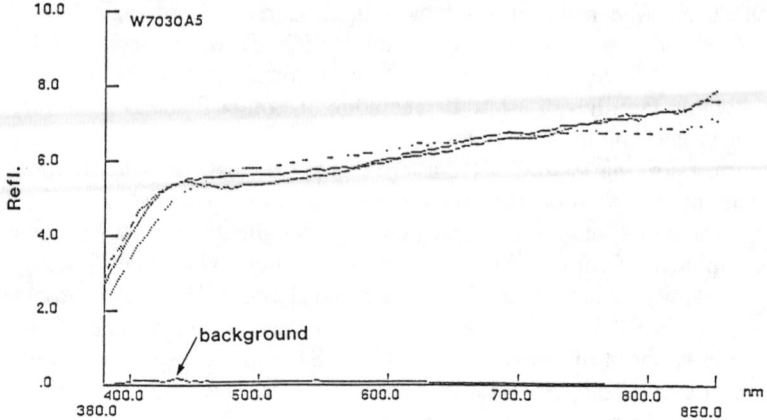

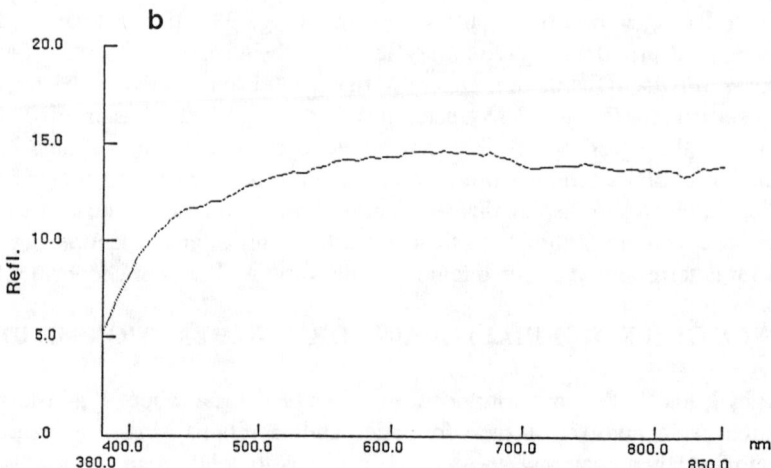

Figure 5. Reflectance spectra collected from two chondritic IDPs (shown in Figure 1) using a microscope photometer. (a) W7035A5 - three spectra collected under different sets of focusing, aperturing, and particle orientation conditions to demonstrate the reproducibility of spectral acquisition. Note the pronounced rise into the near-IR region. The "background" spectrum was collected from the substrate on which the IDP was mounted. (b) W7030A15 - generally flat spectrum above 500 nm.

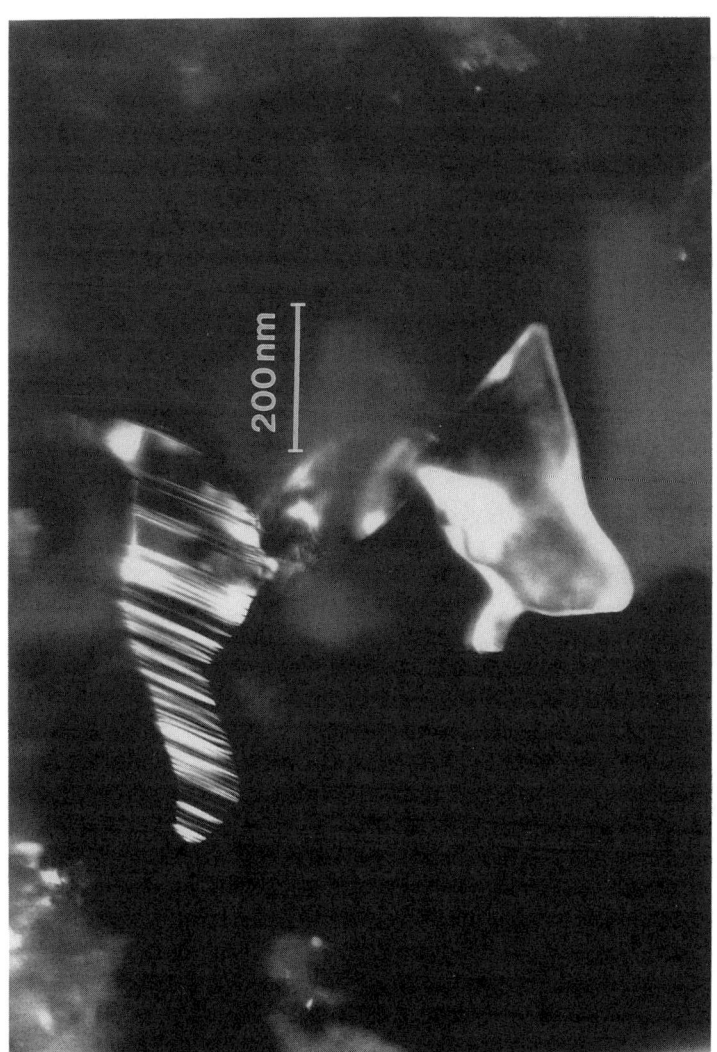

Figure 6. Darkfield micrograph of two enstatite platelets is IDP W7027A11. Platelets exhibiting extreme flattening along (01●), (001) and (100) directions have been observed, together with enstatite whiskers containing axial screw dislocations. These distinctive morphologies are characteristic of crystals formed by gas-to-solid condensation.[48]

pyroxenes in IDPs has shown that some of them contain unusually high levels of Mn (up to 5 wt %). Klöck et al.[5] have suggested that the Mn abundances can be explained in terms of condensation of olivines and pyroxenes from a gas of solar composition.

Anhydrous IDPs also contain ultrafine-grained aggregates, some with bulk compositions that are approximately chondritic. Several types of aggregates have been described by different investigators, resulting in a terminology that is at best confused. Few attempts have been made to correlate the occurrence of specific types of aggregates with other properties of IDPs (e.g., the presence of absence of solar flare tracks). Aggregates have been referred to as "tar balls"[50&51] and granular units (GU's).[52] The term "tar ball" was used to describe aggregates that were assumed to contain carbonaceous material. Although the aggregates were subsequently determined to be carbon-free, there is another class of carbon-rich aggregates that may indeed turn out to be "tar balls".[53&54] Recently the terms unequilibrated, equilibrated, and reduced aggregates (UA's, EA's and RA's) were introduced to describe mineralogically distinct aggregate in track-rich IDPs.[53&54] Since all of these aggregates appear to have undergone an accretional event and/or processing prior to incorporating into IDPs, it is possible that they are the most primordial components of chondritic interplanetary dust. Until a consensus on the types of aggregates in IDPs is reached, adoption of a universal aggregate terminology is premature.

SUMMARY

Electron microscopy and optical spectroscopy have provided significant insight regarding formation mechanisms, post-accretional processes, and possible sources of chondritic IDPs. At least some *layer silicate* IDPs are derived from the main belt of the asteroids, and it may turn out that most, if not all of them, are from hydrous asteroids. Anhydrous IDPs, on the other hand, are more likely to be derived from parent bodies where post-accretional alteration was limited; comets, anhydrous asteroids, and hydrous asteroids beyond the main belt are possible candidates. Evidence of primordial grain-forming reactions appears to be best preserved in track-rich anhydrous (pyroxene and olivine-rich) IDPs, presumably because these IDPs have experienced the least post-accretional alteration and frictional heating during atmospheric entry. Such IDPs are likely to continue to yield information about gas-to-solid condensation and other grain-forming reactions that may have occurred in the solar nebular and presolar interstellar environments. An immediate challenge with anhydrous IDPs lies in understanding the nanometer-scale petrography of the ultrafine-grained aggregates. Whether these aggregates are truly primordial components or merely secondary products (as suggested by Keller et al.[55]), may ultimately be answered by nanometer-scale petrographic studies using analytical electron microscopes. The status of *olivine* IDPs as a discrete class of chondritic objects needs clarification through systematic investigations; some of them have been subjected to heating and equilibration, perhaps during atmospheric entry, while others contain solar flare tracks indicative of minimal frictional heating. Similarly, track-rich pyroxene IDPs may encompass more than one genetic group. *Layer silicate* IDPs have experienced variable degrees post-accretional aqueous alteration, and are likely to continue to provide data on parent body regolith processes.

REFERENCES

1. L. A. Schramm, D. E. Brownlee, and M. M. Wheelock (1989) Major element composition of stratospheric micrometeorites, *Meteoritics* **24**, 99-112.
2. I. D. R. Mackinnon and F. J. M. Rietmeijer (1987) Mineralogy of chondritic interplanetary dust particles, *Rev. Geophys.* **25**, 1527-1553.
3. F. Vilas and M. J. Gaffey (1989) Phyllosilicate absorption features in main-belt and outer asteroid reflectance spectra, *Science* **246**, 790-792.
4. K. D. McKeegan, R. M. Walker, and E. Zinner (1987) Ion microprobe measurements of individual interplanetary dust particles, *Geochim. Cosmochim. Acta* **49**, 1971-1987.
5. W. Klöck, K. L. Thomas, D. S. McKay, and H. Palme (1989) Unusual olivine and pyroxene composition in interplanetary dust and unequilibrated ordinary chondrites, *Nature* **339**, 126-128.
6. G. J. Flynn and S. R. Sutton (1990) Synchotron x-ray fluorescence analyses of stratospheric cosmic dust: new results for chondritic low nickel particles, *Proc. Lunar Planet. Sci. Conf.* (Lunar and Planetary Inst., Houston) pp 335-342.
7. M. S. Germani, J. P. Bradley, and D. E. Brownlee (1990) Automated thin-film analyses of hydrated interplanetary dust particles in the analytical electron microscope, *Earth Planet Sci. Lett.* **101**, 162-179.
8. A. O. Nier and D. J. Schlutter (1990) Helium and neon isotopes in stratospheric particles, Meteoritics **25**, 263-267.
9. L. P. Keller, K. L.. Thomas, and D. S. McKay (1992) An interplanetary dust particle with links to CI chondrites, *Geochim. Coscmochim. Acta* **56**, 1409-1412.
10. S. J. Clemett, C. R. Maechling, and R. N. Zare (1993) Measurement of polycyclic aromatic hydrocarbon (PAH) in interplanetary dust particles, Lunar Planet. Sci., XXIV, 309-310.
11. M. E. Zolensky and R. Barrett (1994) Chondritic interplanetary dust particles: basing their sources on olivine and pyroxene compositions, *Meteoritics*, in press.
12. M. E. Zolensky (1987) Refractory interplanetary dust particles, *Science* **237**, 1466-1468.
13. F. J. M. Rietmeijer (1992) Interplanetary dust particle L2005T12 linked directly to type CM chondrite petrogenisis, *Lunar Planet. Sc. XXIII*, 1153-1154.
14. J. P. Bradley, D. E. Brownlee, and P. Fraundorf (1984) Discovery of nuclear tracks in interplanetary dust, *Science* **226**, 1432-1434.
15. F. J. M. Rietmeijer and I. D. R. Mackinnon (1985) Poorly graphitized carbon as a new cosmothermometer for primitive extraterrestrial materials, *Nature* **316**, 733-736.

16. D. E. Brownlee, D. J. Joswiak, S. G. Love, A. O. Nier, D. J. Schlutter, and J. P. Bradley (1993) Identification of cometary and asteroidal particles in stratospheric IDP collections, *Lunar Planet. Sci. XXIV*, 205-206.
17. D. E. Brownlee, E. Olszewski, and M. M. Wheelock (1982) A working taxonomy for micrometeorites, *Lunar Planet Sci.* XIII, 71-72.
18. I. D. R. Mackinnon, D. S. McKay, G. Nace, and A. M. Isaacs (1982) Classification of the Johnson Space Center stratospheric dust collection, *Proc. 13th Lunar Planet. Sci. Conf., J. Geophys. Res. (Suppl).* **87**, A413-A421,
19. S. A. Sandford and R. M. Walker (1985) Laboratory infrared transmission spectra of individual interplanetary dust particles from 2.5 to 25 microns, *Ap. J.* **291**, 838-851.
20. J. P. Bradley, M. S. Germani, and D. E. Brownlee (1989) Automated thin-film analyses of anhydrous interplanetary dust particles in the analytical electron microscope, *Earth Planet. Sci. Lett.* **93**, 1-13.
21. F. J. M. Rietmeijer (1991) Aqueous alteration in five chondritic porous interplanetary dust particles, *Earth Planet. Sci. Lett.* **102**, 148-157.
22. G.J. Flynn (1994) Cometary dust: A thermal criterion to identify cometary samples among the interplanetary dust collected from the stratosphere. This volume.
23. R. Christoffersen and P. R. Buseck (1986) Mineralogy of interplanetary dust particles from the "olivine" infrared class, *Earth Planet. Sci. Lett.* **78**, 53-66.
24. A.O. Nier (1994) Helium and Neon in Interplanetary Dust Particles. This volume.
25. F. Verniani (1960) *Space Sci. Rev.* **10**, 230.
26. S. G. Love, D. J. Joswiak, and D. E. Brownlee (1993) Densities of 5-15 um interplanetary dust particles, *Lunar Planet. Sci. XXIV*, 901-902.
27. R. P. Binzel (1992) The optical spectrum of 5145 Pholus, *Icarus* **99**, 238-240.
28. J. X. Luu (1993) Spectral diversity among the nuclei of coments, *Icarus* **104**, 138-148.
29. M. Zolensky and H. McSween (1988) In *Meteorites and the Early Solar System.* J. Kerridge and M. Matthews, eds., Univ. Arizona Press, pp. 114-143.
30. M. E. Zolensky, R.A. Barrett and L.B. Browning (1993) *Geochimica et Cosmochimica Acta* **57**, 3123-3148.
31. J. P. Bradley and D. E. Brownlee (1991) An interplanetary dust particle linked directly to type CM meteorites and an asteroidal origin, *Science* **251**, 549-552.
32. M. E. Lawler, D. E. Brownlee, S. Temple, and M. M. Wheelock (1989) Iron, magnesium, and silicon in dust from Comet Halley, *Icarus* **80**, 225-242.
33. J. P. Bradley, H. J. Humecki, and M. S. Germani (1992) Combined infrared and analytical electron microscope studies of interplanetary dust particles, *Ap. J.* **394**, 643-651.
34. E. K. Jessberger, A. Christoforidis, and J. Kissel (1988) Aspects of the major element composition of Halley's dust, *Nature* **332**, 691-695.
35. D. E. Brownlee, M. M. Wheelock, S. Temple, J. P. Bradley, and J. Kissel (1987) A qunatitative comparison of comet halley and carbonaceous chondrites at the submicrometer level, *Lunar Planet. Sci. XVIII*, 133-134.

36. L. M. Muhkin, G. Dolnikov, E. Evianov, M. Fomenkova, O. Pruutsky, and R. Sagdeev (1991) Re-evaluation of the chemistry of dust grains in the coma of comet Halley, *Nature* **350**, 480-481.
37. M. N. Fomenkova, J. F. Kerridge, K. Marti, and L.. A. McFadden (1992) Compositional trends in rock-forming elements of comet Halley dust, *Science* **258**, 266-269.
38. M. Fomenkova and S. Chang (1994) Carbon in comet Halley dust particles. This volume.
39. M. S. Hanner, R. L. Newburn, R. D. Gehrz, T. Harrison, E. P. Ney, and T. L. Hayward (1990) The infrared spectrm of comet Bradfield (1987s) and the silicate emission feature, *Ap. J.* **348**, 312-321.
40. M. J. Gaffey, T. H. Burbine, and R. P. Binzel (1993) Asteroid spectroscopy: progress and perspectives, *Meteoritics* **28**, 335-342.
41. D. J. Tholen (1984) Asteroid taxonomy from cluster analysis of photometry, *PhD Thesis* (University of Arizona).
42. J. D. Bregman, H. Campins, F. C. Witteborn, D. H. Wooden, D. M. Rank, L. J. Allamandolla, M. Cohen, and A. G. G. M. Tielens (1987) Airborne and groundbased spectrophotometry of comet P/Halley from 5-13 micrometers, *Astron. Astrophys.* **187**, 616-620.
43. B. Zellner, D. J. Tholen, and E. F. Tedesco (1985) The eight color asteroid survey: results for 589 minor planets, *Icarus* **61**, 355-416.
44. T. Hiroi, C. M. Pieters, and M. E. Zolensky (1993a) Comparison of reflectance spectra of C asteroids and unique C chondrites Y86720, Y82162, and B7904, *Lunar Planet. Sci. XXIV*, 659-670.
45. T. Hiroi, C. M. Pieters, M. E. Zolensky, and M. E. Lipschutz (1993b) Evidence of thermal metamorphism on the C, G, B, and F asteroids, Science, 261,1016-1018.
46. F. Vilas and B. A. Smith (1985) Reflectance spectrophotometry (0.5-1.0 um) of outer belt asteroids: implications for primitive, organic solar system material, *Icarus* **64**, 503-516.
47. J. Gradie and J. Ververka (1980) The composition of the Trojan asteroids, *Nature* **283**, 840-842.
48. J. P. Bradley, D. R. Veblen, and D. E. Brownlee (1983) Pyroxene whiskers and platelets in interplanetary dust: evidence of vapor phase growth, *Nature* **301**, 473-477.
49. R. Christoffersen and P. R. Buseck (1983) Epsilon carbide: a low temperature component of interplanetary dust particles, *Science* **222**, 1327-1328.
50. J. P. Bradley and D. E. Brownlee (1986) Cometary particles: their sectioning and electron beam analysis, *Science* **231**, 1542-1544.
51. J. P. Bradley (1988) Analysis of chondritic interplanetary dust thin sections, *Geochim. Cosmochim Acta* **52**, 889-900.
52. F. J. M. Rietmeijer (1993) Size distributions in two porous chondritic micrometeorites, *Earth Planet. Sci. Lett.* **117**, 609-617.
53. J. P. Bradley (1993a) Unequilibrated, equilibrated, and reduced aggregates in anhydrous interplanetary dust particles, *Lunar Planet. Sci. XXIV*, 171-172.

54. J. P. Bradley (1994) Nanometer-scale mineralogy and petrography of fine-grained aggregates in interplanetary dust particles, *Geochim. Cosmochim. Acta*, in press.
55. L. P. Keller, K. L. Thomas, and D. S. McKay (1993) Asteroidal agglutinitic particles as a component of anhydrous interplanetary dust, *Meteoritics* **28**, 378-379.

OLIVINE AND PYROXENE COMPOSITIONS OF CHONDRITIC INTERPLANETARY DUST PARTICLES

Michael Zolensky
SN2, NASA, Johnson Space Center, Houston, TX 77058, USA

and
Ruth Barrett
Lockheed ESCO, 2400 NASA RD. 1, Houston, TX 77058, USA

ABSTRACT

We report here analyses of olivines and pyroxenes, and petrofabrics of 27 chondritic IDPs, comparing those from anhydrous and hydrous types. Based on mineralogical evidence, principally the occurrence in some of diopside, we find that hydrous and anhydrous IDPs are, in general, not directly related. The particles containing diopside must have come from parent bodies which experienced extensive low-temperature thermal metamorphism.

INTRODUCTION

A principal goal of meteoritics is to characterize the mineralogy, mineral chemistry, and microstructures of primitive extraterrestrial materials, and to use this information to identify source bodies and (more importantly) characterize the origin and evolution of these bodies.[1,2] It is a commonly held belief that hydrated interplanetary dust particles (IDPs) experienced aqueous alteration on hydrous asteroids while the anhydrous IDPs could be derived from the more primitive asteroids and comets. We wish to discover whether the anhydrous IDPs are actually from the same parent body, being related by aqueous alteration of one to produce the other. We report here analyses of olivines and pyroxenes, and petrofabrics of 27 chondritic IDPs, comparing those from anhydrous and hydrous types. Based on mineralogical evidence, we find that hydrous and anhydrous IDPs are, in general, not directly related.

TECHNIQUES

All of the particles described here were collected in the stratosphere by NASA ER-2 and WB-57 aircraft. All samples were selected on the basis of "chondritic" bulk composition as determined by preliminary energy dispersive X-ray spectrometer (EDX) analyses[3]; chondritic compositions were verified by microparticle instrumental neutron activation analysis (INAA).[4-6] Following the INAA procedure we embedded all particles in EMBED 812 low-viscosity epoxy, and microtomed them into 90 nm thick serial sections. Microtoming was halted approximately halfway through each particle, and the remaining potted butt was then microprobed for major element composition using a CAMECA CAMEBAX microprobe. In this paper we report only particle compositions based on the microprobe analyses; the INAA data will be reported separately. The potted butts were also used to make backscattered electron (BSE) images of the interiors of the IDPs, using a JEOL 35C SEM operating at 15kV, which we found to offer optimum values of resolution vs. electron penetration (and excitation) of the samples. We observed the microtomed sections using a JEOL 2000FX STEM equipped with a LINK EDX analysis system. We used natural mineral standards, and in-house determined k-factors for reduction of compositional data; a Cliff-Lorimer thin-film correction procedure was employed.[7] In general, mineral identifications were made on the basis of *both* composition and electron diffraction data.

CLASSIFICATION

Chondritic IDPs have been classed according to whether they are anhydrous and dominated by either olivine or pyroxene, or contain the hydrous phases saponite or serpentine. We have previously noted that IDPs belonging to the olivine and pyroxene classes have the same basic mineralogy[6], and are mineralogically separable only by consideration of the relative proportion of olivine to pyroxene. In this paper we consider anhydrous IDPs as a single group, simply for convenience. We also consider hydrous IDPs to be those that contain *any* amount of phyllosilicates. An alternate method is to base classification on the dominant crystalline silicate, which would have the effect of moving some hydrous IDPs into the anhydrous group. Unfortunately, the classification of IDPs has not yet been standardized.[6] Examination of the BSE images of chondritic IDPs reveals that some anhydrous chondritic IDPs show low porosity, and that some hydrous ones show relatively high porosity (P.J. Burkett and M.E. Zolensky, unpublished data), as has been noted previously but which may not be fully appreciated (Figure 1).

The vast majority of hydrous chondritic IDPs contain saponite as the dominant phyllosilicate, and are otherwise very similar to the anhydrous IDPs; the only major difference appears to the presence of phyllosilicates. For example, the so called "granular units"[8] (a subset of these structures have been called "tar balls" by Bradley[2]) are observed both in saponite class and anhydrous chondritic IDPs.[6] These granular units are typically composed of sub-micrometer sized aggregates of amorphous phases enclosing many diminutive crystals of olivines, pyroxenes, metals, and sulfides, and probably record the earliest stage of nebular condensation and accretion. Other papers in this volume (by Bradley[7], Klöck[8] and Rietmeijer[9]) discuss the occurrence and mineralogy of these aggregates in much more detail.

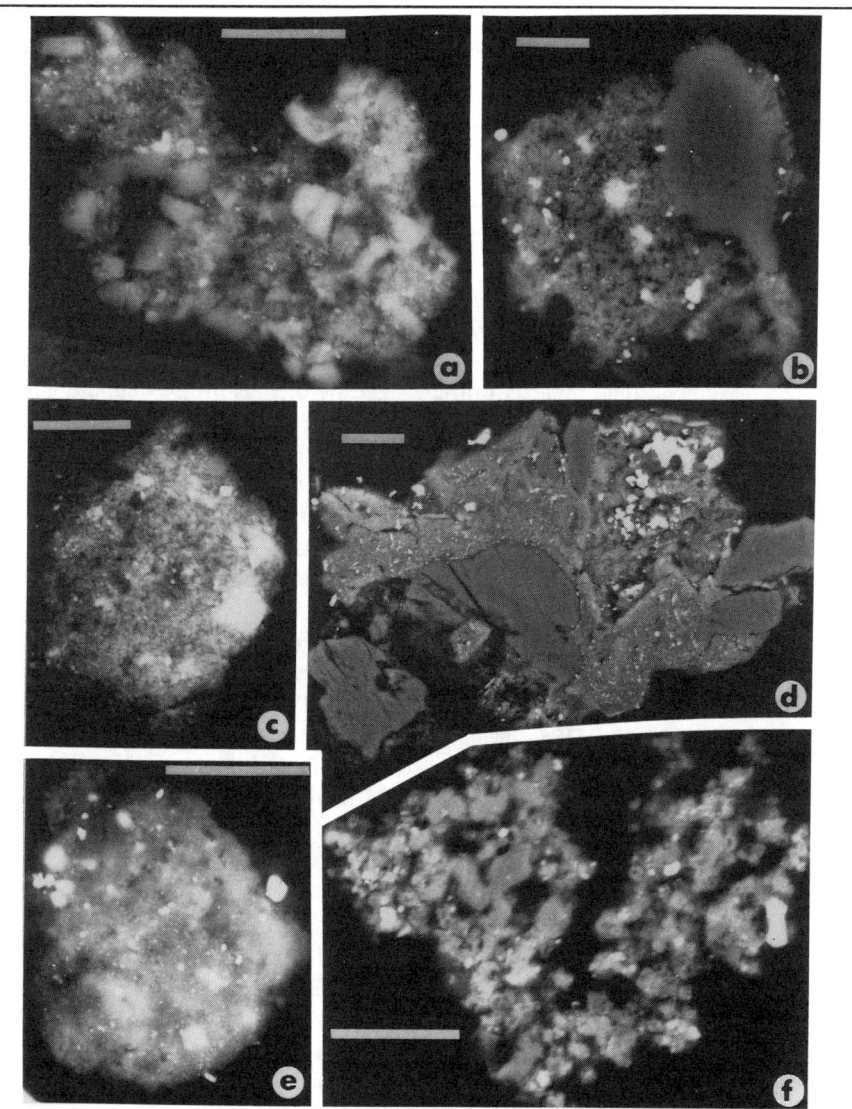

Figure 1. Backscattered electron images of selected chondritic IDPs; all scale bars measure 5 μm. Particles (a)-(d) are anhydrous, while particles (e) & (f) are hydrous. The majority of white grains are Fe-Ni sulfides; most of the black regions are epoxy-filled voids, although some must represent low-Z material. (a) particle #L2005 F39: an anhydrous IDP with numerous subhedral olivine and orthopyroxene crystals, (b) L2005 E36: an anhydrous IDP with a large zoned enstatite crystal, and smaller orthopyroxene and diopside crystals, (c) L2005 C37: a uniformly fine-grained anhydrous IDP, containing abundant olivine and orthopyroxene (d) L2005 Z17: a compact anhydrous IDP with large anhedral orthopyroxene crystals, (e) L2005 Z14: a compact saponite-type IDP, containing small orthopyroxene and diopside crystals, (f) L2011 R3: a highly-porous (~40%), saponite-type IDP, with abundant orthopyroxene.

Serpentine-dominated IDPs are the rarest type, and appear to be mineralogically distinct from other chondritic IDPs. We have not observed granular units within serpentine-class IDPs, and found that these particles rarely contain anhydrous silicates. Serpentine class IDPs have not been found to contain saponite, but IDPs of the saponite class frequently contain a small amount of serpentine.[6] Several workers have concluded that specific serpentine class IDPs are derived from the same hydrous asteroids which produced CM2 carbonaceous chondrite meteorites.[12&13] Only one of the hydrous IDPs which we discuss in this paper is from the serpentine-dominated class.

RESULTS

The compositional ranges of olivines and pyroxenes in IDPs should be useful clues to their histories[14], since these minerals are frequently abundant in these particles, and can be sensitive indicators to the conditions attending grain formation and subsequent metamorphism (if this occurred). To this end we have located and analyzed pyroxenes and olivines in 24 chondritic IDPs (9 hydrous, 15 anhydrous; olivines or pyroxenes were not located in an additional 3 IDPs). Maximum observed grain sizes of olivines and pyroxenes vary from a few hundreds of nanometers (for most particles) to 16 μm in one exceptional case. To these data we have added olivine analyses from three olivine class IDPs, as reported by Christoffersen and Buseck.[15] The results of these analyses are presented in Figures 2 and 3. Figure 2 shows *all* olivine and/or pyroxene analyses that could be obtained from 15 representative chondritic IDPs, through examination of one to two ultramicrotomed sections each. This figure should give the reader an appreciation of the varying abundance and compositional distribution of anhydrous ferromagnesian minerals within chondritic IDPs.

The pyroxenes enstatite, augite, pigeonite and diopside are found in many particles. Their morphologies vary from euhedral to anhedral, with acicular to lath-shaped enstatite crystals being present in some anhydrous IDPs. In the anhydrous IDPs pyroxenes are not commonly intergrown, and where adjacent crystals do occur they do *not* show the 120° crystal intersections typical of recrystallization. While it is possible that some of the augites described here are actually diopside and low-calcium pyroxene in an exsolution relationship, we did not observe evidence of this and exsolution lamellae would have been obvious. We found that there exists no significant difference in the compositions of olivines from olivine- vs. pyroxene-dominated IDPs.[16] The degree of heterogeneity of olivines from anhydrous IDPs is approximately Fo_{52-100} and En_{46-100}. Only a single grain of diopside was found in the 18 anhydrous IDPs examined (see Figure 2).

Olivines in the hydrous IDPs range in composition from Fo_{76-100}, while low-calcium pyroxenes range from En_{76-100}. These ranges are significantly narrower than those from the anhydrous IDPs. Four out of the 9 hydrous IDPs contain diopsides, in contrast to their obvious rarity in anhydrous IDPs. Both anhydrous and hydrous IDPs show a considerable clustering at Mg-rich (enstatite) compositions.

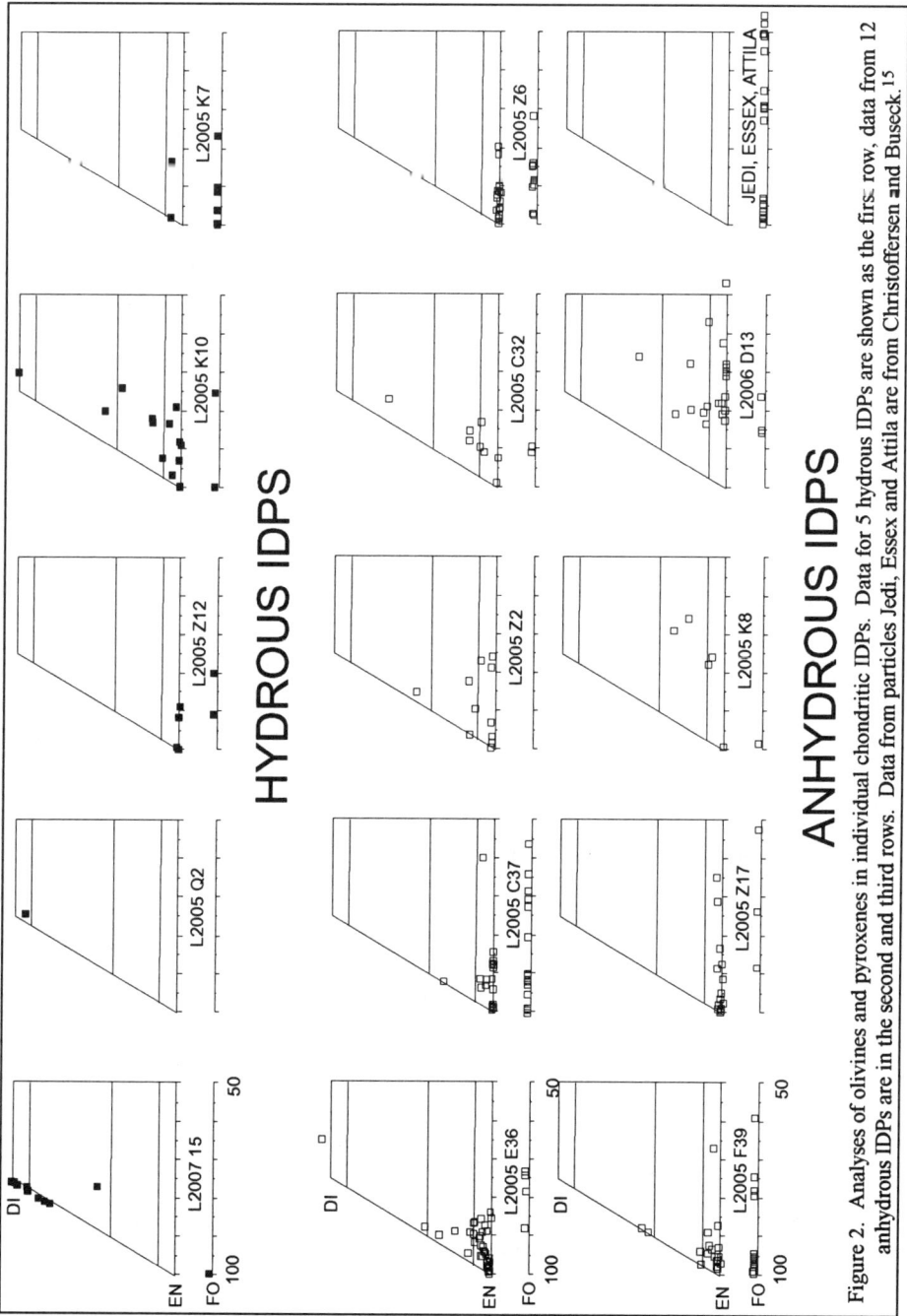

Figure 2. Analyses of olivines and pyroxenes in individual chondritic IDPs. Data for 5 hydrous IDPs are shown as the first row, data from 12 anhydrous IDPs are in the second and third rows. Data from particles Jedi, Essex and Attila are from Christoffersen and Buseck.[15]

110 Olivine and Pyroxene Compositions

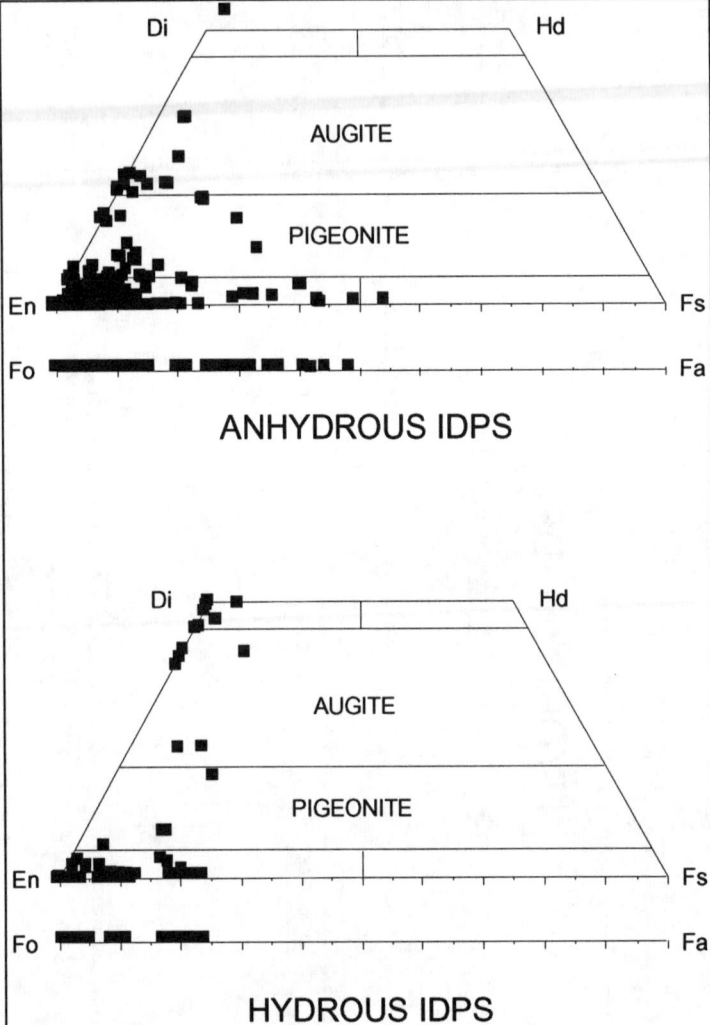

Figure 3. Compositional ranges exhibited by olivines and pyroxenes in chondritic IDPs, with data plotted separately for 9 phyllosilicate-bearing (hydrous) and 15 anhydrous IDPs. Pyroxene analyses are shown projected onto the enstatite (En) - ferrosilite (Fs) - diopside (Di) - hedenbergite (Hd) quadrilateral. Olivines analyses are plotted along the forsterite (Fo) - fayalite (Fa) binary. Data for three olivine-dominated IDPs are from Christoffersen and Buseck.[15] The compositional ranges of olivines and pyroxenes for the hydrous IDPs are more restricted than those for the anhydrous IDPs, although the former contain practically all of the diopside located.

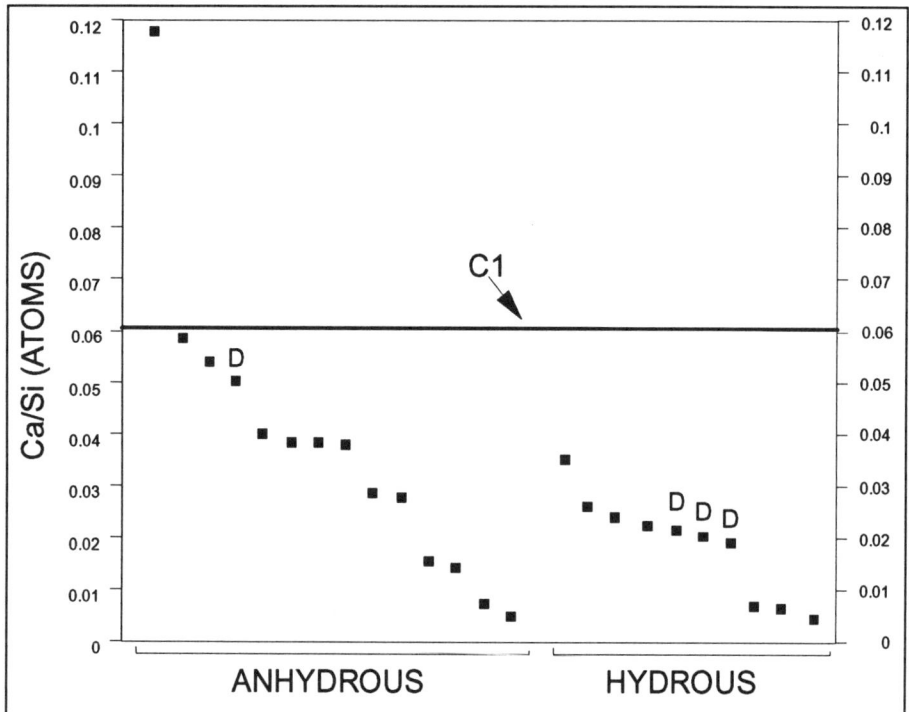

Figure 4. Plot of the Si-normalized Ca concentration of 26 chondritic IDPs. Anhydrous and hydrous IDPs are plotted separately; the C1 value is also indicated.[26] Particles containing diopsides are labeled above with "D".

DISCUSSION

Are anhydrous and hydrous IDPs genetically related by simple hydrolysis of the former material? The presence of primarily Mg-rich (Fe-poor) olivines and pyroxenes in the hydrous IDPs could be due to the preferential early dissolution of Fe-rich olivines and pyroxenes during aqueous alteration, meaning that chondritic IDPs represent a mineralogical continuum. The experimentally-determined rate of fayalite (Fo_6) dissolution in a reducing atmosphere at a pH of 2-7 at 25°C is six times higher than for forsterite (Fo_{91}).[17&18] This relationship probably continues to the pH range 7-12 (the range appropriate to asteroid alteration, according to geochemical modeling)[19], although the absolute rate of olivine dissolution should be adversely affected by this change in pH.[20] While pyroxenes weather at a generally slower rate than olivines under these conditions[21], it is probable that dissolution of Fe-rich pyroxenes is similarly more rapid than Mg-rich varieties. Weathering rates are known to be dependent upon many additional factors, for example, the presence of bidentate organic ligands will increase the dissolution rate of olivine of all compositions by forming metal complexes at the grain surface.[17] Carbon and organic compounds are known to be present in considerable abundance in chondritic IDPs, although their chemistry remains to be fully characterized.[22-24] Nevertheless, if these differential weathering-rate relationships hold for alteration on hydrous asteroids, then they provide a mechanism for the relative destruction of Fe-rich anhydrous silicates, and therefore permit a genetic relationship between hydrous and anhydrous IDPs.

However, this simple mechanism does not explain why diopsides should preferentially be found in hydrous IDPs. Diopside (and other clinopyroxenes) weathers at a generally slower rate than orthopyroxenes, probably due to the development of transport-limiting surface neoformation products on the former, and rapid development of fractures and alteration embayments on the latter.[21&25]

These results suggest that diopside should be a relatively resistant phase during the aqueous alteration of chondritic IDPs, compared to orthopyroxenes and olivines. Accordingly, diopside must never have been present in the vast majority of anhydrous chondritic IDPs, although it is demonstrably present in approximately half of all hydrous IDPs containing remnant olivines and/or pyroxenes. From this we conclude that at least half of the observed hydrous chondritic IDPs have not been derived from bulk of the anhydrous IDPs by aqueous alteration.

For the purpose of the following discussion, the presence or absence of diopside can be used to separate the chondritic IDPs into two separate populations. Population A (22 out of the 27 particles) includes the IDPs without diopside. These materials exhibit a wide range of porosities, from 40% to nearly 0%, and practically the complete range of degree of hydration, although anhydrous particles predominate by a ratio of 14:8. Population B particles (5) contain diopside, show the complete range of degree of hydration, but have a significantly lower range of porosities (15-0%). These particles are dominated by hydrous particles, by a ratio of 4:1.

What does the presence of diopside reveal regarding the origin and history of primitive solar system materials? Could the presence of diopside indicate a more Ca-rich source? To examine this possibility, we have, in Figure 4, plotted the Si-normalized Ca concentrations of the 26 chondritic IDPs for which we have quantitative bulk major element compositions determined by microprobe. On this same diagram we have plotted the C1 value for Ca/Si.[26] Several things are clear from this plot. First, the diopside-containing IDPs are not enriched in Ca relative to the other particles. Secondly, although the hydrous IDPs are depleted in Ca relative to anhydrous particles (as reported earlier[27], the range of Ca concentration in the anhydrous particles is large, encompassing the total range for hydrous IDPs. In fact, most of the chondritic IDPs appear to be depleted in Ca relative to C1 (as would be expected from their more volatile-rich bulk compositions).

A second possibility is that diopside from from low-temperature annealing of amorphous materials. Certainly, chondritic IDPs commonly contain amorphous materials. However, these materials are commonly observed to be either nearly pure silica, feldspathic, or enstatitic in composition[2], none of which are likely to produce diopside upon annealing.

Therefore, the presence of diopside must indicate something else, probably relatively slow cooling, and equilibration with orthopyroxene at a low temperature.[28]

This was probably not a nebular process, and instead required the presence of a parent body. The lower relative porosities of the diopside-containing particles, and their greater relative degree of hydration is consistent with parent-body post-accretion processing.[19] It is noteworthy that the most porous particle we have encountered in this study (see Figure 1) was hydrous (though to a low degree). If we can assume that the sources for hydrous IDPs are hydrous asteroids, this result indicates that asteroid regoliths can today contain materials with very low porosities. We conclude that olivine and pyroxene major-element compositions can be used to help discriminate between IDPs which are (1) purely nebular condensates, and lately resided in anhydrous or icy *(no liquids)*, primitive parent bodies, and (2) more geochemically active parent bodies (probably hydrous and anhydrous asteroids).

ACKNOWLEDGMENTS

We thank Adrian Brearley and David Blake for helpful reviews of an earlier version of this paper. We also thank Patti Jo Burkett for the use of unpublished data on IDP porosities.

REFERENCES

1. Brownlee, D.E. *Ann. Rev. Earth Planet. Sci.* **13**, 34-150 (1985).
2. Bradley, J.P. *Geochim. Cosmochim. Acta* **52**, 889-900 (1988).
3. Zolensky, M.E et al. *Cosmic Dust Catalog 11. Planetary Materials Branch Pub. #83* (NASA Johnson Space Center, Houston) 170p (1990).
4. Lindstrom, D.J., Zolensky, M.E., & Martinez, R.R. *Lunar Planet. Sci. XXI* (Lunar and Planetary Institute, Houston) 700-701, (1990).
5. Lindstrom, D.J. *Nuclear Insts. Methods Phys. Res.* **A299,** 84-588 (1991).
6. Zolensky, M.E. & Lindstrom, D.J. in *Proc. Lunar Planet Sci. Conf. 22nd*, (eds Sharpton, V. & Ryder, G.) (Cambridge University Press) 161-169 (1992).
7. Bradley, J.P. Mechanisms of grain formation, post-accretional alteration, and likely parent body environments of Interplanetary Dust Particles. This volume (1994).
8. Klöck, W. and Stadermann F.J. Mineralogical and chemical relationships of Interplanetary Dust Particles, micrometeorites and meteorites. This volume (1994).
9. Rietmeijer, F.J.M. On the possibility of petrological classification of carbonaceous chondritic micrometeorites. This volume (1994).
10. Goldstein, J.I. in *Introduction to Analytical Electron Microscopy* (eds Hren, J.J. et al.) (Plenum, New York) 813-820 (1979).
11. Rietmeijer, F.J.M. in *Proc. Lunar Planet. Sci. Conf. 19th* (eds Sharpton, V. & Ryder, G.) (Cambridge University Press) 513-521 (1989); .
12. Bradley, J.P. & Brownlee, D.E. *Science* **251**, 549-552 (1991).
13. Keller, L.P., Thomas, K.L. & McKay D.S. *Geochim. Cosmochim. Acta* **56**, 1409-1412 (1992).

14. Klöck, W. et al. *Nature* **339**, 126-128 (1989).
15. Christoffersen, R. & Buseck, P.R. *Earth Planet. Sci. Letts.* **78**, 53-66 (1986).
16. Zolensky, M. & Barrett, R. *Microbeam Analysis*, (1993).
17. Wogelius, R.A. & Walther, J.V. *Chem. Geol.* **97**, 101-112 (1992).
18. Casey, W.H. & Westrich, H.R. *Nature* **355**, 157-159 (1992).
19. Zolensky, M.E., Bourcier, W.L. & Gooding, J.L. *Icarus* **78**, 411-425 (1989).
20. Grandstaff, D.E. *Geochim. Cosmochim. Acta* **41**, 1097-1104 (1977).
21. Eggleton, R.A. in *Rates of Chemical Weathering of Rocks and Minerals* (edsColman S. & Dethier , D.) (Acadenic Press, New York) 21-40 (1986).
22. Clemett, S.J. et al. *Lunar Planet. Sci. XXIV* (Lunar & Planetary Inst., Houston), 309-310.
23 Keller, L.P., Thomas, K.L. & McKay, D.S. *Lunar Planet. Sci. XXIV* (Lunar & Planetary Inst., Houston), 785-786 (1993);
24. Thomas, K.L., Keller, L.P., Blanford, G.E. & McKay, D.S. *Lunar Planet. Sci. XXIV* (Lunar & Planetary Inst., Houston), 1425-1426, (1993)
25. Colin, F., Noack, Y., Trescases, J-J., & Nahon, D. *Clay Minerals* **20**, 93-113 (1985).
26. Anders, E. and Grevesse, N. *Geochim. Cosmochim. Acta* **53**, 197-214 (1989).
27. Scramm, L.S., Brownlee, D.E. and Wheelock, M.M. *Meteoritics* **24**, 99-112 (1989).
28. Lindsley, D.H. *American Mineral.* **68**, 477-493 (1983).

HELIUM AND NEON IN INTERPLANETARY DUST PARTICLES

Alfred O. Nier
School of Physics and Astronomy, University of Minnesota
Minneapolis, MN 55455, USA

ABSTRACT

A summary is given of experiments performed to determine the amounts and isotopic composition of the helium and neon found in individual interplanetary dust particles collected in the Earth's stratosphere. The role of step-heating and pulse-heating techniques for performing the extractions is discussed together with the application of the techniques to determination of the thermal history of the particles. The practicality of using these methods for distinguishing between particles of cometary and asteroidal origin is considered.

INTRODUCTION

For many years it has been recognized that the existence of zodiacal light was strong evidence that dust existed in interplanetary space. It was realized that some of this dust should settle on the Earth, so many attempts were made to collect and classify the particles.[1] Collections were made in the low and in the high atmosphere, on polar ice, in the deep oceans, as well as in less remote places. The results of analyses were not decisive due to the presence of very much larger amounts of dust of terrestrial origin. Small iron-nickel spherules also have been found which appeared to be of extraterrestrial origin. However, it was never demonstrated that they were not merely ablation products of meteorites. In 1964 Merrihue[2] studied the isotopic composition of the noble gases extracted from magnetic separates found in red clay in the deep Pacific and found isotopic and elemental abundance ratios—particularly ^{3}He/^{4}He ratios—distinctly different from those observed in terrestrial samples and more like ratios found in meteorites or lunar soil grains. This gave strong indication that the magnetic separates contained elements of extraterrestrial origin. The complex chemistry experienced over time by elements found in deep sea particles hindered any clear identification of the chemical or mineralogical nature of the extraterrestrial constituents of the separates.

An important breakthrough came when Brownlee et al.[3] showed that some of the particles collected on adhesive surfaces attached underneath the wings of planes flying in the stratosphere had unique characteristics which precluded their being of terrestrial origin. Hence they were truly Interplanetary Dust Particles (IDPs). A summary of studies of their nature[4-7] showed that there are wide differences in composition. Many of the particles have compositions similar to those of CI, CR and CM chondritic meteorites.[8-10] They are irregular in shape, and range from fluffy to solid. In passing through the atmosphere they are heated as they are decelerated, the extent of heating depending upon their size, density, entry velocity and entry angle. See the papers in this volume by Flynn[11&12] for more discussion on this point. As a result, ones larger than 50 to 100 μm in "diameter" are generally melted, or may be altered in other ways. Most of the interest has been in small particles, ranging in size from a few to some tens of microns and having masses measured in nanograms, since they are less likely to have been altered as a result of atmospheric heating.

Because of the scarcity of particles, research has been devoted almost entirely to the investigation of individual IDPs. The small size of the particles has required the use of highly sophisticated instrumentation and considerable ingenuity.

NOBLE GASES IN IDPS

The analysis of noble gases found in materials is a relatively simple and sensitive procedure. Since noble gas elements do not combine with other elements, extraction can easily be accomplished by heating or chemical dissolution. Purification can be by collection on charcoal at reduced temperature, or by the use of various chemical getters. In the case of meteorites, the extraction and isotopic analysis of the noble gases has been a fruitful experience. The gases found may consist of several components with different origins—a primordial component, one due to spallation reactions caused by cosmic ray bombardment over long periods of time, and in some cases elements resulting from radioactive decay.

For IDPs the situation is somewhat different. If the particles have their origin in comets or asteroids, as is generally believed, they may exist as independent particles in interplanetary space for only a relatively short time—of the order of 10,000 years before they settle to the stratosphere, where they are concentrated. There is not sufficient time to accumulate measurable amounts of cosmic ray-produced spallation products during the time the particles existed as independent dust particles. Moreover, the spallation products are created with kinetic energy, and have ranges generally longer than particle diameters, and hence would not be retained. On the other hand, like lunar soil grains, the IDPs are subject to bombardment by the solar wind and solar flares, including the so-called Solar Energetic Particles (SEPs).[13] Noble gas analyses hence are valuable, but in a way different from the corresponding analyses of meteorites.

The first noble gas analyses on IDPs were made by Rajan et al.[14] In a study of 10 individual particles, having masses in the range of 0.21 to 24 ng, ^{4}He concentrations, comparable to those observed in lunar soil grains, were found. These were attributed to solar wind accumulation. Calculations have shown[15] that in the 10,000 year time span when the particles may have been free in interplanetary space, solar wind bombardment could readily account for all of the helium observed. In an investigation by Hudson et al.[16] 13 particles were combined, and an attempt made to measure the ^{20}Ne/^{22}Ne ratio. The investigators concluded that the ratio observed, 13 ± 3, fell in the range expected for solar wind neon, and this provided further evidence that the particles were of extraterrestrial origin.

Additional evidence for an extraterrestrial origin was obtained by Bradley et al.[17] who, in a study of several IDPs collected in the stratosphere, observed solar flare tracks. The densities of the tracks were consistent with the particles being exposed to solar cosmic rays in the interplanetary medium for a period of roughly 10^4 years. Moreover, the fact that the tracks existed, and had not been erased by temperature annealing, indicated that the particles had not been heated above 500 to 600°C in their deceleration in the atmosphere. Sandford[18], in a comprehensive study, considered the general problem of interpreting solar flare track densities in IDPs.

The cosmological importance of ^{3}He led Nier and Schlutter to undertake an investigation of ^{3}He/^{4}He, as well as ^{20}Ne/^{22}Ne ratios, in individual IDPs. In an initial study, 16 particles, having diameters of approximately 15 μm, were analyzed.[19] The results are given as line A in Table 1. With the assumption that the average density of the particles was 2 g/cm^3 the average helium content was (0.027 ± 0.01) cm^3 STP/g, a value about 1/6 that found for typical lunar grains of the same approximate size, and in the same range observed by Rajan et al.[14] Except for one particle for which the ^{3}He/^{4}He was (1.45 ± 0.05) x 10^{-3}, the average for the remaining fifteen particles was (2.4 ± 0.3) x 10^{-4}. For the 10 cases where neon measurements could be made, the average value for ^{20}Ne/^{22}Ne was found to be 12.0 ± 0.5; for the three cases where ^{21}Ne/^{22}Ne measurements could be made, an average value determined was 0.035 ± 0.006.

Table 1. Results of noble gas extractions from IDPs

Investigation	Heating	# of Part.	Avg. Dia. mm	^{4}He cm^3 STP/g x 10^3	^{3}He/^{4}He x 10^4	^{4}He/^{20}Ne	^{20}Ne/^{22}Ne
A [19]	total	16	~15	27±2	2.4±0.3	33±7	12.0±0.5
B [29]	step	20	~15	38±10	2.8±0.2		
C [33]	pulse	24	~20	5.6±2.4	5.3±0.9[c]	35±9	
D [34]	pulse	22	5-15	170±50	5.3±0.7	58±10	11.8±0.8
SW[a]					4.3±0.2	~600	13.7±0.3
Lunar fines[b]				150±20	3.7±0.1	62±5	12.6±0.1

[a]Geiss J. et al. (1972) Apollo 16 Prel. Sci. Rep. NASA SP-315, 14.1-14.10.
[b]Average of Eberhardt P. et al. (1970) Proc. Apollo XI Lunar Sci. Conf. 2, 1037-1070; Eberhardt P. et al. (1972) PLSC 3rd, 1821-1856; Hintenberger H et al. (1971) PLSC 2nd, 1607-1625; Pepin R.O. et al. (1974) PLSC 5th, 2149-2184; Pepin R.O. et al. (1975) PLSC 6th, 2027-2055.
[c]Average for 8 IDPs which appeared "normal;" average for 14 others was (60±13) x 10^{-4}.

Following the success of the first study, in which ^{3}He/^{4}He ratios were determined for the first time for IDPs, it was realized that more sophisticated experiments could be undertaken to learn more about the nature and source of the particles. The results from these appear as lines B, C, and D in Table 1, and will be discussed later.

SOURCES OF INTERPLANETARY DUST PARTICLES

The likely source of interplanetary dust particles has been a subject of interest for many years. Until relatively recently, it was generally believed that most of the dust came from comets as they were heated in their approach to the Sun. Detailed calculations of the dynamics of IDPs and the probability of their capture by the Earth's gravitational field have suggested that a significant fraction of the particles collected in the Earth's atmosphere originate in asteroids. The dynamic computations are based on the calculations of Whipple[20], who considered the general problem of deceleration heating of particles which entered the Earth's atmosphere. His computations were extended by a number of investigators, making use of additional experimental data which were becoming available.[21-26] This greatly expanded our understanding of the overall problem. In addition, there is now compelling mineralogical and petrographic evidence linking several chondritic IDPs to CI, CR and CM meteorite petrogenesis, and hence a common asteroidal origin.[8-10]

The net result of the calculations may be summarized in an approximate form as follows: a particle outside of the Earth's atmosphere picks up speed toward the Earth's center as it experiences the Earth's gravitational pull. When it reaches a sufficiently dense part of the atmosphere, below 100 km, but above the stratosphere, it is decelerated by the drag force of the atmosphere. The kinetic energy of the particle is converted to heat, warming the particle. The extent of the heating depends primarily on the speed of the particle and its diameter and mass density.[11&12]

IDPs which have their origin in main-belt asteroids, and evolve to Earth-crossing orbits under Poynting-Robertson drag, have little velocity relative to the earth when they begin to feel the Earth's gravitational pull and hence, in their fall, reach speeds of only approximately 12 km/s, the same as the escape velocity of objects leaving the Earth's gravitational attraction. On the other hand, IDPs having their origin in comets carry with them the velocity of the comets, and hence have a total velocity higher than that of typical asteroids when they begin their deceleration. In the case of low periapsis comets, the total velocity may reach 15 to 20 km/s, or more. Hence, other things being equal, one would expect comet particles to be heated several hundred degrees more than those from asteroids during deceleration. The actual situation is somewhat more complex since particles reach the earth coming from various directions and with a range of velocities. The orbital dynamics problem has been discussed in some detail by Jackson and Zook.[27] Also, the deceleration heating of a particle depends not only upon its velocity but also on the diameter and mass density of the particle. This phenomenon has been explored by Flynn and Sutton[28], who, following detailed computations, point out that even assuming similar densities and diameters, a small fraction of the dust particles from a fast comet may be heated less during atmospheric deceleration than a small fraction of particles from main-belt asteroids.[12]

LABORATORY STEP- AND PULSE-HEATING OF IDPS

Step-Heating Extractions

The fact that generally a particle of a given size and density, which originates from a low periapsis comet, is heated more in its deceleration in the Earth's atmosphere than a similar particle of asteroidal origin, suggests that there should be a difference in the amount of degassing during deceleration. This should be reflected in the amount of gas which remains in the particle and the release temperature of the gas. It follows that if step-heating is employed in laboratory extraction of the gas, useful information *might* be obtained which could help decide between cometary and asteroidal origin of the particles. This observation led to the study of 20 individual particles in which gas was extracted by step-heating.[29] The results are summarized in line B in Table 1.

In the experiment performed, the power to the small ovens holding the particles was increased in one-fourth watt steps until all of the gas was removed. A plot was made showing the amount of gas removed as a function of the oven power. Fig. 1 shows the release curves for ^{4}He for the 12 particles of the 20 which contained the largest amount of gas. The curves have been normalized to facilitate comparisons. The shaded area covers the ranges of curves found in a companion study of 14 lunar soil grains which had been subject to the same heating sequence. While there are differences in the IDP curves, no doubt related to the mineralogy and porosity of the particles, it appears that the release patterns of the 12 IDPs did not differ much from those of the lunar grains, indicating that the 12 IDPs did not suffer any extreme heating during atmospheric entry.

Four other particles in the group of 20 contained less helium and the release curves were shifted to higher power, i.e., higher temperature, suggesting that they had suffered more heating during entry. The remaining four of the 20 particles contained too little helium to permit an analysis. A tentative conclusion from the results is that 12 of the particles were probably of asteroidal origin. One cannot rule out the possibility a few could have been of cometary origin and entered the atmosphere at a grazing angle and consequently were not heated as much as if they had had a more typical average entry direction.

The study was interesting from another aspect. The particles placed at our disposal by the NASA curatorial facility at the Johnson Space Center were actually fragments of larger parent particles which shattered when they struck the collection flags under the collecting plane's wing. Companion fragments were distributed to other laboratories for elemental and mineralogical analysis. While the studies have not all been completed, results to date indicate that fragments showing low amounts of helium have companion fragments low in zinc, a volatile element, and have an abundance of magnetite.[30] These observations are consistent with the view that the parent particles suffered heating during atmospheric entry.

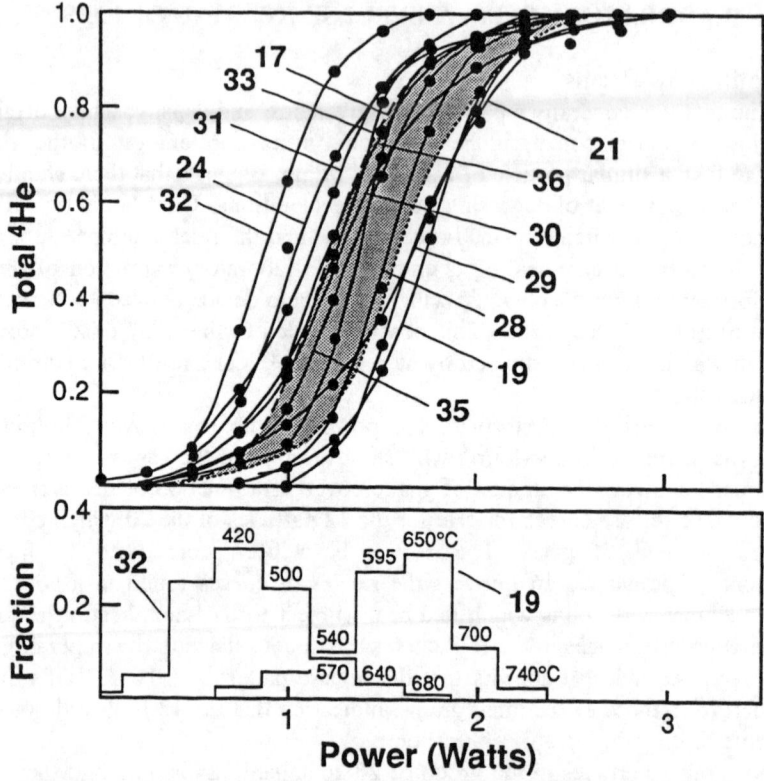

Figure 1. Helium release by step-heating from fragments of 12 IDPs. The curves have been normalized to facilitate comparisons. Plotted is the cumulative release versus the power delivered to the ovens. The power was increased in 1/4 watt steps until all of the gas was extracted. The steps were five minutes in duration. The shaded area shows the range of similar curves found for 14 lunar soil grains of comparable size studied for comparison. The lower plots give the fractional amounts and the corresponding temperatures for the two IDP fragments, #32 and #19, which had the lowest and highest release temperature, respectively. The purpose of the lower plot is to illustrate the range of release temperatures encountered for the twelve particles in the group. For details see reference 29.

Pulse-Heating Extractions

The theoretical analyses of Love and Brownlee[31], as well as earlier calculations, show that the heat pulse felt by IDPs during their deceleration in the atmosphere lasts only about two seconds. This suggests that if a comparison is to be made between this heating and subsequent laboratory step-heating, it would be more realistic to extract gas from IDPs by employing a succession of heat pulses of increasing temperature. These pulses should have the approximate shape of atmospheric deceleration heating pulses. In practice this was accomplished by applying constant power pulses of five second duration. Because of the small masses of the ovens used, they were near peak temperature for approximately two seconds. The temperature profile matched the Love and Brownlee calculated values[31] remarkably well.[32&33] Gas release curves such as shown in Fig. 2 were obtained for each particle.

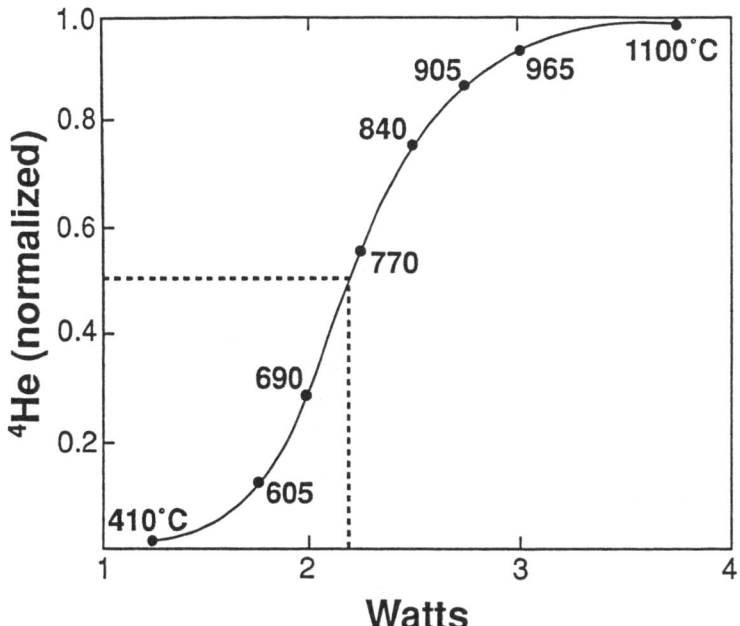

Figure 2. He release from a fragment of an IDP which was heated by pulse-heating. The power was increased in 1/4 watt increments, and the pulses were five seconds in duration. As noted in the figure, for this particular particle, 2.2 watts were required to extract 50% of the helium. The corresponding peak pulse temperature was 750°C. See reference 33 for details.

To see how such a curve would relate to the result of a previous heating, such as might have occurred during atmospheric entry, tests were made in which the heating sequence was stopped at some arbitrary value and started over again from zero power. It was found that in the second heating sequence no gas was released until the power level was reached at which the first heating was halted. Here approximately 20% as much gas was released as was during the first heating sequence. The second heating was then continued until all of the gas was extracted. As expected, the curve for the second heating is shifted to higher power (i.e., temperature). Had this been an IDP entering the earth's atmosphere, the first heating would have corresponded to what happened to the particle in its deceleration in the atmosphere; the second heating would have corresponded to what we would have observed in our laboratory extraction. A little reflection will show that the amount of shift for the second heating sequence will depend upon the fraction of the total gas removed in the first extraction. In other words, in a laboratory extraction performed on an *actual IDP, the position of the extraction curve, along with a determination of the total amount of gas extracted, is a measure of the particle's previous thermal history.*

To test the validity of the procedure[33], interrupted heating sequences were employed for six individual IDPs and seven individual lunar grains with uniformly consistent results. In the second heating sequence no gas was released until the power level was reached at which the first heating was stopped. Here, on the average, approximately 20% as much gas was extracted as at the point where the first heating was halted.

DISCUSSION

The Isotopic Ratio $^3He/^4He$

The summaries of the several studies given in Table 1 include some puzzling results. First of all, we note that the average $^3He/^4He$ ratios for the first two investigations, A and B, are only about half those of the last two, C and D. It would be tempting to assume the cause was some error in calibration. This is highly unlikely. In the A and B groups combined, 36 particles, there was only one particle whose $^3He/^4He$ did not fall between 1.5 and 6.5 x 10^{-4}. In the case of the C study, the particles came from three different collection flags, six from L2005, six from L2006 and twelve from L2011. Eleven of those from flag L2011, along with three from flag L2006, had $^3He/^4He$ values in excess of 1.6 x 10^{-3}. For the remaining 8 particles for which $^3He/^4He$ ratios could be determined, the average value was (5.3 ± 0.7) x 10^{-4}, as noted in Table 1. In the case of the line D investigation of 22 particles, in only one case did the $^3He/^4He$ ratio exceed 10^{-3}.

The four studies of 82 particles listed in Table 1 included the only $^3He/^4He$ ratio determinations which have ever been made for stratospheric-collected IDPs. There clearly are variations among particles—some rather dramatic. The ratios appear to fall in three general groups: (1) values less than 3 x 10^{-4}, (2) ones near 5 x 10^{-4}, and (3) ones above 2 x 10^{-3}. If one assumes that the helium found in IDPs should be due primarily to solar wind implantation with a ratio between 4 and 5 x 10^{-4} one might explain the low values in the (1) group as caused by diffusion loss of the lighter isotope. On the other hand, the low values found might be associated with Solar Energetic Particles (SEPs), observed along with solar wind gas, in lunar soil grains.[13] This explanation would require that the IDPs lost their lightly held surface solar wind implantation, probably during entry. Such an assumption suffers from the contradiction that IDPs having low $^3He/^4He$ ratios sometimes have relatively high total helium contents (studies A and B). The (2) group might be explained as merely solar wind. The (3) group is a puzzle. The high ratios might be attributed to a contribution of cosmic ray produced spallation helium. Since the group includes a fairly large number of IDPs, it will be interesting to see if a similar proportion is maintained as additional IDPs are studied. The explanation suffers from the fact that in the few cases where neon isotopic measurements could be made, the ^{21}Ne/^{22}Ne fell near 0.03 rather than a higher value which would be expected if spallation products are present.

The analysis of the ^{3}He/^{4}He ratios is complicated by the fact that in two of the investigations, lines B and C, the particles studied were actually fragments of "parent" IDPs, which were believed to have fragmented when they struck the collection flags mounted on the collecting plane's wings. Could it be that the ^{3}He/^{4}He ratio in IDPs exhibits a depth effect and that the variations observed in study C are related to the position occupied by the fragment studied in the parent particle? If so, why were there not large variations in the line B study where the particles investigated also were fragments of parent particles?

The ^{20}Ne/^{22}Ne ratios, where measured, fell near 12.0, a value close to that found in bulk analyses of lunar soil grains and for SEPs.[13] The ratios observed are clearly below that attributed to the solar wind. The ^{4}He/^{20}Ne ratios, where measured, fell in the general range found in the bulk analyses of lunar soil grains and well below that found in the solar wind.

The Temperature Release Patterns

As discussed earlier, the position of temperature release curves such as shown in Figs. 1 and 2 are a measure of the thermal history of an IDP. Referring to curves such as shown in Fig. 2, where a pulse-heating sequence is employed, the question arises as to the appropriate power (i.e., temperature) to assign to a curve when comparisons are made with other particles. Should it be the onset temperature, the leveling-off temperature, or some other point such as the 50% point, as shown in the figure? Both the onset and leveling-off temperatures are hard to define, particularly for particles with little gas, where there are appreciable random fluctuations in the mass spectrometer ion currents. Normalized curves for the various particles all have approximately the same slope and shape, and differ mainly in their position along the power (temperature) scale. For this reason the 50% extraction point, almost always the steepest part of the curve, has arbitrarily been chosen in associating a temperature with a curve. In discussing a group of particles, it serves merely as a convenient means for comparing the "ease" with which helium is released when the particles are heated.

Ideally, we would like to determine the temperature for which gas is first released in the laboratory extractions, since this temperature falls near the maximum temperature reached in the atmospheric deceleration heating of the particle before it is collected. Unfortunately, because of the small amount of gas present in an IDP, determination of this point is not easy to achieve. In an unpublished work just completed of 12 IDPs having diameters between 10 and 15 mm, it was possible to show that the temperature for release of 10% of the helium averaged about 150°C below the temperature for 50% release. From this it was possible to conclude that the average peak temperature reached by the particles was approximately 760°C, a plausible amount for the particles studied.

The IDP Size and Fragmentation Problem

After the first isotopic ratios for helium and neon were successfully completed[19], it became apparent that it would be highly desirable to conduct comprehensive experiments on selected IDPs, with different laboratories applying the methods in which they had special competence. At the time it was felt that this could best be accomplished by taking advantage of the fact that some of the larger IDPs shatter upon striking the collection flags on which they accumulate. Accordingly, the fragments of "parent" IDPs collected on the flags were distributed to the different laboratories which could provide useful information about the particles—elemental and mineralogical analyses, morphology, gas content, etc. At the time, the step- and pulse-heating gas extraction techniques had not been developed, and it was not fully appreciated that using the techniques for deciding between a cometary and an asteroidal source of the particles would be seriously compromised

The problem arises, as pointed out by Love and Brownlee[31] and others, because the deceleration heating of a particle entering the Earth's atmosphere depends not only upon its energy but also upon its density and "diameter." Hence a larger particle is heated more, merely because of its size, independent of whether it is of cometary or asteroidal origin. This extra heating can mask the difference in heating associated with the difference in the origin of the particles. The effect almost certainly degraded the usefulness of the study summarized in line C of Table 1. Here the fragments placed at our disposal all had "diameters" estimated to be close to 20 μm. On the other hand, the parent particles had "diameters" of approximately 40 μm. It is this larger diameter which determined the deceleration heating experience. For the 24 IDP fragments which formed the basis of the study, the average peak pulse temperature for extraction of 50% of the helium was $798 \pm 24°C$.

The data presented in line D of Table 1 illustrate the case where the IDPs were all relatively small—5 to 15 μm in diameter, and were *not* believed to have been fragments of larger parent IDPs. We note that in spite of the small size of the particles, on the average they contained more total helium, and the amount per gram of material was dramatically higher than for the particles in other studies—particularly those of the fragments listed in line C of Table 1. The substantially lower temperature, $629 \pm 28°C$, for the removal of 50% of the gas points strongly to the difference being due to the heating effect discussed in the previous paragraph. Brownlee et al.[34], in a careful analysis of release curves of the individual particles similar to the curve shown in Fig. 2, concluded that most of the IDPs in the study had an asteroidal origin but that in excess of 20% had a cometary origin.

CONCLUSIONS

It has been demonstrated that it is possible to determine quantitatively the amounts of and isotopic composition of the helium and neon extracted from individual interplanetary dust particles collected in the Earth's stratosphere. By employing step- or pulse-heating to accomplish the extraction, it is possible to learn about the thermal history of the particles, and through this, in appropriate cases, to distinguish between IDPs of cometary and asteroidal origin. However, to exploit the potential of the measurements fully, a thorough knowledge of particle sizes and densities is needed.

ACKNOWLEDGMENTS

The research discussed was supported by grants to the University of Minnesota by the Planetary Materials and Geochemistry Program of the National Aeronautics and Space Administration. The particles studied were provided by the Curatorial Branch of the Johnson Space Center in Houston, TX. The author is indebted to Dennis Schlutter for assistance with the many experiments on which the work was based. He is further indebted to Donald Brownlee and Michael Zolensky for helpful discussions, and to George Flynn and two other reviewers of the manuscript.

REFERENCES

1. Hodge P. W. (1981) *Interplanetary Dust*. Gordon and Breach, New York.
2. Merrihue C. (1964) *Ann. N. Y. Acad. Sci.* **119**, 351-367.
3. Brownlee D. E., Tomandl D. A. and Olszewski E. (1977) *Proc. Lunar Sci. Conf. 8th*, 149-160.
4. Fraundorf P., Brownlee D.E. and Walker R. M. (1982) in *Comets* (Ed. L.L. Wilkening), Univ. of Ariz. Press, Tucson, pp. 383-409.
5. Bradley J. P., Sandford S. A. and Walker R. M. (1988) in *Meteorites and the Early Solar System* (Ed. J. F. Kerridge and M. S. Matthews), Univ. of Ariz. Press, Tucson, pp. 861-895.
6. Brownlee D. E. (1985) *Ann. Rev. Earth Planet. Sci.* **13**, 147-173.
7. Mackinnon I.D.R. and Rietmeijer F.J.M. (1987) *Rev. Geophys.* **25**, 1527-1553.
8. Bradley J.P. and Brownlee D.E. (1991) *Science* **251**, 549-552.
9. Keller L.P., Thomas K.L. and McKay D.S. (1992) *Geochim. Cosmochim. Acta* **56**, 1409-1412.
10. Rietmeijer F.J.M. (1992) *Lunar Planet. Sci XXIII* (Abstract) 1153-1154.
11. Flynn G. (1994) Changes to IDP composition and mineralogy by terrestrial encounters. This volume.
12. Flynn G. (1994) Cometary dust: A thermal criterion to identify cometary samples among the Interplanetary Dust collected from the stratosphere. This volume.
13. Wieler R., Baur H., and Signer P. (1986) *Geochim. Cosmochim. Acta* **50**, 1997-2017.
14. Rajan R. S., Brownlee D. E., Tomandl D., Hodge P. W., Farrar H. and Britten R. A. (1977) *Nature* **267**, 133-134.
15. Pillinger C. T. (1979) *Rep. Prog. Phys.* **42**, 897-961.
16 Hudson B., Flynn G. J., Fraundorf P., Hohenberg C. M. and Shirck J. (1981) *Science* **211**, 383-386.
17. Bradley J. P., Brownlee D. E. and Fraundorf P. (1984) *Science* **226**, 1432-1434.
18. Sandford S.A. (1986) *Icarus* **68**, 377-394.

19. Nier A. O. and Schlutter D. J. (1990) *Meteoritics* **25**, 263-267.
20. Whipple F. L. (1950) *Proc. Natl. Acad. Sci. U. S. A.* **36**, 687-695; (1951) **37**, 19-30.
21. Fraundorf, P. (1980) *Geophys. Res. Letts.* **10**, 765-768.
22. Zook H. A. and McKay D. S. (1986) *LPSC 17th* (Abstract), 977-978.
23. Flynn G.J. (1986) *Meteoritics* **21** (Abstract) 362-363.
24. Sandford S.A. (1986) *Meteoritics* **21** (Abstract) 501.
25. Flynn G. J. (1989) *Icarus* **77**, 287-310.
26. Sandford S.A. and Bradley J.P. (1989) *Icarus* **82**, 146-166.
27. Jackson A. A. and Zook H. A. (1992) *Icarus* **97**, 70-84.
28. Flynn G.J. and Sutton S.R. (1991) *Proc. Lunar Planet. Sci. Conf.* **21**, 541-547.
29. Nier A. O. and Schlutter D. J. (1992) *Meteoritics* **27**, 166-173.
30. Klöck W., Flynn G. J., Sutton S. R. and Nier A. O. (1992) *Meteoritics* **27**, 243-244.
31. Love S. G. and Brownlee D. E. (1991) *Icarus* **89**, 26-43.
32. Nier A. O. and Schlutter, D. J. (1993) *LPSC 24th* (Abstract), 1075-1076.
33. Nier A. O. and Schlutter D. J. (1993) *Meteoritics* **28**, 675-681.
34. Brownlee D. E., Joswiak D. J., Love S. G., Nier A. O., Schlutter D. J. and Bradley J. P. (1993) *LPSC 24th* (Abstract), 205-206.

CHANGES TO THE COMPOSITION AND MINERALOGY OF INTERPLANETARY DUST PARTICLES BY TERRESTRIAL ENCOUNTERS

G. J. Flynn,
Dept. of Physics, SUNY-Plattsburgh,
Plattsburgh, NY 12901, USA

ABSTRACT

Individual interplanetary dust particles can experience alteration or contamination by a variety of mechanisms during Earth encounter. Velocity selection effects produce a stratospheric dust population different from the interplanetary dust population. The thermal pulse experienced during Earth atmospheric entry is, in some cases, sufficient to alter the mineralogical and chemical properties of the interplanetary dust recovered from the Earth's stratosphere. Stratospheric residence or the collection and curation process is known to produce a surface contamination of S and possibly other elements. A thorough understanding of the nature and magnitude of these contamination and alteration processes is essential if the properties of the interplanetary dust, and of the parent bodies of the dust and of the primitive solar nebula, are to be inferred.

INTRODUCTION

Major objectives of the study of interplanetary dust particles (IDPs) are to constrain the physical and chemical conditions in the early Solar System, to characterize the particles making up the Zodiacal Cloud, and to infer the physical, chemical, mineralogical and isotopic properties of the IDP parent bodies -- the comets and the asteroids. However, terrestrial interactions alter the properties of some IDPs from those of the Zodiacal Cloud particles. It is important to understand the nature and extent of these terrestrial alterations, and to identify those particles which have experienced alterations, in order to infer the properties of the interplanetary dust, the asteroidal and cometary parent bodies, and the early solar nebula. The interactions which produce these alterations can be separated into four distinct phases: near-Earth gravitational segregation, atmospheric entry deceleration, stratospheric residence, and the collection/curation process.

NEAR-EARTH GRAVITATIONAL SEGREGATION

Öpik[1] showed that the effective radius (R_{eff}) for capture of a particle by a planet having a radius of R_{act} depends on the relative velocity of that particle (v_g) with respect to the planet:

$$R_{eff} = R_{act}(1 + v_e^2/v_g^2)^{1/2}$$

where v_e is the planetary escape velocity.

Flynn[2] calculated the geocentric velocities at the Earth collection opportunity for IDPs derived from main-belt asteroids and from comets. These calculations showed that IDPs derived from main-belt asteroidal sources whose orbits evolved under the influence of solar gravity and Poynting-Robertson radiation drag have significantly lower geocentric velocities at the Earth-collection opportunity than IDPs derived from comets. Jackson and Zook[3] extended this modeling to include the effects of planetary gravitational perturbations, and confirmed that the main-belt asteroidal IDPs generally have significant lower geocentric velocities at the Earth-capture opportunity than do the cometary IDPs. The velocity ranges determined in both studies for main-belt asteroidal and cometary IDPs are compared in Table 1.

Table 1
Velocity Range at Nodal Crossing for 10 μm Radius IDPs

Parent Body	Velocity Range at 1 AU Nodal Crossing (km/s)	
	Flynn[2]	Jackson and Zook[3]
Main-Belt Asteroids	0 to 5.5	1 to 7.5*
Comets		
Perihelia >1.5 AU (or >1 AU)+	5.5 to 11	5.3 to 17@
Perihelia ≤1.5 AU (or ≤1 AU)+	>11	>11.8

* Except for IDPs derived from the Hungaria and Phocaea families, which have velocities up to 13.5 km/s at nodal crossing.
+ Jackson and Zook[3] separate the comet groups at a perihelion of 1 AU, while Flynn[2] separates them at 1.5 AU.
@ Except IDPs from one comet having 25.3 km/s at nodal crossing.

Jackson and Zook[3] found that some 10 μm radius IDPs from main-belt asteroids have geocentric velocities as low as 1 km/s (prior to acceleration by Earth's gravity) at the Earth capture opportunity, while the particles derived from the most favorable known comets all have geocentric velocities higher than 5.3 km/s.

The Earth capture cross-section is shown as a function of IDP geocentric velocity in Figure 1. For an IDP with a 1 km/s encounter velocity the effective Earth capture cross-section is 126 times the physical cross-section of the Earth[4]. In comparison, the Earth capture cross-section for a particle derived from Comet Encke is only 1.15 times the physical cross-section. If IDPs of both types were equally prevalent in the Zodiacal Cloud, the low-velocity, asteroidal IDPs would outnumber the IDPs from Encke by a factor of 109 to 1 in Earth collections.

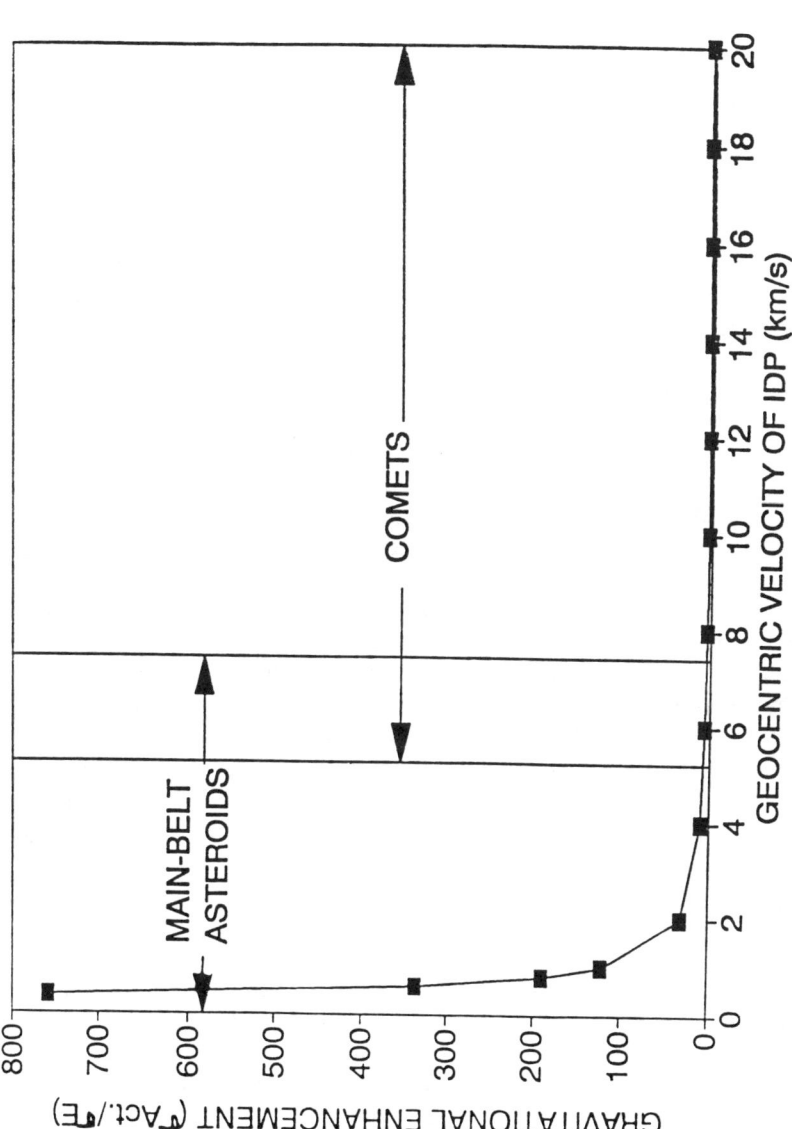

Figure 1: Gravitational enhancement, the ratio of the effective cross-sectional area for Earth capture of an IDP to the physical cross-section, as a function of the geocentric velocity of the IDP prior to Earth gravitational acceleration. The velocity ranges for IDPs from main-belt asteroids and from comets are taken from Jackson and Zook[3] as discussed in the text.

Thus the ratio of asteroidal to cometary dust in all near-Earth collections from the stratosphere, the polar ices, the sea floor, and any Earth-orbiting satellites will be different from that ratio in interplanetary space. Gravitational focusing strongly biases all near-Earth collections in favor of IDPs derived from main-belt asteroids over those comets.[4]

ATMOSPHERIC DECELERATION

IDPs which encounter the Earth are decelerated by collisions with air molecules and subsequently radiate away the energy acquired in these collisions.[5] Flynn[6] has shown that noticeable deceleration of an incoming 20 μm diameter IDP occurs at altitudes above 175 km, and atmospheric deceleration continues until the IDP has slowed to atmospheric settling speeds (a few meters/sec or less) at about 80 km altitude. The duration of the atmospheric deceleration phase is generally less than a minute, though, in rare cases, IDPs can be captured into Earth orbit and experience repeated gentle decelerations of much longer durations.[7]

Atmospheric Entry Heating

A procedure to model the heating pulse experienced by an IDP on atmospheric entry was developed by Whipple[5], and extended by Fraundorf.[8] The peak temperature reached on atmospheric entry can be calculated for an individual IDP given the size, density, entry velocity, and entry angle.

Flynn[6] developed a computer simulation of the atmospheric entry heating and deceleration process, allowing the actual atmospheric density profile to be incorporated. This model predicts substantially less heating for very small particles (<5 μm) and those encountering the Earth at grazing incidence, than does the Fraundorf[8] model. Love and Brownlee[7] have extended these calculations with a computer simulation that includes the effect of partial vaporization. The simulations allow calculation of the time-temperature profile experienced during atmospheric deceleration for a particle of a given size, density, entry velocity, and entry angle.

However, these entry heating calculations all rest on assumed values for three parameters never measured experimentally: the IDP emissivity at temperatures from a few hundred degrees to the silicate melting temperature (~1600 K), the IDP drag parameter at gas densities appropriate for the upper atmosphere, and the degree of elasticity of the collisions. Following the arguments of Whipple[5], Flynn[6] has described the rationale for selecting appropriate values for these parameters.

However, Fraundorf[8] suggested the assumption that IDPs radiate like a blackbody (emissivity = 1) might break down for small particles. Rizk et al.[9] have repeated the entry heating calculations using both the blackbody assumption and a theoretical expression for the emissivity appropriate for olivine spheres of selected sizes. For spheres of forsteritic olivine ≥10 μm in diameter they found peak temperatures they calculated for the olivine spheres agreed to within 100 K with the predictions using a blackbody emissivity.[9] Since most IDPs are dark, their true emissivities are likely to be somewhere between those calculated for spheres of olivine and the blackbody emissivity. Thus, the results of Rizk et al.[9] indicate that the blackbody emissivity assumption produces errors of <100 K in the entry heating calculations for IDPs ≥10 μm in diameter.

Rizk et al.[9] found the peak temperatures they calculated for spheres of olivine <10 μm in diameter to be several hundred degrees higher than those calculated using the blackbody assumption. The olivine spheres provide a lower bound on the emissivity of chondritic IDPs, however the porous, fine grained IDPs are likely to be better emitters than are the olivine spheres. The validity of entry heating calculations using the blackbody assumption for IDPs substantially smaller than 10 μm in diameter has yet to be established.

The entry heating model determines only the surface temperature versus time function of an IDP. Interior temperatures, which are used to constrain the entry velocities of individual IDPs[6], are generally assumed to be roughly equal to the surface temperatures because of the small sizes of the IDPs. Szydlik and Flynn[10&11] have explicitly modeled the internal temperature profiles of homogeneous, spherical IDPs up to 100 μm in diameter, and shown that for thermal conductivities as low as the value measured for lunar soil in a vacuum the particles have equilibration times short compared to the duration of the thermal pulse experienced on atmospheric entry. Thus, at least up to the temperature at which IDPs undergo significant phase changes, the particles will have an approximately uniform temperature. A consequence of the Szydlik and Flynn[10&11] results is that the rotation state of the IDPs on atmospheric entry is unimportant to the thermal calculations, since both a rapidly rotating IDP and one entering in a fixed orientation will be heated uniformly.

Mineral Transformations

Other then melting, the most obvious mineralogical alteration of IDPs on heating is the production of magnetite or maghemite, which has previously been observed in natural and artificially produced meteorite fusion crusts.[12] Magnetite rims are observed on both anhydrous and hydrated IDPs.[13] Partial rims are also observed.[14] Fine grained magnetite in the interior or on the surface of individual olivine crystals is also observed.[8]

Studies on IDPs show the presence of a magnetite rim correlates qualitatively with other indicators of atmospheric entry heating.[15] However, pulse heating studies have not yet been conducted on IDPs to determine the magnetite formation temperature, or the variation of this temperature with the initial IDP mineralogy. In studies on analog materials, Flynn et al.[16] have reported that magnetite crystals form within 25 seconds on some layer-silicates heated to 1100°C.

Sandford[17] has shown that a layer-silicate type IDP, pulse heated in steps from 190° to 1200°C, showed pronounced changes in the depth of the 3.0 μm water band in the particle's infrared absorption spectrum. This feature disappeared completely at 560°C. However, the alteration temperature measured by Sandford[17] has not been correlated directly with mineralogical changes, such as the alteration of the basal lattice spacing, visible in the Transmission Electron Microscope (TEM).

Zolensky and Lindstrom[18] have observed serpentine in L2005P17 exhibiting corrugated basal layers and which dehydrates more easily than typical serpentine under the TEM electron beam. They have previously observed similar features in serpentine heated to approximately 300° C, and they suggest the damage to L2005P17 serpentine may have been induced by atmospheric entry heating.

Klöck and Zolensky (Pers. comm., 1993) have independently observed magnesiowüstite, (Mg,Fe)O and periclase (Mg,Ca)O, in heated IDPs, and suggest that these minerals formed from the thermal breakdown of Mg-Fe-Ca carbonates.

Although olivine and pyroxene, the dominant minerals in the anhydrous IDPs, are more stable than are the layer-silicates, Keller et al.[15] and Rietmeijer[19] have reported the alteration of olivine to laihunite in at least two IDPs, and they attribute this to high-temperature oxidation of the olivine. Rietmeijer[19] has also reported the alteration of a rare, Ca-rich olivine to garnet in another IDP.

Volatile Element Loss

Laboratory simulation experiments have demonstrated that volatile elements are lost from meteorites upon heating (see the review by Lipschutz and Woolum[20]), but the time scales of these meteorite experiments are generally hours or days. Wulf and Palme[21] have reported similar experiments on Murchison and Allende heated to temperatures from 1050°C to 1300°C for durations of 4 days. However, the duration of the entry heating pulse experienced by IDPs is only a few seconds.[6]

The effect of atmospheric entry heating on the element abundances in IDPs may depend on factors such as grain size of the mineral(s) containing the volatile element, the host phase of the volatile element, as well as the peak temperature and duration of the thermal pulse. Sulfur loss, for example, may depend on grain size of Fe-Ni-S because of diffusion rate limitations.[22]

Fraundorf et al.[23] showed that S was lost from chondritic IDPs heated for 20 seconds, to simulate the atmospheric entry thermal pulse. The S depletion became severe above 800°C, and they suggested a chondritic S/Si ratio was probably incompatible with an IDP having been heated above 800°C for even a few seconds.[23] However, some chondritic IDPs might have low intrinsic S/Si ratios (i.e., the distribution of S-bearing phases might be inhomogeneous in the parent body), thus low S/Si should not necessarily be interpreted as indicating severe heating.[23]

Flynn[2] suggested that Zn, which is generally more volatile than S in meteorite heating experiments, could also serve as a monitor of atmospheric entry heating for the IDPs. Flynn and Sutton[24] have identified a subset of the chondritic IDPs which exhibit Zn depletions to less than 0.1 x CI. Flynn et al.[25], Thomas et al.[13], Klöck et al.[26] and Keller et al.[15] have demonstrated that Zn depletions correlate well with mineralogical evidence (magnetite rims and heated textures) and He release temperature inferences of substantial atmospheric entry heating. However, as Fraundorf et al.[23] cautioned for S, it is possible that Zn is not homogeneously distributed within the parent body, and low Zn/Fe could, in some cases, result from an inhomogeneous Zn distribution. Pulse heating experiments on IDPs have not yet been performed to establish the Zn mobility temperature.

Fraundorf et al.[23] suggested that C, not measured in their pulse heating experiments, might be lost at low temperature. However, Thomas et al.[13] have identified both anhydrous and hydrated IDPs which have low-Zn abundances and well-developed magnetite rims, but carbon contents of about 2 times the CI abundance. The mobility of C in IDPs has yet to be established, and is likely to depend critically host phase(s) of the carbon, hydrocarbons, carbonates, and/or poorly graphitized carbon.[27]

Flynn et al.[28&29] have compared the abundances of the elements from Cr to Br in a set of IDPs which experienced severe atmospheric entry heating and a second set which were less heated (see Figure 2). They suggest that the element mobility sequence is Zn, followed by Br, followed by Ge. However this mobility sequence rests on the assumption that both the severely heated and the less heated IDP groups started out with the same average chemical compositions. Since cometary IDPs are expected to be heated to significantly higher temperatures, on the average, than main-belt asteroidal IDPs[2], compositional differences between the severely heated and the less heated IDPs might reflect differences in the source compositions as well as the effects of element mobilities.

Solar Flare Track Annealing

During their residence in space the IDPs are exposed to the solar wind and solar flare ions. The crystals in the IDPs accumulate radiation damage tracks from the passage of heavy ions (principally Fe) from the solar flares, and these tracks have been observed in TEM images of individual crystals within the IDPs.[30] Fraundorf et al.[23] have shown that fresh Fe ion tracks are annealed in olivine by a 25 second heating pulse at 572°C, while tracks in enstatite were annealed by heating to 705°C. Thus solar flare tracks are likely to be annealed in particles which experience significant thermal pulses during atmospheric entry. Track annealing experiments have not been performed on actual solar flare tracks in IDP silicates, though these tracks are known to be less stable in the electron beam of the TEM than fresh Fe tracks.

Loss of Solar Implanted Noble Gases

Solar wind ions, including the noble gases, are implanted in IDPs during their space residence. Rajan[32] and Hudson et al. (1981) have measured concentrations of He, Ne, Ar, and Xe in IDPs at comparable concentrations to those in the gas-rich lunar soils. Nier and Schlutter[33] have demonstrated that implanted solar He is released from lunar grains and IDPs during heating pulses comparable to those experienced on atmospheric entry, and they suggest the release profile indicates the peak temperature reached by the particle during atmospheric entry. See also the paper by Nier[34] in this volume.

Consistency of the Thermal Indicators

Klöck et al.[26] have compared the heating inferences from a chemical indicator (degree of Zn depletion), a mineralogical indicator (the formation of a magnetite rim), and a noble gas indicator (the abundance and minimum release temperature of He) in a set of IDPs each of which was examined by three different techniques. Generally the mineralogical and chemical measurements were performed sequentially on the same fragment, but the He measurements were made on a different fragment an IDP which broke up on collection. Klöck et al.[26] found these three indicators provided a consistent picture of the degree of heating experienced by individual IDPs. However, Flynn et al.[35] found that two chemical indicators, Zn and S, gave contradictory indications (both Zn normal but S depleted, and S normal but Zn depleted) of the degree of heating in 14 out of 30 IDPs for which both elements were measured, and that the Zn content agreed with other indicators of heating (magnetite and/or He content) in 6 of the 7 cases for which independent evidence of heating (magnetite rims or He content) was available.

Particle Fragmentation

Particle fragmentation during entry could produce more extreme heating than calculated for the resulting fragments since a large particle penetrates more deeply into the atmosphere prior to deceleration. Fragmentation can also alter the distribution of densities, since particles are likely to fragment into their strongest subunits. However, Love and Brownlee[7] have calculated that few IDPs experience fragmentation because even fragile cometary particles melt before encountering a dynamic pressure sufficient for fragmentation.

Isotopic Fractionation

Significant mass fractionations (up to 32‰·amu) have been observed for Ni and Fe in magnetite-wüstite spherules recovered from the sea floor. Davis and Brownlee[36] infer the pre-atmospheric mass of these spherules from the fractionation.

The large Zn depletions in some IDPs (≤ 0.01 x CI in one IDP[13]) might suggest vaporization of a substantial fraction (99% or more) of the starting Zn. If so, then large mass fractionations would be expected in the residual Zn due to Rayleigh fractionation. Detection of isotopic mass fractionations among the volatile elements in IDPs thought to have been severely heated would confirm this heating, and could be used to estimate the fractional mass loss of the element measured.

Contamination During Entry

Jessberger et al.[37] have suggested that IDPs acquire substantial contamination during atmospheric entry or residence, and that this contamination may substantially perturb the minor element content of the IDPs. The possibility of contamination during atmospheric entry can be assessed from direct measurements of the atmospheric content of meteoric elements.

The E-layer of the terrestrial mesosphere, between 80 and 110 km altitude, is derived from meteoric ablation.[38] Concentrations of Na and Fe, contributed by meteoric vapor, have been monitored in the mesosphere, and both individual meteors and average concentration profiles have been measured.[38]

Individual IDPs entering the Earth's atmosphere experience most of their deceleration in the mesospheric layers rich in meteoric volatile elements. Limits on the extent to which individual IDPs can be contaminated by meteoric volatile elements during deceleration in the upper atmosphere can be established by considering the extreme cases: the direct passage of an IDP through a meteoric vapor trail, or the passage of an IDP through the mesospheric layer rich in meteoric volatiles.

The worst case for contamination would occur if an IDP were to pass directly through a meteoric vapor trail. In this case the local concentration of meteoric material is maximized. Peak densities of 10^4 atoms/cm^3 for Na and 10^5 atoms/cm^3 for Fe have been measured in meteor vapor trails.[38] However these trails are small (~100 m wide) and persist for only a short period of time (typically a few minutes or less).[38]

During the direct passage of a 10 μm diameter IDP through a 100 m diameter vapor trail having a Na ion density of 10^4 atoms/cm^3 and an Fe density of 10^5 atoms/cm^3, the IDP would directly collide with only 80 Na ions and only 800 Fe atoms. This level of potential contamination is far below the Na and Fe contents of IDPs having a chondritic composition (i.e., the composition of the CI meteorite Orgueil), suggesting that the contamination for Na and Fe acquired by chondritic IDPs through this mechanism is negligible. Since the other elements in the E-layer are also derived from meteors, this suggests that the contamination by other elements is similarly small.

Kane and Gardner[38] have used Lidar to monitor an area of the sky for meteor vapor trails. They detected 89 meteor events in 450 hours of data acquisition. Assuming an average trail duration of 1 minute, a meteor trail appeared within the field of view of their instrument for about 0.0033 of the observing time. Thus events in which a decelerating IDP passes through a meteor trail must be quite rare.

Alternatively, contamination could occur as the IDP passed through the mesospheric layer of vapor accumulated from many individual meteors. The effectiveness of this mechanism to produce IDP contamination can be modeled using the terrestrial accretion rate of meteors and the mesospheric residence time of their debris. The terrestrial accretion rate of meteoritic particles from 10^{-13} to 10^6 grams in mass is estimated to be 16 x 10^9 grams/year.[39] For example, if each particle were to initially have a Zn content equal to the CI meteorite content (~300 ppm), than the terrestrial accretion rate of meteoritic Zn would be 4.8 x 10^6 grams/year. Although some particles in this size range are known to survive atmospheric entry with their Zn intact, we can obtain an upper limit on the addition to this meteoric vapor layer by assuming that all 4.8 x 10^6 grams/year of Zn is deposited into the upper atmosphere. The surface density of Zn in this atmospheric (i.e., the total mass density along a line from the top of the atmosphere to the surface) can then be calculated as the product of the Zn residence time (t_{Zn}) and the Zn accretion rate divided by the surface area of the atmosphere:

$$4\pi R^2 = 4\pi (6.4 \times 10^6 \text{ m})^2 = 5.1 \times 10^{14} \text{ m}^2 \quad .$$

Thus the maximum surface density of Zn in this atmospheric layer would be:

$$9.4 \times 10^{-9} \text{ grams/m}^2 \cdot \text{year} \times t_{Zn} \quad .$$

If we then allow a 10 μm diameter IDP to pass through this layer during atmospheric deceleration, the particle could accrete all the Zn encountered by its cross-sectional area:

$$\pi r^2 = 8 \times 10^{-11} \text{ m}^2$$

Thus the total Zn accretion possible in this model is

$$8 \times 10^{-19} \text{ grams/IDP·year} \times t_{Zn}$$

To acquire a chondritic Zn content (300 ppm) by this mechanism requires the addition of 1.5×10^{-13} grams of Zn to a 10 μm diameter, density 1 gm/cm^3, spherical IDP. Thus the value of the Zn residence time (t_{Zn}) would have to be 2×10^5 years for the meteoric layer to have a high enough Zn content for this mechanism to be viable. However, Hunten et al.[40] have computer modeled the deposition of meteoric material into the Earth's upper atmosphere. They suggest that meteoric vaporization products probably recondense into "smoke" particles with a mean size of about 1 nm at an altitude of about 80 km. Even employing a viscous settling model, which results in much longer residence times than for models permitting coagulation and eddy diffusion used by Hunten et al.[40], Sutton and Flynn[41] calculate the descent of the 1 nm meteoric smoke from 80 km to 40 km takes no longer than 50 years, far short of the 2×10^5 year Zn residence time required for the IDP to encounter a chondritic amount of Zn during deceleration in the mesosphere. Nonetheless, the true behavior of meteoric vapor in the Earth's atmosphere has not been observationally verified, and only model calculations indicate these effects are negligible.

Alternatively, we could allow the IDP to collect Zn over a much larger area than its physical cross-section. However, even in the 50 year Zn residence time is assumed, each particle would require a effective Zn accretion cross-section 4000 times its physical cross-section. No mechanism by which a hot, high-speed IDP decelerating through the upper atmosphere could collect Zn (or other volatiles) over a cross-section so much larger than the physical cross-section of the particle has been suggested.

It appears the interaction of IDPs with meteoric vapor during deceleration in the upper atmosphere does not produce significant contamination of IDPs as they decelerate in the upper atmosphere.

ION IMPLANTATION

During the initial stage of the deceleration the individual air molecules strike the surface of the IDP with energies ranging from about 4 eV/amu at a 30 km/sec velocity to about 1 eV/amu at a 10 km/sec velocity when viewed in the rest frame of the IDP. Thus the air molecules may penetrate several atomic layers into the IDP surface before stopping.[5] The low N content of IDPs suggests few if any of these molecules are retained.[32]

STRATOSPHERIC RESIDENCE

Modeling of the settling of small particles through the Earth's atmosphere suggests that IDPs spend weeks to months in the stratosphere, with typical settling rates of a few cm/s in the lower stratosphere.[23] From measurements of the concentrations of particles of various sizes in the stratosphere Zolensky and Mackinnon[42] calculate the time between collisions for particles >1 µm is 5×10^9 years, and for particles >0.1 µm is 10^8 years, suggesting contamination through collision by solid particles during stratospheric residence is small.

Nonetheless, some IDPs are observed to have acquired contamination either during stratospheric residence or collection/curation. Several of the chondritic IDPs (e.g., U2015D8, and W7066A5) documented in the JSC Cosmic Dust catalogs are rigidly attached to aluminum oxide spheres, presumably solid rocket exhaust, as shown in Figure 3. Other chondritic IDPs (such as U2015D4) are found rigidly attached to fragments classified as terrestrial contaminants based on their element abundance patterns. Other IDPs show surface correlated increases in S, suggestive of contamination by stratospheric sulfate aerosols.[42]

Chemical Contamination

Aerosol particles, predominantly sulfuric acid droplets, are abundant in the Earth's stratosphere. Three stratospheric particles analyzed for surface chemistry by Auger spectroscopy showed a thin (~150 Å thick) sulfur-rich layer, suggesting contamination by stratospheric sulfuric acid aerosols. Because the contamination layer is thin, its effects on bulk particle chemistry are small, <1 weight % for S.[43]

Halogens (particularly Br and Cl) are abundant in the stratosphere. Rietmeijer[14&44] has shown a correlation of particle settling time with Br abundance, and Flynn et al.[35] have reported enrichments of Br in IDPs with otherwise CI-like volatile element abundances, both suggesting the possibility of Br contamination.

Jessberger et al.[37] suggest that all the volatile enrichments above the CI concentrations detected in the IDPs may result from contamination by meteoric material residing in the Earth's atmosphere. However, Stephan et al.[45] have examined the spatial distributions of Zn and Br in an IDP using Time-of-Flight SIMS, and concluded that the Zn appeared to be indigenous. Flynn et al.[28] suggest that the absence of heated IDPs with high contents of Zn and other volatiles indicates that once volatile elements are depleted by atmospheric entry heating the particles are not subsequently contaminated by significant (i.e. CI-level) amounts of those volatile elements. This indicates that the level of stratospheric contamination experienced by severely heated IDPs (identified by their mineralogy and texture in the TEM, not by their trace element content) is small or nonexistent. However, the extent of chemical contamination of individual IDPs has not be convincingly assessed.

Mineralogical Contamination

Rietmeijer[46&47] has detected fragments of silica-rich glass and tridymite in an IDP. He suggests these fragments are terrestrial volcanic debris incorporated into the IDP during stratospheric residence.

COLLECTION AND CURATION

During the collection and curation process the IDPs are exposed to a variety of potential contamination sources, although considerable efforts are made to minimize these effects. See the paper in this volume by Warren and Zolensky.[48]

Silicone Oil

IDPs are collected from the stratosphere by impact into a layer of silicone oil. Despite efforts to remove this oil, Sandford and Walker[49], using infrared absorption, and Rietmeijer[50], using Analytical Transmission Electron Microscopy, have demonstrated that silicone oil remains on the highly porous particles after cleaning.

Overlapping Impacts

Contamination could occur by the chance impact of an IDP onto a region of the collector already occupied by another particle. The probability of these overlapping impacts can be assessed from the fraction of the collector surface covered by particles. Zolensky and Mackinnon[42] have examined the material washed from the surface of the W7017 collector, which sampled the stratosphere over the western United States for 45 hours during the summer of 1981. This was prior to the eruption of the El Chichón volcano in the spring of 1982, so the concentration of volcanic debris in the stratosphere was expected to be low. They estimated that 650 particles >5 µm in diameter and 25,000 particles >1 µm were present on the W7017 collector.[42] Preliminary examination suggested that submicron particles occurred with about the same frequency as particles >1µm.[42]

The fractional area of the W7017 collector covered by particles in each size range is estimated in Table 2. With only 3.9×10^{-5} of the collector surface area covered by particles, the probability that any individual chondritic IDP will overlap with another particle on the collector is quite low. If all 25,650 particles >1 µm in size were spread uniformly over the 30 cm² collector surface they would occur at the intersections of a grid with a 340 µm linear spacing. The probability that the last 10 µm diameter IDP would overlap with one particle >1 µm in size is ~0.003. Thus, even with 65 chondritic IDPs >5 µm on the W7017 surface[42] the probability of a chance overlap between a chondritic IDP and a stratospheric contaminant is low.

The situation is quite different for the chance overlap of a "cluster particle" with a previously collected stratospheric particle. "Cluster particles" are a group of fragments occurring within a localized region, generally 100 to several hundred micrometers, on the collector surface. All particles within this region are designated as members of the cluster, and are frequently treated as if they were all fragments of the same parent particle which broke up on or shortly before impact with the collector. The probability of a chance overlap between a cluster distribution and a single fragment of the uniformly distributed particles approaches 1 as the size of the cluster distribution approaches 340 µm. Thus most large clusters are likely to include at least one unrelated overlap fragment >1 µm in size if the particle number density determined by Zolensky and Mackinnon[42] for the W7017 collector is representative of other collector surfaces.

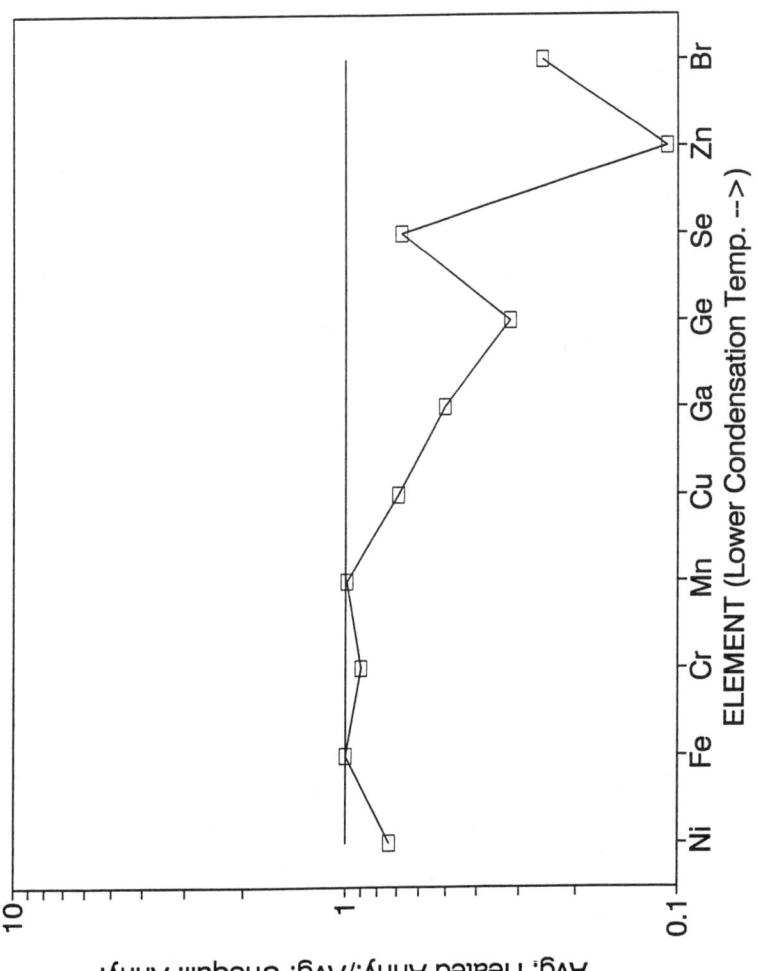

Figure 2: Average element abundances in 8 severely heated anhydrous IDPs ratioed to the average element abundances in 5 unequilibrated anhydrous IDPs show the severely heated IDPs have significantly lower contents of Zn, Br, and Ge, as well as smaller depletions in Ga, Cu, and Ni, than the unequilibrated IDPs. (Data from Flynn et al.[29]).

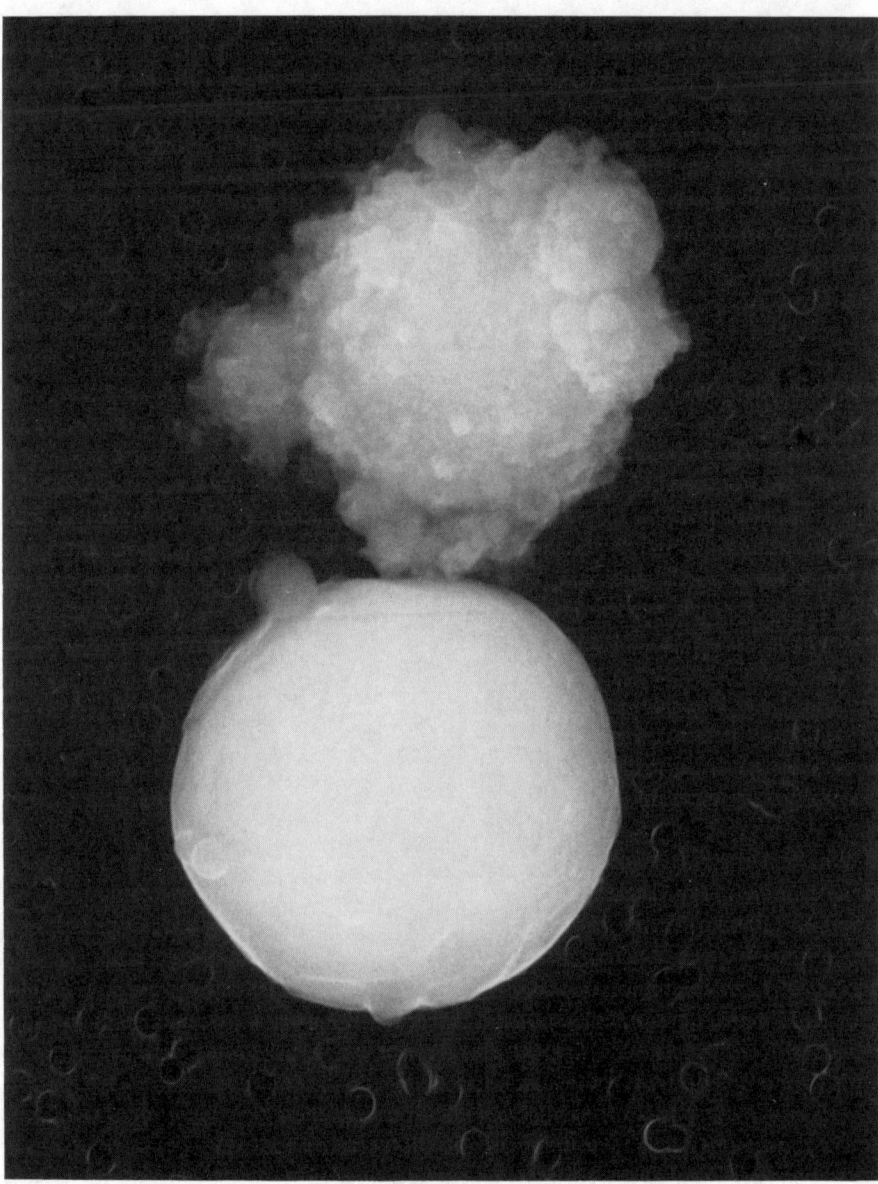

Figure 3: Johnson Space Center catalog photo showing chondritic IDP U2015D8 rigidly attached to an aluminum oxide sphere (U2015D9). This suggests some mechanism exists for the contamination of IDPs with man-made material found in the Earth's stratosphere. Each individual grain in this image measures 6 μm in diameter. (NASA JSC Photo S-84-41340.)

Table 2:
Stratospheric Particles on the W7017 Collector

Size Range	Number on W7017 Collector[*]	Minimum Area Covered (μm^2)[+]	Fraction of Surface Covered[**]
>5 μm	650	3.3×10^4	1.1×10^{-5}
1 to 5 μm	25,000	7.9×10^4	2.6×10^{-5}
<1 μm	25,000	4.9×10^3	1.6×10^{-6}
Total	50,650	11.7×10^4	3.9×10^{-5}

[*] Data from Zolensky and Mackinnon[42]
[+] Assumes a diameter of 8 μm for the >5μm particles, 2 μm or the 1 to 5 μm particles, and 0.5 μm for the < 1 μm particles.
[**] Total collector surface area ~30 cm^2.

Examination of the W7066*A cluster supports the idea that clusters may experience significant overlap. Most fragments of the W7066*A cluster fit the "cosmic" classification, but fragment #5 is a porous aggregate attached to an aluminum oxide sphere (AOS), likely to be solid rocket exhaust, and a second AOS occurs within the field of the W7066*A cluster (see Cosmic Dust Courier, Number 5, June 1984, p.2 - available from NASA JSC - see Warren and Zolensky[48]).

Collector surfaces with higher particle concentrations than found on the surfaces examined by Zolensky and Mackinnon[42] would have higher probabilities for chance overlap. Similar measurements have never been performed on the Large Area Collectors, so we can only guess at the probability for overlaps among the numerous cluster particles identified on these collectors. Researchers examining fragments of cluster particles need to be particularly alert for the possibility of overlap particles unrelated to the cluster parent which may occur within the cluster field.

Interactions on the Collector

Interactions on the collector surface may also cause contamination. Stratospheric sulfuric acid droplets and terrestrial particles, collected along with the IDPs, may interact with the IDPs during impact onto the collector, residence on the collector surface, or cleaning. This might explain the unusual chemical contents, such as high Cd, measured in rare cases in otherwise chondritic IDPs. Rietmeijer[22] has suggested that droplets of sulfuric acid may be mobilized and deposited onto IDPs during the particle cleaning process.

CONCLUSIONS

Many IDPs recovered from the Earth's stratosphere have experienced thermal pulses during atmospheric deceleration sufficient to alter their mineralogies, volatile element contents, and noble gas abundances.[26] This thermal pulse may also cause track annealing, and could result in isotopic fractionation. Other IDPs may have experienced contamination during atmospheric entry, stratospheric residence, or the collection/curation process. The degree to which any individual IDP has been altered during Earth encounter requires detailed examination of that particle, and comparison to other IDPs which have experienced documented alterations.

ACKNOWLEDGMENTS

Constructive reviews by W. Klöck and F. J. M. Rietmeijer substantially improved this paper. This work was supported by NASA Grant # NGT3587.

REFERENCES

1. Öpik, E. J. (1951) *Proc. R. Irish Acad.* **54**, Sect. A, 165-199.
2. Flynn, G.J. (1989) *Icarus,* **77**, 287-310.
3. Jackson, A. A. and Zook, H. A. (1992) *Icarus* **97**, 70-84.
4. Flynn, G.J. (1990) *Proc. Lunar Planet. Sci. Conf. 20th*, 363-371.
5. Whipple, F. L. (1950) *Proc. Nat. Acad. Sci.* **36**, 687-695.
6. Flynn, G.J. (1989) *Proc. 19th Lunar Planet. Sci. Conf.*, 673-682.
7. Love, S. G. and Brownlee D, E. (1991) *Icarus* **89**, 26-43.
8. Fraundorf, P. (1980) *Geophys. Rev. Lett.* **10**, 765-768.
9. Rizk, B. et al. (1991) *J. Geophys. Res.* **96**, A2, 1303-1314.
10. Szydlik, P. P. and Flynn, G.J. (1992) *Lunar Planet Sci. XXIII*, 1391-1392.
11. Szydlik, P. P. and Flynn, G.J. (1992) *Meteoritics* **27**, 294-295.
12. Brownlee, D. E. et al. (1975) *J. Geophys Res.* **80**, 4917-4924.
13. Thomas, K.L. et al. (1992) *Lunar Planet. Sci. XXIII*, 1427-28.
14. Rietmeijer, F.J.M. (1992) *Meteoritics* **27**, 280-281.
15. Keller, L. P. (1992) *Lunar Planet. Sci. XXIII*, 675-676.
16. Flynn, G.J. et al. (1993) *Proc. NIPR Symp. Ant. Met.* in press.
17. Sandford, S. A. (1986) *Lunar Planet. Sci. XVII*, 754-755.
18. Zolensky, M.E. and Lindstrom, D.J. (1992) *Proc. Lunar Planetary Science* **22**, 161-169.
19. Rietmeijer, F.J.M. (1992) *Lunar Planet. Sci. XXIII*, 1151-1152.
20. Lipschutz, M.E. and Woolum, D.S. (1988) in *Meteorites and the Early Solar System* (eds. J.F. Kerridge and M.S. Matthews), Univ. Arizona Press, 462-487.
21. Wulf, A. V. and Plame, H. (1991) *Lunar Planet. Sci. XXII*, 1527-1528.
22. Reitmeijer, F.J.M. (1993) *Earth Planet. Sci. Lett.* **117**, 609.
23. Fraundorf, P. et al. (1982) in *Comets,* U. of Aizona Press, 383-409.
24. Flynn, G.J. and Sutton, S.R. (1992) *Proc. Lunar Planet. Sci. Conf.* **22**, 171-184.
25. Flynn, G.J. et al. (1992) *Lunar Planet Sci. XXIII*, 375-376.

26. Klöck, W. et al. (1992) *Meteoritics* **27**, 243-244.
27. Mackinnon, I.D.R. and Rietmeijer, F.J.M. (1987) *Rev. Geophys.* **25**, 1527-1553.
28. Flynn, G.J. et al. (1993) *Lunar Planet. Sci. XXIV*, 495-496.
29. Flynn, G.J. et al. (1993) *Meteoritics* **28**, 349-350.
30. Bradley, J. P. et al. (1984) *Science* **226**, 1432.
31. Rajan, R. S. et al. (1977) *Nature* **267**, 133-134.
32. Hudson, B. et al. (1981) *Science* **211**, 383-386.
33. Nier A.O. & Schlutter D.J.(1993) *Lunar Planet. Sci XXIV*, 1075.
34. Nier, A.O. (1994) Helium and Neon in Interplanetary Dust Particles. This volume.
35. Flynn, G.J. et al. (1993) *Lunar Planet. Sci. XXIV*, 497-498.
36. Davis A.M. & Brownlee, D.E. (1993) *Lunar Planet. Sci XXIV.*, 373-74.
37. Jessberger, E. K. et al. (1992) *Earth and Planet. Sci. Lett.* **112**, 91-99.
38. Kane, T. J. and Gardne,r C. S. (1993) *Science* **259**, 1297-1299.
39. Hughes, D. W. (1978) in *Cosmic Dust*, Wiley, New York, 123-185.
40. Hunten, D. M. et al. (1980) *J. Atmos. Sci,* **37**, 1342-1357.
41. Sutton, S.R. and Flynn, G.J. (1990) *Proc. 20th Lunar Planet. Sci. Conf.*, 357-361.
42. Zolensky, M.E. and Mackinnon, I. D. R. (1985) *J. Geophys. Res.* **90**, 5801-08.
43. Mackinnon, I.D.R. & Mogk, D.W.(1985) *Geophys Res. Lett.* **12**, 93.
44. Rietmeijer, F.J.M. (1993b) *Lunar Planet. Sci. XXIV*, 1261-1262.
45. Stephan, T. et al. (1993) *Lunar Planet. Sci. XXIV*, 1349-1350.
46. Rietmeijer, F.J.M. (1986) *Lunar Planet. Sci. XVII*, 708-709.
47. Rietmeijer, F.J.M. (1987) *Lunar Planet. Sci. XVIII*, 836-837.
48. Warren, J. and Zolensky, M.E. (1994) Collection and Curation of Interplanetary Dust Particles Recovered From the Stratosphere by NASA. This volume.
49. Sandford, S.A. and Walker, R.M. (1985) *Astrophys. J.* **291**, 838.
50. Rietmeijer, F.J.M. (1988) *Journal of Volcanology and Geothermal Research.* **34**, 173-184.

NASA WB-57F stratospheric aircraft carrying particle collection devices (arrowed) under the wings. These collectors are exposed only in the stratosphere (above the majority of terrestrial particles) to maximize the percentage of extraterrestrial particles on the collectors. These devices also collect a wealth of terrestrial and space debris particles, of considerable interest to atmospheric scientists. (NASA photo S81-31582)

CHEMICAL COMPOSITIONS OF PRIMITIVE SOLAR SYSTEM PARTICLES

S. R. Sutton
Department of the Geophysical Sciences
and
Center for Advanced Radiation Sources
The University of Chicago
Chicago, IL 60637, USA

ABSTRACT

Bulk chemical compositions for several hundred stratospheric micrometeorites have been determined using electron beam X-ray analysis, proton-induced X-ray emission, synchrotron X-ray fluorescence, secondary ion mass spectrometry, or instrumental neutron activation analysis. The *chondrite-like* particles are chemically similar to carbonaceous chondrites but exhibit enrichments and depletions in some volatile elements that distinguish them from members of any of the conventional meteorite classes. One possibility is that the stratospheric micrometeorites are samples of a new, more primitive class of chondritic material.

INTRODUCTION

Chemical studies of micrometeorites are of fundamental importance to studies of the early solar system for two principal reasons. First, micrometeorites are distinct samples from conventional meteorites (either material from different parent bodies or previously unsampled portions of the meteorite parent bodies). Second, atmospheric entry selection effects (such as destruction of friable objects and melting) are less significant than those for conventional meteorites.[1] As a result, particles which have experienced little post-accretional processing have a significant chance of surviving the Earth encounter and subsequent collection. Chemical analyses of these relatively unaltered micrometeorites may lead to a better understanding of the compositions of the most primitive materials in the solar system and thereby constrain the conditions (physical and chemical) that existed in the early solar nebula.

Micrometeorites have been collected from the stratosphere, polar ices and ocean sediments, but the stratospheric collection is the best source of the most unaltered material because they are less massive and are typically not heated to their melting points during atmospheric entry, and their terrestrial residence times (both stratospheric and curatorial times) are relatively very small. Despite the fact that the stratospheric micrometeorites have masses in the nanogram range, a variety of microanalytical techniques have been applied to bulk chemical analyses with part-per-million sensitivity. In some cases, multidisciplinary studies (e.g., chemistry and mineralogy) have been performed on individual particles.

The purpose of this paper is to review the status of efforts to establish the chemical compositions of primitive interplanetary dust particles through analyses of the stratospheric samples. This review concentrates on the *bulk* compositions of these particles and, as such, the research on mineral compositions (obtained largely by electron beam techniques) will not be summarized. A review of available microanalytic techniques and their capabilities is included as well as results to date. The first order conclusion is that the *chondrite-like* particles are chemically similar to carbonaceous chondrites but in detail are distinct from members of any of the conventional meteorite classes.

SAMPLING CONSIDERATIONS

Several hundred stratospheric particles have been analyzed for bulk composition by various techniques with differing sensitivities and elements. Since each particle is typically 1 nanogram in mass, this translates into a total analyzed mass of the order of one microgram. There are several important implications of this sampling situation.

o   The large number of analyzed particles is advantageous for identifying rare members of the interplanetary dust population, a situation somewhat akin to the discovery of rare achondrites in the Antarctic Meteorite Collection.

o   The large number of particles may make it possible to identify chemical subclasses in the collection.

o   The small mass of each particle is problematic from a pairing standpoint since few chemical heterogeneity studies have been conducted on meteoritic material at the nanogram scale.[2&3]

o   Attempts to compare average compositions of the micrometeorites with the bulk compositions of the various meteorite classes are plagued by the relatively low total mass analyzed.

o   Conclusive evidence of the extraterrestrial nature of *individual* particles is difficult to obtain and commonly is lacking. Most of the particles analyzed for bulk composition (and other properties for that matter) are inferred to be extraterrestrial based on circumstantial evidence such as major element composition, mineralogy and textural similarity to particles in which solar flare tracks have been observed.

o   The high surface to volume ratio of these small, often porous, samples makes contamination issues paramount.

Despite these complications, the bulk chemical compositions of the stratospheric particles is of great interest in providing better constraints on the nature of interplanetary dust and primitive Solar System bodies.

ANALYTICAL TECHNIQUES

Chemical analytical techniques with high sensitivity are required for the analysis of IDPs because of their small size (typically in the nanogram range). A variety of methods have been applied.

Electron Beam X-ray Analysis (EBXA)

Electron beam X-ray analysis is mainly applicable to major elements from the third and fourth rows with concentrations in excess of about 0.1%. Typically, these are elements with atomic numbers between those of Na and Ni. This approach has been applied typically by the attachment of a solid state X-ray detector to a scanning electron microscope (SEM) or electron microprobe. The electron beams used in this work (~15-40 kV) typically penetrate to depths on the order of a few micrometers. This dimension is the maximum sampling depth of the technique except for elements with atomic numbers below that of Mg where sampling is more shallow due to self absorption effects of the characteristic fluorescent X-rays. As a first step in the chemical characterization process, qualitative EBXA is obtained on each particle as part of the curatorial process at NASA-JSC (see the Cosmic Dust Catalogs produced by the JSC Curatorial Facility for a description of these procedures). These results provide semi-quantitative information on the relative abundances of elements with atomic numbers typically between those of Na and Ni. EBXA has also been used to obtain quantitative concentrations of Na, Mg, Al, Si, P, S, K, Ca, Ti, Cr, Mn, Fe, and Ni.[3&4] Carbon and oxygen have also been analyzed with EBXA using windowless, energy-dispersive X-ray detectors.[5]

Proton-Induced X-ray Emission (PIXE)

PIXE is a trace element technique applicable to elements in the third and fourth rows with concentrations above about 10 ppm. Results have been reported for most elements between Al and Br.[6] The proton beams used in this work (~3 MeV) typically penetrate IDPs with little attenuation and the sampling depth is therefore constrained primarily by the self absorption effects of the fluorescent X-rays. Such effects become important for elements with atomic number below about that of sulfur.

Synchrotron X-ray Fluorescence (SXRF)

SXRF is a trace element technique applicable to elements in the third and fourth rows with concentrations above about 1 ppm. Results have been reported for most elements between S and Mo.[6] The X-ray beams used in this work (a continuum between about 3 and 30 keV) typically penetrate IDPs with little absorption and the sampling depth is therefore constrained primarily by the self absorption effects of the fluorescent X-rays. As with the other fluorescence-based techniques, such effects become important for elements with atomic number below about that of sulfur. The energy deposition by the X-ray microbeam is several orders of magnitude lower than that of charged-particle beam techniques making it well suited for volatile element analyses.

Secondary Ion Mass Spectrometry (SIMS)

Unlike the 3 previously discussed fluorescence-based techniques, SIMS uses intense ion beams to sputter material from the sample and elemental abundances are derived from mass spectrometry of the resulting ionized species.[8] (Isotopic compositions determined with this technique are not discussed in this review.) Elemental sensitivity varies dramatically across the periodic table. SIMS is especially sensitive for elements with high ion yield. Results represent the bulk composition of the sputtered fraction of each sample. SIMS is obviously a destructive technique.

Instrumental Neutron Activation Analysis (INAA)

INAA is based on spectrometry of gamma rays emitted by neutron-activated species in the sample.[9&10] Both the irradiating neutrons and emitted gamma rays have long absorption lengths in IDPs and therefore INAA is truly a bulk technique in these samples. Although extremely sensitive for some elements, INAA is generally limited to the relatively large IDPs in the collection ≥ 20 µm).

Other Techniques

Scanning auger microprobe (SAM) has been used for analyses of particle surfaces particularly in the search for contaminating aerosol species.[11] The laser microprobe/mass spectrometer (LM/MS; see reference 12) and double laser mass spectrometry (mL^{2}MS; see reference 13) techniques have provided analyses of volatile/hydrocarbon species in stratospheric particles. The work on hydrocarbons will not be discussed here but is described by Gibson and Bustin in this volume.[14]

RESULTS

Analytical Capabilities

Bulk chemical microanalysis of stratospheric micrometeorites using these techniques is summarized in Figure 1. Shown for each element are those techniques for which at least one, non-limit (i.e., upper limits were omitted) result has been published. No attempt has been made to indicate the number of particles analyzed and this varies dramatically across the table. A number of pieces of information can be derived from this figure.

o The coverage is fairly remarkable considering the mass of these particles. At least one analysis exists for 51 out of the 89 naturally-occurring elements (i.e., about 60%).

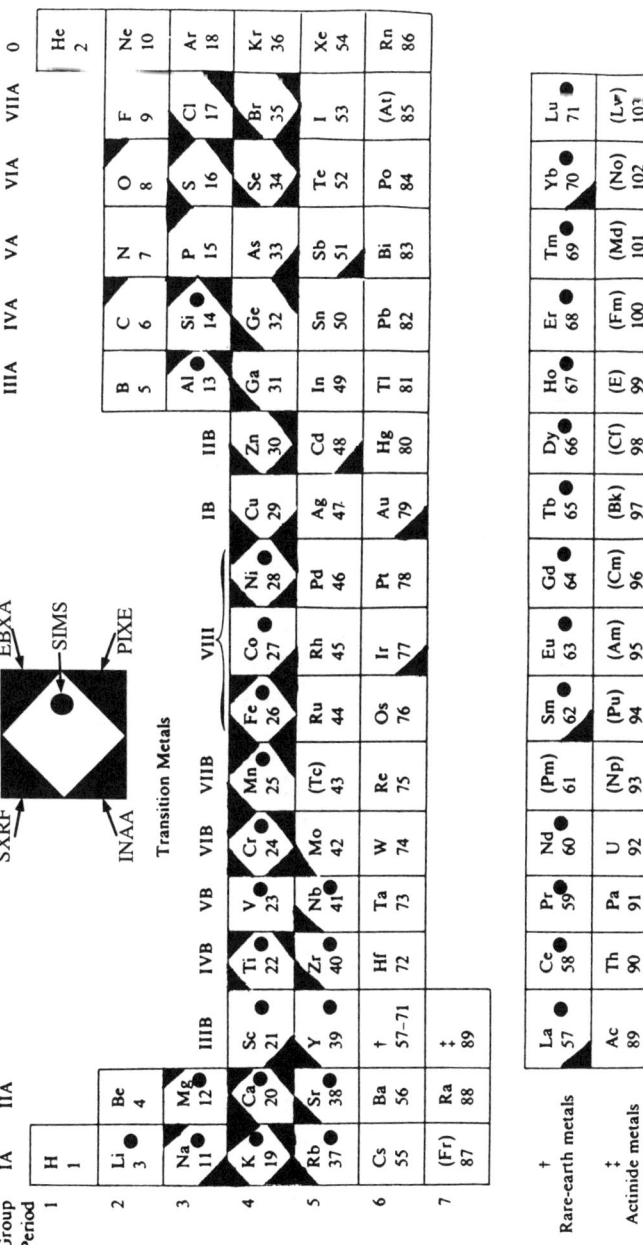

Figure 1. This annotated periodic table summarizes published bulk chemical microanalyses of stratospheric micrometeorites. Shown for each element are those techniques for which at least one, non-limit result has been published. No attempt has been made to indicate the number of analyses reported in each case.

o   Considerable overlap exists between the various techniques, most notable for the first row transition elements. This overlap suggests that intercomparisons might be done on individual particles although few such studies have actually been carried out.

o   There are many elements for which no data exists. As indicated above, roughly 40% of the periodic table is unrepresented and this is primarily a sensitivity phenomenon since the majority of the results reside in the upper rows.

o   One can easily see from this figure which technique(s) are applicable for particular elements of interest to a particular study.

Major Element Results

In the most extensive study of major elements in *chondrite-like* micrometeorites, Schramm et al.[3] analyzed 200 stratospheric micrometeorites by EBXA for Mg, Al, Si, S, Ca, Cr, Mn, Fe, and Ni (Figure 2). The particles were grouped morphologically into *porous* and *smooth* types. The porous particles were found to have compositions close to that of CI meteorites but the smooth ones were depleted in Ca and Mg analogous to depletions in CI and CM matrix attributed to parent body leaching (most likely on an asteroidal parent body). Schramm et al. concluded that the porous, most chemically primitive particles should be considered a new type of carbonaceous chondrite unrepresented in the meteorite collection. Thomas et al.[5] showed that *chondrite-like* IDPs tend to be enriched in carbon relative to C-rich meteorites. INAA results for *chondrite-like* IDPs led to the conclusion that Na and K contents are within a factor of two of CI abundance. Flynn and Sutton[7] suggested that Ni determination by SXRF was a quick screening method for identifying *chondrite-like* particles among the particles classified as C-type in the preliminary examination procedure. Thus, from a major element standpoint, the IDP compositions are generally comparable to those of carbonaceous meteorites but several elements are anomalous. Enrichment *and* depletion effects have been observed.

Trace Element Results

The value of trace elements in IDPs was first demonstrated by Ganapathy and Brownlee using INAA.[10] Trace elements for which there are quantitative data for IDPs comprise a surprisingly long list, a direct result of the complementary nature of the various analytical techniques. Unfortunately, there are few cases where these trace element techniques have been applied to the same particle. The multitude of data on elements of different cosmochemical character coupled with the scarcity of analytical intercomparisons on individual particles makes the task of summarizing the average composition of IDPs complex. In general, the conclusion of the trace element analyses on *chondrite-like* particles agrees with that of the major element analyses, i.e., particles which have *chondrite-like* major element compositions also have *chondrite-like* trace element compositions. The most intriguing deviations from this conclusion concern the volatile elements.

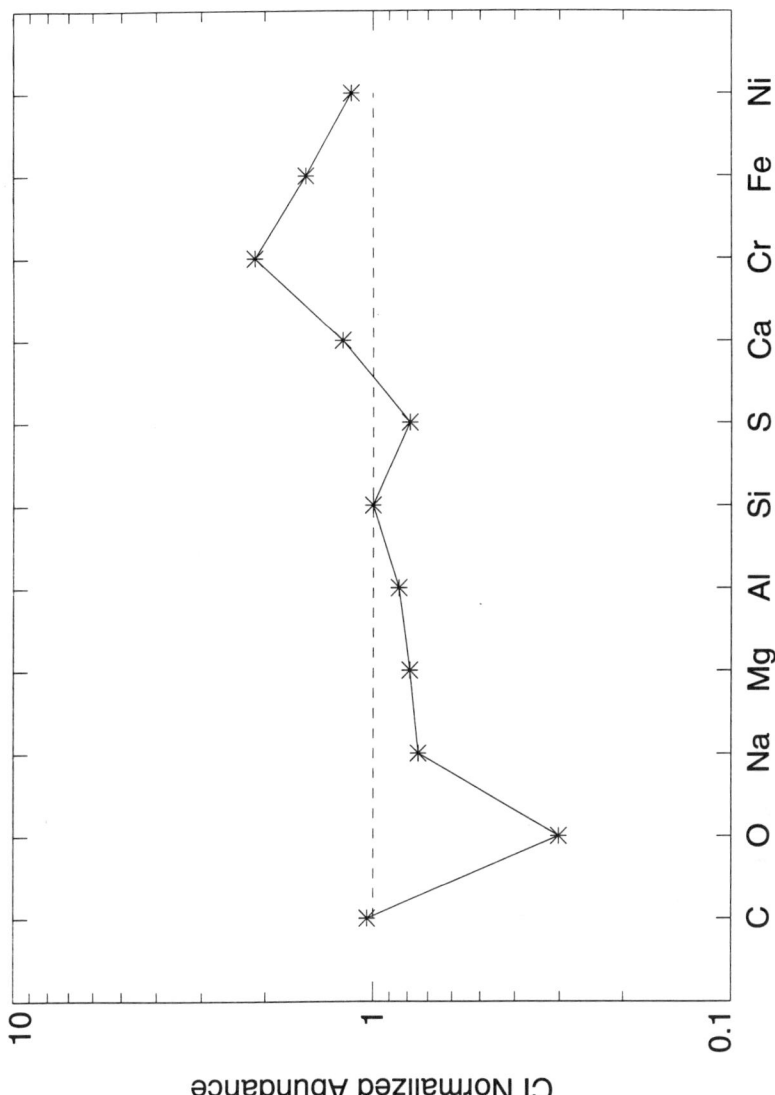

Figure 2. Summary of major element compositions (Si normalized) of 200 IDPs by electron beam X-ray analysis (from reference 3).

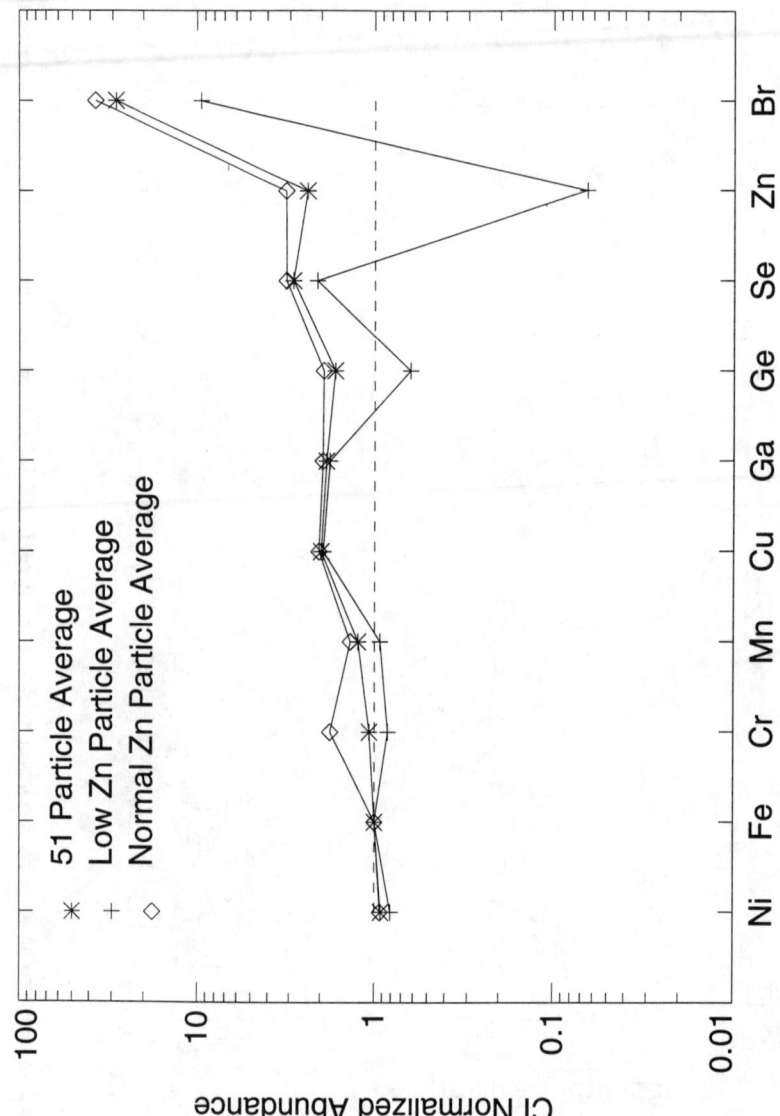

Figure 3. Average composition of 51 IDPs by synchrotron X-ray fluorescence (SXRF) analysis (from reference 15).

Volatile trace element enrichments

The deviations from the nominal chondritic abundance pattern that have received the greatest attention are the general enrichments in volatile elements first documented with PIXE by Van der Stap et al.[15] Subsequently, similar enrichments in many particles have been observed [references 16&17, and references therein]. The average compositions of 51 chondritic IDPs determined by SXRF show an abundance pattern enriched from CI in volatiles and complementary to the depletions in the non-CI carbonaceous meteorites (Figure 3; reference 17).

Volatile trace element depletions

Trace element abundance patterns for some IDPs are similar to those of most *chondrite-like* particles but exhibit varying degrees of Zn depletion.[18&19] The magnitude of the Zn depletion is well-correlated with the thickness of the magnetite rim suggesting that an atmospheric heating mechanism is responsible.

Sc-Co Provenance Signature

Lindstrom[20] showed that Sc and Co (normalized to Fe) were particularly useful in determining provenance of particles since terrestrial contaminants, igneous material and chondrites have unique values of these ratios.

Rare Earth Elements (REE)

Stadermann[21] demonstrated the capability of SIMS to produce full REE patterns on individual IDPs. His results for four particles showed a flat *chondrite-like* pattern. One IDP yielded a refractory trace element pattern including a ten-fold enriched REE pattern and negative Eu anomaly.

DISCUSSION

The bulk chemical compositions of *chondrite-like* stratospheric particles are similar to those of carbonaceous meteorites but some compositional anomalies exist. In particular, volatile elements tend to be enriched relative to CI meteorites. The differences between the compositions of stratospheric particles and carbonaceous meteorites can be interpreted in basically three ways:

o   Stratospheric micrometeorites are samples of primitive material that has different origins from those of conventional meteorites.

o   The stratospheric micrometeorite collection is a biased sample of meteoritic material.

o   The compositions of stratospheric micrometeorites have been altered prior to analysis.

If stratospheric micrometeorites are samples of primitive material from the early solar system, their volatile element enrichments have important cosmochemical implications. One possibility is that the IDPs are more chemically primitive than any of the members of the carbonaceous chondrite class and they represent the only samples of this primitive materials available for study. It is intriguing to note that the IDP volatile trace element abundance pattern is complementary to those of the non-CI carbonaceous meteorites.

There are potential sources of sampling bias. IDPs might simply be the fine-grained fraction of the meteorite parent bodies, although chemical analyses on chondritic matrices and mineralogical comparisons suggest a distinction.[5] Atmospheric entry may also produce a sampling bias in several ways. First, low velocity-in plane trajectory particles may preferentially survive unaltered. Second, the meteorite-sized siblings of IDPs may be too friable to survive entry as large objects. The particle collection system may also favor particles of particular physical characteristics.

There are potential sources of contamination (discussed in detail by Flynn in this volume[22]). IDPs may be contaminated by stratospheric aerosols during atmospheric entry and/or collection. Considerable debate continues about whether the volatile trace element enrichments are in fact aerosol contamination (see below and reference 16). Si contamination clearly exists from residual silicone oil used on the collectors.[23] Curation and transfer to investigators may also introduce contamination but analyses of non-meteoritic, stratospheric particles (e.g., aluminum oxide spheres) have failed to reveal any obvious contamination problems in these procedures.[7] Exposure of the particles to the energetic beams used in the microanalytical techniques have the potential for causing volatile element loss, but, again, the studies that have been done on these effects (primarily via repetitive analyses) have shown few cases where such losses are significant. Conceivably, adsorption of volatiles during the long space exposure as small objects might be significant considering their high surface to volume ratios but this effect has not been explored.

Determination of volatile trace element depletions in partially heated IDPs may prove to be a powerful indicator of atmospheric heating. A quantitative method to determine thermal history of the particles would be useful in inferring the proportions of cometary and asteroidal dust based on orbital differences between these two populations.[24]

Thus, bulk compositional analyses of stratospheric IDPs have provided important new insights on the nature of interplanetary dust and the relationships between IDPs and members of the meteorite classes. Yet, these bulk analyses provide only an overview of the cosmochemical systematics, as is true in all areas of geochemistry and cosmochemistry. Improvements in analytical capabilities are needed to allow compositions (coupled with mineralogy, etc.) to be obtained for individual lithologies within individual IDPs. At the moment, part-per-million sensitivity is achievable within nanogram masses (femtogram sensitivity). This next technological step will require analyses that are capable of ppm sensitivity within 0.01 nanogram components (10 attogram sensitivity). With such a capability one could perform a detailed trace element comparison of mineral compositions between IDPs and meteorite matrices and identify the host phases of key volatile elements.

FUTURE STUDIES

Primitive stratospheric IDPs have chemical compositions similar to CI carbonaceous meteorites but clearly most of them are not fragments of CI meteorites. suggesting that they represent a new (and perhaps more primitive) class of carbonaceous material. Future studies need to be done in the following areas: (i) *Consortia Studies:* Multiple techniques have been applied to some IDPs but true consortia studies are needed to provide complete characterization of individual particles analogous to the consortia currently being organized for uncommon meteorite types, such as SNCs. This research will be particularly fruitful in understanding abundance anomalies and atmospheric heating effects by comparing composition with mineralogy. Desperately needed are trace element microanalyses with spatial resolutions (sub-µm) comparable to those of the TEM mineralogical studies. (ii) *Meteorite Associations:* The use of elemental ratios for determining associations with meteorites appears promising. Lindstrom[20] showed the potential value of Sc/Fe and Co/Fe ratios and additional work is needed to explore the suitability of other elemental ratios. (iii) *Contamination:* Contamination is a major concern in samples of this size. Additional systematic studies need to be done to quantify these effects. (iv) *Orbital Collections:* Orbital collections are desperately needed to allow the chemical analyses described here to be performed on particles of known orbital parameters. Although such collection activities are presently very destructive to particles, resulting in significant chemical fractionation[25], future orbital collection techniques should address this problem.

ACKNOWLEDGMENTS

The preparation of this review article was supported by NASA NAG9-106 and NAGW-3651.

REFERENCES

1. Love, S. G., and D. E. Brownlee (1991) Heating and thermal transformation of micrometeoroids entering the Earth's atmosphere. *Icarus* **97**, 70-84.
2. Lindstrom, D. J., and M. E. Zolensky (1990) Compositional variations in cosmic dust-sized pieces of Murchison matrix. *Lunar Planet. Sci. XXI*, 698-699.
3. Schramm, L. C., D. E. Brownlee, and M. M. Wheelock (1989) Major element composition of stratospheric micrometeorites. *Meteoritics* **24**, 99-112.
4. Zolensky, M.E. and Lindstrom, D. (1992) Mineralogy of 12 large "chondritic" interplanetary dust particles. *Proc. 22nd Lunar Planet. Sci. Conf.*, 161-169.
5. Thomas, K. L., G. E. Blanford, L. P. Keller, W. Klöck, and D. S. McKay, Carbon abundance and silicate mineralogy of anhydrous interplanetary dust particles (1993). *Geochim. Cosmochim. Acta* **57**, 1551-1566.
6. Wallenwein, R., H. Blank, E. K. Jessberger and K. Traxel (1987) Proton microprobe analysis of interplanetary dust particles. *Anal. Chim. Acta* **195**, 317-322.

7. Flynn, G. J., and S. R. Sutton (1990) Synchrotron x-ray fluorescence analyses of stratospheric cosmic dust: New results for chondritic and low-nickel particles. *Proc. 20th Lunar Planet. Sci. Conf.*, 335-342.
8. Stephan, T., W. Klöck, E. K. Jessberger, H. Rulle, and J. Zehnpfenning (1993) Multielement analysis of interplanetary dust particles using TOF-SIMS. *Lunar Planet. Sci. XXIV*, 1349-1350.
9. Lindstrom, D. J. (1990) INAA of cosmic dust particles from the large area collector. *Lunar Planet. Sci. XXI*, 700-701.
10. Ganapathy, R., and D. E. Brownlee (1979) Interplanetary dust: trace element analyses of individual particles by neutron activation. *Science* 206, 1075-77.
11. Mackinnon, I. D. R., and D. W. Mogk (1985) Surface sulphur measurements on stratospheric particles. *Geophys. Res. Lett.* 12, 93-96.
12. Gibson, E. K., Jr. (1992) Volatiles in interplanetary dust particles: A review. *J. Geophys. Res.-Planets* E3, 3865-3875.
13. Clemett, S. J., C. R. Maechling, R. N. Zare, P. D. Swan and R. M. Walker (1993) Identification of complex aromatic-molecules in individual interplanetary dust particles. *Science* 262, 721-725.
14. Gibson, E. and Bustin, R. (1994) Volatiles in Interplanetary Dust Particles: A comparison with volatile-rich meteorites. This volume.
15. Van der Stap, C. C. A. H., R. D. Vis, and H. Verheul (1986) Interplanetary dust: Arguments in favour of a late stage nebular origin. *Lunar Planet. Sci. XVII*, 1013.
16. Jessberger, E. K., J. Bohsung, S. Chakaveh and K. Traxel (1992) The volatile element enrichment of chondritic interplanetary dust particles. *Earth Planet. Sci. Letters* 112, 91-99.
17. Flynn, G. J., S. R. Sutton and S. Bajt (1993) Trace element content of chondritic cosmic dust: Volatile enrichments, thermal alterations, and the possibility of contamination. *Lunar Planet. Sci. XXIV*, 495-496.
18. Flynn, G. J. and S. R. Sutton (1992) Trace elements in chondritic stratospheric particles: Zinc depletion as a possible indicator of atmospheric entry heating. *Proc. 22nd Lunar Planet. Sci. Conf.*, 171-184.
19. Thomas, K. L., L. P. Keller, G. J. Flynn, S. R. Sutton, K. Takatori, D. S. McKay (1992) Bulk compositions, mineralogy, and trace element abundances of six interplanetary dust particles. *Lunar Planet. Sci. XXIII*, 1427-1428.
20. Lindstrom, D. J. (1992) Scandium/iron and cobalt/iron ratios as indicators of the sources of stratospheric dust particles. *Lunar Planet. Sci. XXIII*, 779-780.
21. Stadermann, F. J. (1991) Rare earth and trace element abundances in individual IDPs. *Lunar Planet. Sci. XXII*, 1311-1312.
22. Flynn, G. J. (1993) Changes to IDP composition and mineralogy by terrestrial encounters. This volume.
23. Rietmeijer, F. J. M. (1987) Silicone oil: A persistent contaminant in chemical and spectral microanalyses of interplanetary dust particles. *Lunar Planet. Sci. XVIII*, 836-837.

24. Flynn, G. J. (1989) Atmospheric entry heating: A criterion to distinguish between asteroidal and cometary sources of interplanetary dust. *Icarus* 77, 287-310.
25. Amari, S., J. Foote, C. Simon, P. Swan, R. M. Walker, E. Zinner, E. K. Jessberger, G. Lange, and F. Stadermann (1992) SIMS chemical analysis of extended impacts on the leading and trailing edges of LDEF experiment A0187-2. *Lunar Planet. Sci. XXIII*, 25-26.

158 Primitive Solar System Particles

The Solar Maximum (Solar Max) Satellite. In April of 1984 the Space Shuttle Challenger (after deploying the LDEF satellite) repaired the Solar Max, in the course of which several thermal control blankets and louvers from the main electronic box were permanently removed. Subsequent analyses of these materials back on Earth provided an early look at the particulate environment in low-Earth orbit. (NASA photo S13-34-1381)

CARBON IN PRIMITIVE INTERPLANETARY DUST PARTICLES.

Lindsay P. Keller
MVA, Inc., 5500 Oakbrook Parkway, Norcross, GA 30093, USA

Kathie L. Thomas
Lockheed, 2400 NASA Rd. 1, Houston, TX 77058, USA

and
David S. McKay
SN, NASA Johnson Space Center, Houston, TX 77058, USA.

INTRODUCTION

A major source of information regarding the nature and formation of carbonaceous materials in the early solar system is the study of primitive interplanetary dust particles (IDPs). Carbon is a significant component of most chondritic IDPs, and the nature of the carbon-rich phases bears on the chemical and physical processes that have affected carbon from its nucleosynthesis to its incorporation into primitive solar system bodies. In this abstract, we review the data regarding carbon in IDPs since approximately 1987. Brownlee[1] summarized the state of carbon in IDPs in a workshop held at Ames in 1987; other recent reviews have summarized the formation mechanisms that have been proposed for carbonaceous materials in primitive solar system materials.[2&3] We discuss here the abundance of carbon in IDPs, the nature and distribution of carbon, and topics and strategies for future work.

ABUNDANCE OF CARBON IN IDPS

Blanford et al.[4] published the first quantitative bulk carbon analyses of IDPs and subsequent work has greatly expanded the initial observations to include analyses of ~20 anhydrous particles along with their detailed mineralogy.[5] Analyses of anhydrous IDPs show a range of carbon abundances from ~4- to 45-wt.% with an average of ~13 wt.% (Figure 1). Carbon abundance correlates with mineralogy such that pyroxene-rich IDPs tend to have the highest carbon contents, whereas olivine-rich particles have chondritic levels of carbon (<3 x CI).[5] The correlation of carbon abundance with the mineralogy of anhydrous IDPs may provide evidence for potential sources. Pyroxene-rich IDPs have been proposed as strong candidates for cometary particles based on their unusual mineralogy and high carbon contents.[5] The major element compositions of anhydrous IDPs *as a group* are more primitive than the CI chondrites; and their general lack of hydrated phases indicates that they are also physically more primitive than CI chondrites.

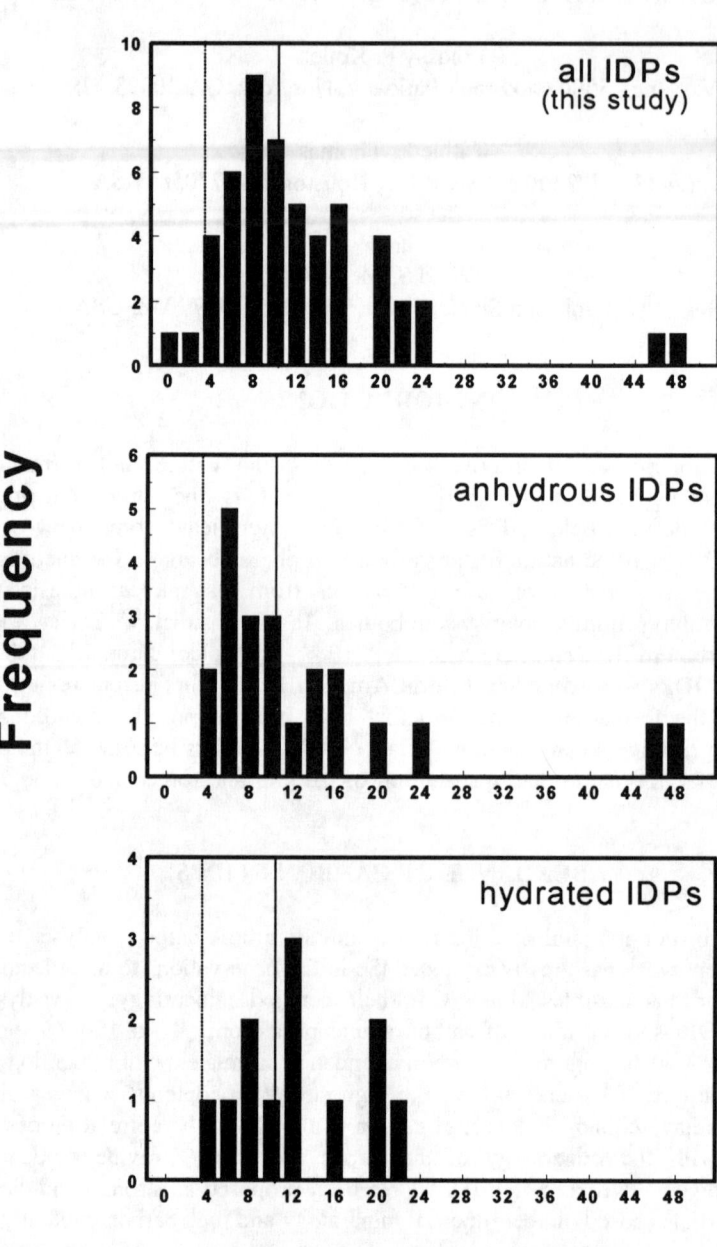

Figure 1. Carbon abundance in chondritic interplanetary dust particles. Vertical dashed line at 3.5 wt.% C is the CI carbon abundance. Vertical solid line at 10.5 wt.% C indicates 3 x CI. Bulk carbon <3 x CI is considered chondritic (see Thomas et al.[7] in this volume for a detailed discussion).

Hydrated IDPs on average are also strongly enriched in carbon relative to CI chondrites. Carbon abundances in hydrated IDPs range from 5 to over 20 wt.% with an average of 13 wt.% (Figure 1), although nearly 1/2 of the hydrated IDPs analyzed have chondritic (<3 x CI) carbon abundances.[6] The strong enrichment in carbon in hydrated IDPs as a group relative to the carbonaceous chondrites suggests that most hydrated IDPs are derived from parent bodies that are more carbon-rich than the known carbonaceous chondrites, although the extent of aqueous alteration effects in hydrated IDPs is comparable to that in the altered carbonaceous chondrites.

The limitations of current analytical techniques (SEM/EDX analysis of small particles) used for the determination of bulk carbon are discussed in this volume by Thomas et al.[7] A new method where IDPs are prepared with noncarbonaceous embedding media and TEM support substrates allows for the determination of bulk carbon contents of individual thin sections using thin-window EDX spectrometers.[8] Advances in other microbeam techniques such as Proton Induced X-ray Emission (PIXE) also hold promise for bulk carbon determinations.

NATURE OF THE C-BEARING PHASES

A host of carbon-bearing materials are known from IDPs. The most abundant carbon-rich material is poorly-graphitized or amorphous carbon that occurs as grain coatings, as a matrix in certain fine-grained aggregates (in anhydrous IDPs)[3], and as distinct "clumps" in anhydrous IDPs.[5] In high-carbon anhydrous IDPs, the amorphous carbon can be readily recognized by TEM examination of microtome thin sections, and the distribution of carbonaceous material can be documented.[5] Amorphous carbon is distributed throughout anhydrous IDPs and is not concentrated on the surfaces of the particles.[5]

From the analyses of hydrated IDPs, it is now known that carbonates are only a minor component of most hydrated IDPs, and that the high carbon abundances in hydrated IDPs indicates that additional carbon-bearing phases are present in significant concentrations probably in the form of amorphous or poorly-graphitized carbon.[6] The distribution of these carbon phases in hydrated IDPs is not known because it is difficult to differentiate poorly-ordered fine-grained carbonaceous materials from fine-grained phyllosilicates, but is assumed to be homogeneously distributed. TOF-SIMS (Time-of-flight, secondary ion mass spectrometry) analysis with high spatial resolution has been used for mapping the distribution of carbon in fine-grained hydrated IDPs and has the capability to address the problem of silicone oil contamination.[9]

From meteorite studies it is known that a large number of other organic compounds are present at individual concentrations typically <100 ppm. One might expect IDPs to contain a suite of organic compounds similar to those in carbonaceous chondrites, although the proportions may be dramatically different. Thomas et al.[7] note that in some anhydrous particles, the carbonaceous material is vesicular; a texture they interpret as a loss of hydrocarbons during heating. In other IDPs, the carbonaceous material is featureless. Although is has been long been suspected that IDPs contain indigenous polyaromatic hydrocarbons (PAHs), the direct analysis of

PAHs in IDPs has only recently been performed.[10&11] A variety of PAHs were observed in two IDPs, but apparently the mass distribution patterns are different from other analyzed extraterrestrial materials.[11] Additionally, it was inferred from the odd-mass peaks in the spectra that nitrogen-bearing functional groups were present.[11] A careful evaluation of heating during atmospheric entry needs to be considered in these studies (see below).

The occurrence and concentration of refractory carbons such as silicon carbides and diamonds in IDPs has not been demonstrated. FeNi-carbides are a minor constituent of anhydrous IDPs, and are believed to have formed by catalytic reactions between metal grains and CO gas.[3]

There is intriguing evidence that the carbonaceous material in IDPs is not all the same. Wopenka[12], using Raman spectroscopy, noted distinct differences in the degree of order exhibited by carbonaceous material in a group of 20 IDPs. She recognized that the interpretation of the degree of order is a complex function of the types of precursor organics and the metamorphic history of the particle, including atmospheric entry heating. However, solar flare tracks were observed in two of the IDPs that showed different degrees of order, indicating that they were probably not strongly heated during atmospheric entry. Wopenka speculated that the differences in these two particles reflected different organic precursor material. These differences could also indicate different levels of parent body thermal metamorphism. More measurements of this type in conjunction with detailed mineralogical studies to determine the atmospheric entry heating effects in a suite of well-characterized IDPs would be desirable. Major differences may indicate multiple sources and provide new information about the formation processes. The carbonaceous material in IDPs probably contains the main carrier phase for the large deuterium enrichments, but the identity of the specific carrier phase(s) is not known.

Secondary processes such as aqueous alteration and atmospheric entry heating have modified the mineralogy and chemistry of many IDPs, and these processes may have modified the carbon phases in IDPs as well. Although the range of carbon abundances are similar in the anhydrous and hydrated IDPs, the aqueous alteration process has probably changed the carbon phases in hydrated IDPs (Are there primary and secondary organics just like there are primary and secondary minerals?). There are basic differences in the amount and variety of organic compounds in anhydrous carbonaceous chondrites (e.g., Allende) and altered chondrites (e.g., Murchison)[2], and thermodynamic modelling suggests that reactions converting certain PAHs to amino and carboxylic acids are energetically favorable at the the conditions inferred from the mineral assemblages.[13] Thus, we would expect to see similar effects in cosmic dust particles.

STRATEGIES FOR FUTURE WORK

Studies involving the structure of the carbonaceous component of IDPs or the molecular compounds in IDPs must select particles that are pristine and that have not been strongly heated during atmospheric entry (although this may induce a bias towards particles derived from objects in asteroidal orbits). A major consideration in

IDPs studies is understanding how atmospheric entry heating has modified the carbon chemistry of IDPs. Analyses of carbon in hydrated IDPs shows that carbon abundance does not correlate with the degree of atmospheric entry heating estimated from mineralogical changes, even though ~75% of hydrated IDPs analyzed show evidence for entry heating.[6] There are indications that the form of carbon may change with entry heating including the loss of low-mass PAHs[10], increasing order in poorly-graphitized carbon[12&14], and decarbonation of Mg-Fe carbonates.[6] We need to be able to interpret results from heated particles, thus, laboratory pulse-heating experiments should be performed to understand mineralogical changes and changes in the carbon chemistry of IDPs as a function of heating.

One of the approaches to the unanswered questions regarding carbon in IDPs is to organize consortium studies on cluster particles. Cluster particles were originally 30- to 50-μm sized particles which broke into several fragments on the collection surfaces. Cluster particles will be extremely valuable for multdisciplinary study because there are usually many ~10-μm sized fragments available for a wide variety of investigations including mineralogical, chemical, isotopic, and spectroscopic measurements (although the siginificant heterogeneities exhibited by cluster particles at the 10 μm scale[15] may limit extrapolations to other fragments of the same cluster). Continued advances in analytical techniques, sample preparation methodologies, and in the number of well-characterized IDPs will help to address many of the important questions that still remain regarding the nature of carbon-bearing materials in IDPs.

ACKNOWLEDGEMENTS

This work was supported by NASA RTOP 152-17-40-23 (PMG) and 199-52-11-02 (Exobiology). We thank J. Bradley and the editors for constructive reviews.

REFERENCES

1. Brownlee, D. E. (1990) Carbon in comet dust. *NASA Conf. Publ. 3061*, 21-25.
2. Cronin, J. R. et al. (1988) Organic Matter in Carbonaceous Chondrites, Planetary Satellites, Asteroids and Comets. In *Meteorites and the Early Solar System*, pp. 819-860.
3. Bradley, J. P. et al. (1988) Interplanetary Dust Particles. In *Meteorites and the Early Solar System*, pp. 861-898.
4. Blanford, G. et al. (1988) Microbeam analyses of four chondritic interplanetary dust particles for major elements, carbon, and oxygen. *Meteoritics* 23, 113-121.
5. Thomas, K. L. et al. (1993) Carbon abundance and silicate mineralogy of anhydrous interplanetary dust particles. *Geochimica et Cosmochimica Acta* 57, 1551-1566.
6. Keller, L. P. et al. (1993) Carbon abundances, major element chemistry and mineralogy of hydrated interplanetary dust particles. *Lunar Planet. Sci. XXIV*, 785-786.

7. Thomas, K. L. et al. (1993) Quantitative analyses of carbon in anhydrous and hydrated interplanetary dust particles. This volume.
8. Bradley, J. P. et al. (1993) Carbon analyses of IDPs sectioned in sulfur and supported on beryllium films. *Lunar Planet. Science XXIV*, 173-174.
9. Stephan, T. et al. (1993) Multielement analysis of carbon-rich interplanetary dust particles with TOF-SIMS. *Meteoritics* **28**, 443-444.
10. Clemett, S. J. et al. (1993) Measurement of polycyclic aromatic hydrocarbon (PAHs) in interplanetary dust particles. *Lunar Planet. Science XXIV*, 309-310.
11. Clemett, S. J. et al. (1993) Identification of complex aromatic molecules in individual interplanetary dust particles. *Science* **262**, 721-725.
12. Wopenka, B. (1988) Raman observations on individual interplanetary dust particles. *Earth and Planetary Science Letters* **88**, 221.
13. Shock, E. and Schulte, M. D. (1990) Amino-acid synthesis in carbonaceous meteorites by aqueous alteration of polycyclic aromatic hydrocarbons. *Nature* **343**, 728.
14. Rietmeijer, F. (1992) Pregraphitic and poorly graphitised carbons in porous chondritic micrometeorites. *Geochimica et Cosmochimica Acta* **56**, 1665.
15. Thomas, K. L. et al. (1993) Analysis of fragments from cluster particles: Carbon abundances, bulk chemistry, and mineralogy. *Meteoritics* **28**, 448-449.

QUANTITATIVE ANALYSES OF CARBON IN ANHYDROUS AND HYDRATED INTERPLANETARY DUST PARTICLES

K.L. Thomas and L.P. Keller
Lockheed, NASA Rd.1,
Houston, TX 77058 USA

G.E. Blanford
University of Houston-Clear Lake,
Houston, TX 77058 USA

and
D.S. McKay
NASA/JSC SN, Houston, TX 77058 USA

INTRODUCTION

Carbon is an important and significant component of most anhydrous and hydrated IDPs. We have analyzed ~40 anhydrous and hydrated chondritic interplanetary dust particles (IDPs) for major and minor elements, including carbon and oxygen.[1-5] Quantitative analyses of light elements in small particles are difficult and require careful procedures in order to obtain reliable results. In our work, we have completed extensive analytical checks to verify the accuracy and precision of carbon abundances in IDPs.[1] In our present work, additional methods are used to verify carbon abundances in IDPs including analysis of ultramicrotomed thin sections of IDPs embedded in sulfur, and direct observation of carbonaceous material in these thin sections. Our work shows conclusively that carbon is significantly enriched in chondritic IDPs relative to CI abundances.

CARBON ANALYSIS PROCEDURES

Uncoated IDPs supported on beryllium substrates were analyzed using a thin window energy-dispersive X-ray detector on a scanning electron microscope equipped with a turbomolecular pump. This procedure eliminates carbon contamination and contributions from carbon-bearing substrates. Spectra were processed through the PGT bulk particle data reduction (BSAM) program, which performs peak overlap corrections, background subtractions for elements Z>6, and calculates the ratios of X-ray counts from the unknown to counts from a pure, flat element standard (k-ratio). Carbon k-ratios were determined manually for each spectrum. After manual background subtraction for carbon, we recorded counts in a window between 170-340 eV. These counts were divided by the average of measurements of a cleaved diamond standard to obtain carbon k-ratios. All k-ratios were used as input to the CITPIC (Ver. 2.03) matrix and $\phi(\rho z)$ correction procedure developed by Armstrong and Buseck.[6]

The CITPIC program was designed for analyzing small, irregularly-shaped particles. We evaluated our analytical procedure by analyzing particle standards from three sources under the same conditions as those for IDPs: small particles with homogeneously distributed carbon (calcite), and IDP analog materials including small particles of the Orgueil CI chondrite and the Allende CV3 chondrite. The detailed results of these studies (given in reference 1) show that our carbon abundances are accurate and have relative errors of ~10%. For elements ranging from carbon to nickel, we determined that the overall agreement of our analyses of individual grains of calcite, Orgueil, and Allende particles with bulk values is excellent.[1] Carbon abundances in 29 Orgueil particles ranged from 0-7.4 wt.% with a mean value of 3.4 wt.%, in excellent agreement with literature values (Orgueil bulk carbon is 3.4 wt.% as shown in reference 7).

Since IDPs are contaminated with silicone oil (see Warren and Zolensky[8] and Flynn[9] in this volume), which contains ~30% C by wt., Orgueil and Allende particles were saturated with silicone oil, washed in hexane, and analyzed for major elements, carbon, and oxygen (results for Allende are given in reference 1). Results from these particles show that there is a real contamination problem from the silicone oil which causes an ~10% enhancement in silicon abundances. Although silicone oil contains carbon and oxygen, we are not detecting excesses of these elements. The hexane rinse may be effective in removing silicone oil from the particle surfaces, beyond the depth of detection for carbon and oxygen. The actual contamination of IDPs from silicone oil is based on particle parameters such as porosity and the extent to which the particle was rinsed. We believe IDP carbon abundances are not significantly affected even though the silicon abundances are overestimated.

A frequent criticism of our technique is that energy dispersive X-ray analysis is essentially a surface analysis technique for light elements and we must assume a homogeneous carbon distribution within the sample volume. To address this criticism, we are now quantitatively analyzing thin sections of IDPs embedded in sulfur for carbon and are directly observing carbon distribution in the TEM. IDPs are initially embedded in glassy sulfur, thin sectioned, and placed on beryllium support films.[10] Analysis of carbon in sulfur-embedded thin sections of one IDP shows excellent agreement with that determined by our bulk particle method. The average bulk carbon from IDP thin sections is 8.6 wt.% compared with 8 wt.% from our bulk particle technique (results described in reference 10).

Carbon can also be directly observed in anhydrous IDPs by point counting of individual particle thin sections in the TEM.[1] Point counting of IDP thin sections allows us to demonstrate where the carbon is located, estimate the volume of the carbonaceous material, and show a correlation of carbon in the thin section with the bulk carbon abundance. Our results indicate that the distribution of carbon in anhydrous IDPs appears to be rather homogeneous; carbon surrounds the internal grains and is distributed unevenly on the particle surface. The carbon seems to act as a matrix holding the individual grains together.

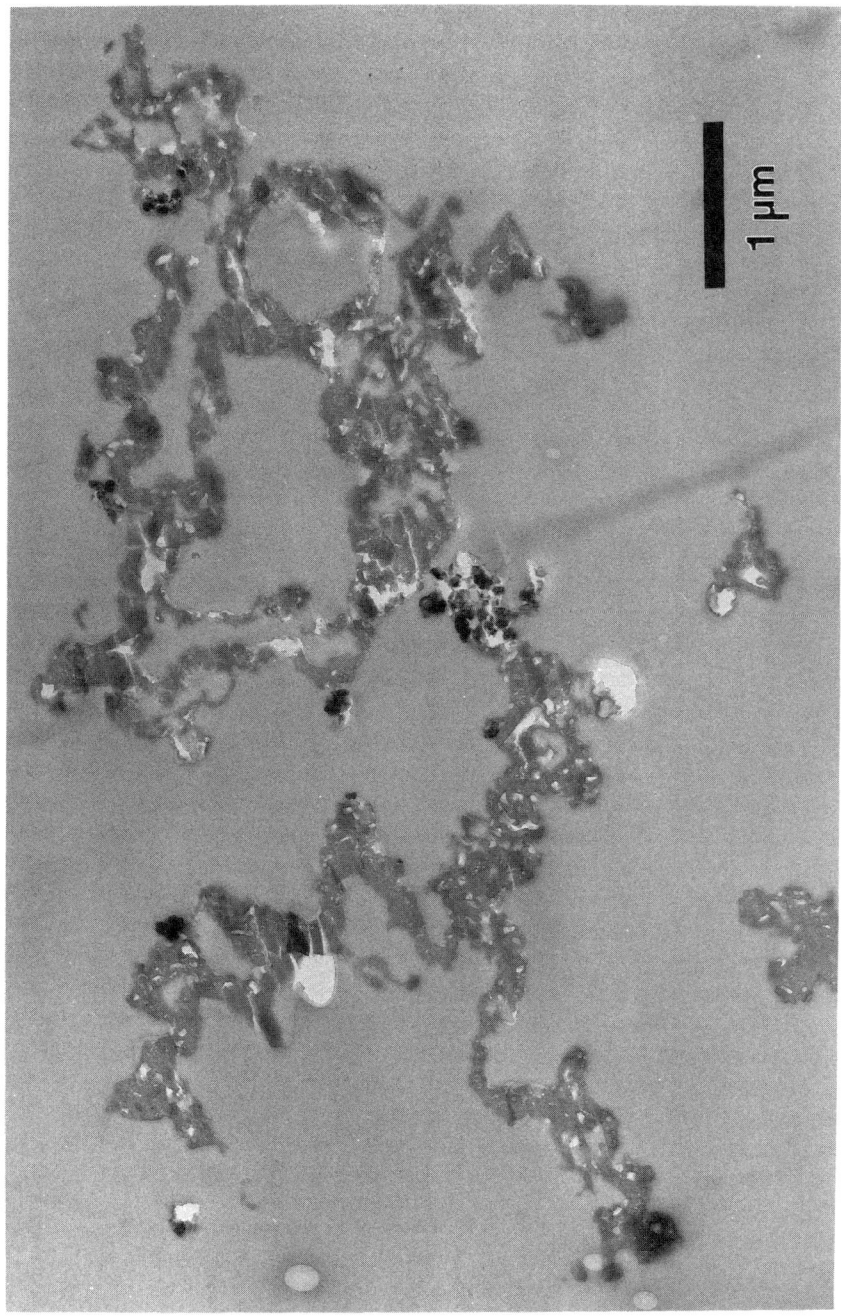

Figure 1. TEM photomicrograph of IDP L2006B23 after embedding in epoxy and thin-sectioning.

168 Quantitative Analyses of Carbon

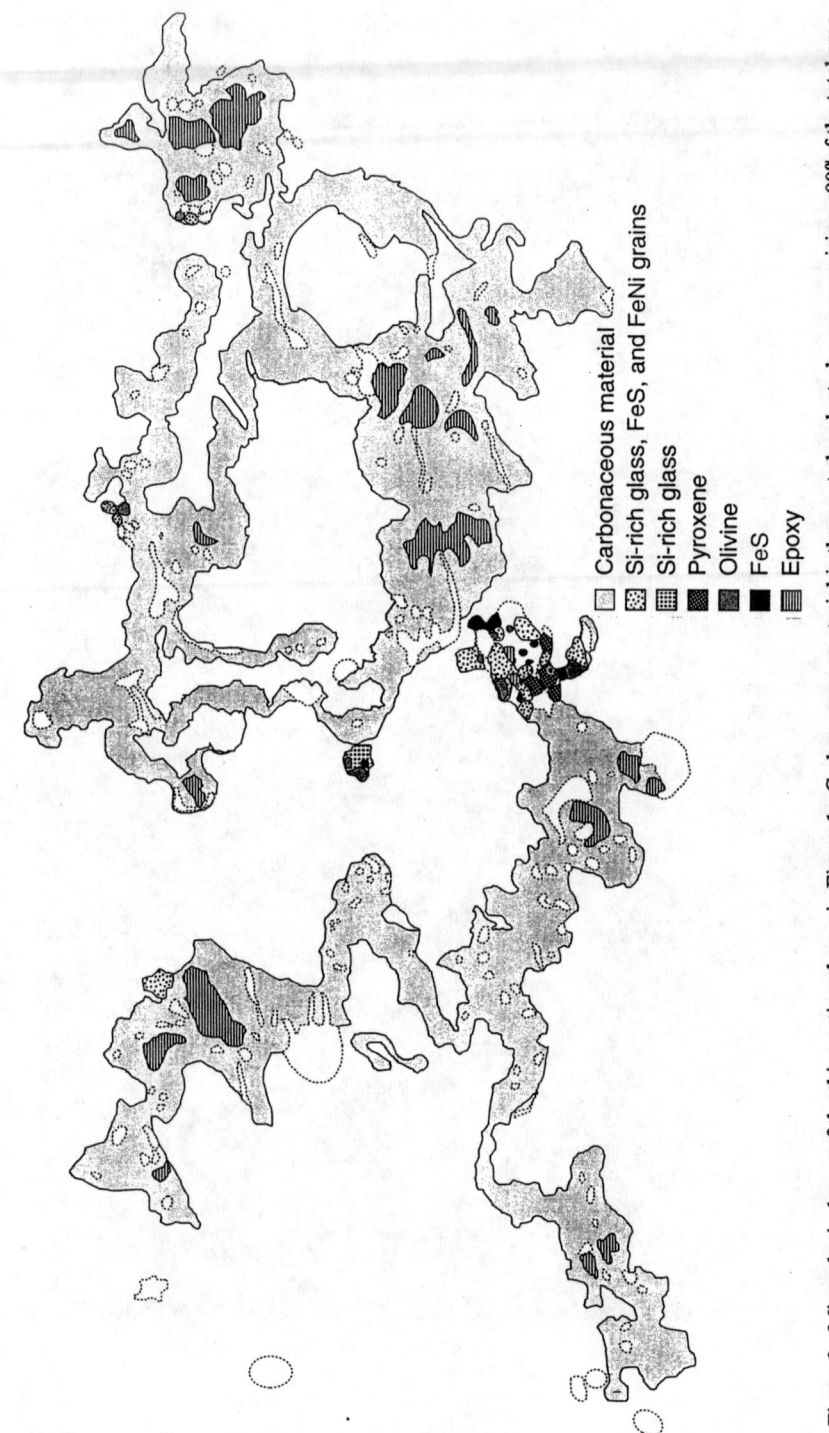

Figure 2. Mineralogical map of the thin section shown in Figure 1. Carbonaceous material is the most abundant phase, comprising ~90% of the total area of this particle.

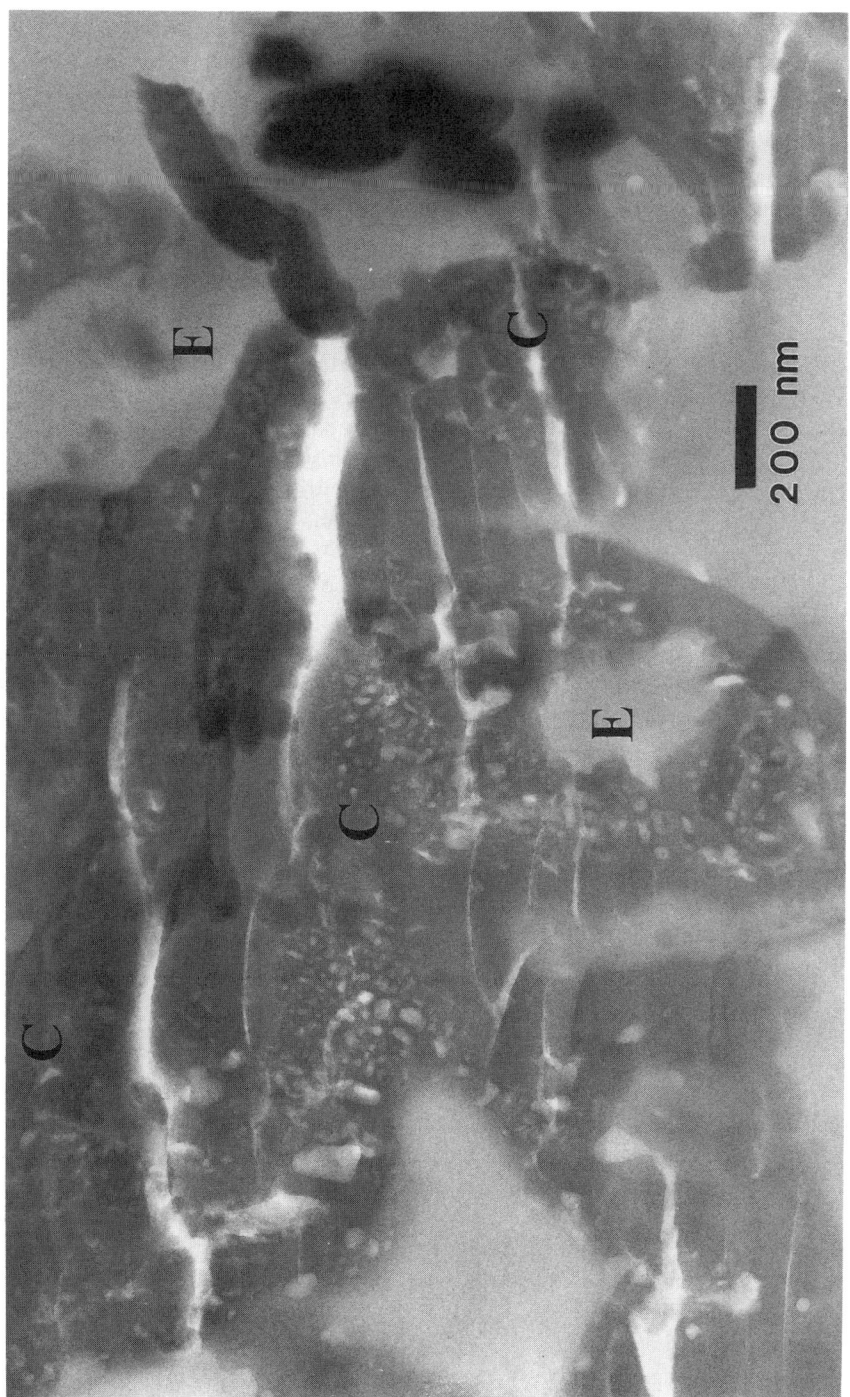

Figure 3. TEM photomicrograph showing the vesicular texture of carbonaceous material (C) in a thin section of L2006B23 (E indicates epoxy).

ANHYDROUS IDPS

Our results[1] show that anhydrous IDPs have a chondritic composition within a factor of 2 of CI chondrites for most major and minor elements with the exception of carbon, which ranges from ~0.6 to 13xCI. We have identified a relationship between carbon abundance and silicate mineralogy which, in general, shows that particles dominated by pyroxenes have higher carbon abundances (C>3xCI) than those dominated by olivines (C<3xCI). Particles containing equal amounts of pyroxene and olivine can be grouped with either the pyroxene or olivine-dominated IDPs based on carbon abundances.

Carbon can be directly observed in all our pyroxene-dominated IDPs because of their high carbon abundances, but it is more difficult to observe carbon in thin sections when C<3xCI (e.g., olivine-dominated IDPs). We performed point count analyses of thin sections of two pyroxene-dominated IDPs: W7027H14 and L2006B23. Results show that W7027H14 contains 40 to 50 volume percent carbonaceous material which is in good agreement with an estimate of ~40 volume percent, assuming a particle diameter of 10 µm, a mean particle density of 1 gm/cm^3 (see reference 9), and ~23 wt.% C (bulk). L2006B23 has ~45 wt.% C, the highest reported bulk carbon of any IDP (Figures 1&2). The volume percent of C is ~90, determined by point counting, and agrees with the theoretical estimate of 90 volume percent based on a particle diameter of 15 µm, mean density of 1 gm/cm^3 (ref. 11), 45 wt.% C (bulk), and ~50% porosity. (Editors: Gibson and Bustin report small amounts of hydrous material within another portion of particle L2006B23.[12])

The nature of the carbonaceous material in anhydrous IDPs is poorly known. We have not observed graphitized carbon (i.e., 0.34 nm spacings) in any particles nor have we observed carbon in the form of carbonates. Rather, the carbonaceous material could be poorly graphitized or amorphous. The carbon-rich phases in L2006B23 have a vesicular texture (Figure 3), indicating the loss of volatiles, probably hydrocarbons. Since all IDPs have been heated during atmospheric entry[13], carbon-rich compounds (e.g., hydrocarbons) could be volatilized or pyrolized. It seems plausible that several carbon phases could co-exist in anhydrous IDPs.[12,14,15]

HYDRATED IDPS

Our results show that 12 hydrated IDPs have a chondritic composition within a factor of 2 of CI for analyzed elements with the exception of carbon which ranges from ~2 to 6xCI.[3,4,16] Carbon abundances are <3xCI in six of the hydrated IDPs. The range of carbon abundances in hydrated IDPs overlaps that of anhydrous IDPs. No relationship has been observed between carbon abundance and the presence of any particular silicate phase. Carbonaceous material may be distributed in hydrated IDPs in a similar manner to that in anhydrous IDPs but is difficult to differentiate carbonaceous material from fine-grained, poorly crystalline phyllosilicate matrix, which dominates hydrated IDPs. Therefore, it has been impossible to directly observe amorphous or poorly-graphitized carbon in IDP thin sections. Fine-grained carbonates have been found in three hydrated IDPs with 8, 15 and 20 wt.% carbon.

Carbonates can account for some, but not all, of the high carbon in these IDPs. Two IDPs, L2005P9 and L2006J14, have 20 and 22 wt.% carbon respectively, and lack carbonates. Therefore, we suggest that poorly graphitized carbon, amorphous carbon, or hydrocarbons are present in these carbon-rich IDPs.

All IDPs are heated to some degree during atmospheric entry[13], and our subset of hydrated IDPs have magnetite rims on the exterior particle surface which is evidence of significant heating.[3&16] These heated IDPs have carbon abundances which range up to 6xCI. If some carbon was present as a volatile species (e.g., hydrocarbons), then some of this low-Z material could be volatilized or pyrolized due to atmospheric heating. Therefore, the actual carbon abundances of these heated, hydrated IDPs may be higher than we report.

CONCLUSIONS

1. We are able to quantify carbon abundances accurately in small particles.
2. Point counting and analysis of IDP thin sections embedded in sulfur are two additional techniques used to validate carbon abundances in IDPs.
3. Most anhydrous and hydrated IDPs have carbon abundances much higher than those of any known meteorite. Anhydrous IDPs have carbon abundances which range from ~0.6 to 13xCI. They can be classified into 3 groups based on carbon abundance and mineralogy: pyroxene-rich IDP with C>3xCI, olivine-rich with C<3xCI, and mixed mineralogy with carbon abundances ranging between olivine- and pyroxene-rich IDPs. One anhydrous IDP has the highest reported bulk carbon (45 wt.%) of any IDP. Carbonaceous material acts as a matrix holding individual mineral grains together.
4. Hydrated IDPs have carbon abundances which range from 2 to 6xCI. Carbon abundance is not apparently correlated with any silicate phase. Although some IDPs have abundant carbonates, other forms of carbonaceous material must be present to account for high carbon abundances. The distribution of carbon in hydrated IDPs is poorly known.

ACKNOWLEDGMENTS

This work was supported by NASA RTOPS 152-17-40-23 and 199-52-11-02.

REFERENCES

1. Thomas K.L. et al. (1993) *Geochimica et Cosmochimica Acta* **57**, 1551-1566.
2. Thomas K.L. et al. (1993) *Lunar Planet. Sci. XXIV*, 1425-1426.
3. Keller L.P. et al. (1993) *Lunar Planet. Sci. XXIV*, 785-786.
4. Thomas K.L. et al. (1992) *Meteoritics* **27**, 296-297.
5. Keller L.P. et al. (1994) Carbon in primitive Interplanetary Dust Particles. This volume.
6. Armstrong J.T. and Buseck P.R. (1975) *Anal.Chem.* **47**, 2178-2192.
7. Anders E. and Grevesse N. (1989) *Geochim. Cosmochim. Acta* **53**, 197-214.

8. Warren J. and Zolensky M.E. (1994) This volume.
9. Flynn G. (1994) Changes to IDP composition and mineralogy by terrestrial encounters. This volume.
10. Bradley J. P. et al. (1993) *Lunar Planet. Sci. XXIV*, 173-174.
11. Flynn G.J. and Sutton S.R. (1991) *Proc. Lunar Sci. Conf. 21st*, 541-547.
12. Gibson E. and Bustin R. (1994) Volatiles in Interplanetary Dust Particles: A comparison with volatile-rich meteorites. This volume.
13. Flynn G.J. et al. (1993) *Lunar Planet. Sci. XXIV*, 497-498.
14. Reitmeijer, F.J.M. (1992) *Trends in Mineral.* **1**, 23-41.
15. Clemett, S.J. et al. (1993) *Science* **262**, 721-725.
16. Keller L.P. et al. (1992) *Lunar Planet. Sci VVIII*, 675-676.

VOLATILES IN INTERPLANETARY DUST PARTICLES: A COMPARISON WITH VOLATILE-RICH METEORITES

Everett K. Gibson, Jr.
SN4, Planetary Sciences Branch
NASA Johnson Space Center
Houston, Texas 77058, USA

and

Roberta Bustin
Chemistry Department, Arkansas College
Batesville, Arkansas 72501, USA

ABSTRACT

Fourteen interplanetary dust particles (IDPs) have been analyzed for their volatiles using a laser microprobe/mass spectrometer technique. For comparison, a suite of 10 to 20 micron sized samples from the Orgueil (CI) and Murchison (CM) carbonaceous chondrites was analyzed. Comparison of the abundance patterns for the IDPs and meteorites suggests that the volatiles found within the IDPs are similar to those observed for the CM chondrites groundmass.

INTRODUCTION

During the past several years we have been studying and classifying the nature of volatiles within interplanetary dust particles (IDPs) using a laser microprobe/mass spectrometer (LM/MS) technique. Volatile abundances and distributions found for the IDPs were compared with those measured for chondritic meteorites in order to determine if the IDPs are related to the parent bodies of these primitive meteorites. It is clear that the IDPs studied may be small non-representative samples of their asteroidal or cometary parent bodies. Because of the heterogeneity and wide variety of volatile-rich and volatile-depleted components present within primitive and chondritic meteorites, it is very unlikely that the analysis of a 5 to 10 micron size region within these meteorites would provide information which would allow one to know the composition of the whole meteorite. Clearly, only an analysis of all of the components present within the meteorite and knowledge about their abundances and distributions within the specimen can provide a complete understanding of the nature of the parent body and its origin. A similar situation exists with analysis of IDPs. Before a complete and rational understanding of the nature of the IDPs' parent bodies and hence, origins for the IDPs can be established, analysis of sufficient IDPs must be carried out in order to obtain a statistical and accurate sampling of their sources.

Analysis of volatile species within extraterrestrial materials must take into account the nature of any possible heating which the particles might undergo during their deceleration in the Earth's atmosphere. The heating might result in the loss of volatile species present within the IDPs if the entry velocities were great enough for the particles to undergo melting or extensive alteration because of the thermal effects. However, there is a battery of evidence that suggests not all particles underwent the heating to temperatures above 500° to 600° C. Evidence for the lack of substantial heating (for some IDPs) can be seen from the following observations on IDPs: (a) abundance of organic compounds (i.e. polycyclic aromatic hydrocarbons and other low molecular weight hydrocarbons)[1-5]; (b) noble gas helium release temperatures[6]; (c) low-temperature clay minerals[7]; (d) presence of carbonates[8]; (e) cosmic ray tracks in silicate minerals[9], and (f) zinc abundance data.[10] From the above list, there is ample evidence that there is a population of IDPs which have not undergone substantial heating, and meaningful data on the nature of volatiles present within IDPs can be obtained.

Table 1. EDX/SEM Elemental Analysis of IDPs

Particle	Major and (Minor) Components[1]
L2005B21	Si, Fe, Mg, O, (Al)
L2005C21	Si, S, Mg, O, Fe, Ca, Na, Al
L2005C24	Si, Mg, Fe, O, Ca, (Al)
L2005C26	Si, Mg, Fe, O, (Ca), (Al), (Ni)
L2005C28	S, Si, Fe, Mg, O, (Al), (Ni)
L2005C30	Si, Mg, Fe, O, (Ca), (Ti), (Al), (Na)
L2005D27	Si, Mg, O, S, Fe, (Na), (Al), (Ca)
L2005D34	Si, Mg, O, Na, Fe
L2005E38	Si, Mg, Fe, O, (Na)
L2005E39	Si, Mg, C, Fe, O, (Na)
L2006A6,7	C, Si, Na, (O)
L2006A12	Si, Mg, O, Fe, (Ca), (C), (Na), (Al)
L2006A26	Si, Ca, Mg, O, Fe, C
L2006B16	Si, Mg, O, Fe, C

[1]Elemental abundances listed in order of abundance with minor amounts placed in parentheses. Sulfur abundances are difficult to know accurately because of the overlap of the sulfur peak with the strong background gold peak from the sample mount.

Table 2. Indigenous Volatile Species Present in IDPs

Particle	Volatile Components Measured within IDP
L2005B21	OH
L2005C21	C, C_2H_5, O_2 or S, SO_2, C_5H_6, C_6H_6, C_6H_7, $C_6H_5CH_3$
L2005C24	OH, C_2H_5, C_2H_6, C_4 or SO, C_4H, C_5H_8, C_5H_9, C_6H_6 $C_6H_5CH_3$
L2005C26	C, CH, and higher molecular weight hydrocarbons
L2005C28	CH, OH, C_5H_7, CS_2, C_6H_6, C_6H_7, C_7H_{11}
L2005C30	OH, C_2H_5, C_2H_6, CO_2, C_5H_8, C_5H_9
L2005D27	C, C_2H_5, CO_2, SOH?, SO_2, CS_2, C_6H_6, C_6H_7, $C_6H_5CH_3$, C_7H_9, C_7H_{11}, C_7H_{16}
L2005D34	CH, OH, C_6H_6
L2005E38	OH, O_2 or S, SO
L2005E39	C, C_2H_5, C_5H_5, C_5H_6, C_5H_7
L2006A6,7	C, OH, C_2H_5, CO_2, C_5H_5, C_5H_6, C_6H_6
L2006A12	C, C_2H_5, CO_2, C_5H_5, C_5H_7, C_6H_6
L2006A26	NONE
L2006B16	CO, SO

EXPERIMENTAL

We analyzed IDPs and 10 to 20 micron size regions within carbonaceous chondrites using the techniques of Hartmetz et al.[3,4,11] Individual IDPs and pieces of CI, CM and CV carbonaceous chondrites similar in size were mounted on gold foils; the IDPs were cleaned by rinsing with hexane and Freon to remove silicone oil prior to mounting on the gold foil. The gold foils were previously cleaned by treatment with an atomic oxygen plasma. Each particle was analyzed with EDX-SEM to determine the elemental composition (Table 1); the majority of the IDP particles studied contained the same elements as chondritic meteorites, based on the similarity of their elemental compositions and those of chondritic meteorites, particularly the carbonaceous chondrites. Individual samples were vaporized under vacuum using a Q-switched Nd/YAG laser (1.06 microns), and the released volatiles directly analyzed with a quadrupole mass spectrometer.[12] Spectral information obtained was processed using the procedures of Hartmetz et al.[3,4,11] Blanks were obtained by analyzing regions of the gold foils 30 to 50 microns from where the particles had been mounted. The background data was subtracted from the signals obtained from analysis of the IDP. The resulting spectrum was considered to be from the IDP. As shown by Hartmetz et al.[3,4,11], selected IDPs contain trace amounts of the silicone oil from the collection procedures employed. The IDPs used in our studies were from the Large Area Collectors (LAC) L2005 and L2006 along with particles collected prior to the utilization of the LACs flown aboard a NASA ER-2 aircraft during a series of flights that were made within west-central North America.

RESULTS

An overview of the indigenous volatiles observed for the analysis of 14 chondritic IDPs is given in Table 2. The volatile abundances observed within the IDPs varied dramatically from almost no volatiles (L2005B21) (Fig. 1) to what is considered a volatile-rich IDP (L2006A12) (Fig. 2). IDP particle L2006A12 contained an abundance of several hydrocarbon "families" in addition to species such as C, O, CO, CO_2, and COS. Tables 1 and 2 show the variety of elemental compositions and indigenous volatile species found for 14 IDPs. As noted by Hartmetz et al.[3&4], IDPs are porous and may retain silicone oil and/or Freon residues from the collection devices or hexane used to remove the silicone oil from the surface. In our studies a volatile component is considered to be indigenous only if it does not occur in the mass spectrum of silicone oil, Freon, or hexane. The 14 particles shown in Tables 1 and 2 include both hydrated and anhydrous IDPs. Several of the particles which we analyzed are from clusters of IDP material on the collection plate. The exact classification of the particle into either anhydrous or hydrated category is difficult because of heterogeneous nature of the particles. Our analytical technique may detect the presence of hydrated components (e.g. OH) in trace amounts within the particle. One might be tempted to classify the particle as hydrated on this basis but in reality the bulk of the IDP is composed of anhydrous components. It is clear that there are inadequacies in the classification scheme for IDPs when dealing with particles that contain some hydrous phases, but are not dominated by them. Zolensky discusses this point further in this volume.[13] Table 3 lists the major classes of volatiles which we have observed for all of the IDPs analyzed to date in our laboratory. The particles have been separated into six categories based upon the abundances and types of volatiles observed from the IDPs. Fourteen of the particles contained sulfides whereas only four particles contained carbonates.

One of the more unusual fine-grained aggregate (FGA) particles analyzed is L2006B16 which was observed to contain unusually large amounts of carbon in its EDX spectrum. Thomas et al.[14&15] noted a fragment of this sample contained 46 wt. % carbon. Our analysis of the particle showed it contained a form of carbon which released essentially only CO during vaporization (Fig. 3). We do not know if the carbon was graphite or another form of carbon. Analysis of the particle showed the presence of a water-bearing component (most likely a clay phase) along with trace amounts of sulfur-related volatiles. Particle L2006B16 is an unusual IDP, as our analysis and those of Thomas et al.[14&15] have shown it to be enriched in carbon. It is worth noting that Thomas also consider this particle to be anhydrous.[14&15]

TABLE 3. OVERVIEW OF VOLATILE SPECIES IDENTIFIED IN 28 IDPs

Little or No Indigenous Volatiles	Large Amount of Indigenous Volatiles	Carbonaceous Material	Carbonates	Water or Hydroxyl	Sulfur Species
L2001D3	L2002C4	L2002C4	L2003D2	L2005D21	L2002C4
L2004D3	L2003D2	L2003D2	L2006A6,7	L2005C24	L2003D2
L2005B21	L2003E3	L2004C3	L2006A12	L2005C28	L2004C3
L2005C21	L2004C3	L2003E3	U2034D7	L2005C30	L2004D3
L2005C24	L2006A12	L2005C21		L2005D34	L2005C21
L2005C26	U2017A4	L2005C24		L2005E38	L2005C24
L2005C28	U2017A5	L2005C26		L2006A6,7	L2005C26
L2005C30	U2022G13	L2005D34		L2006B16	L2005E38
L2005D27	U2034D7	L2005E39		U2022F5	L2006A12
L2005D34		L2006A6,7		U2022F20	L2006B16
L2005E38		U2034D7		U2034D7	U2017A4
L2006A26					U2017A5
L2006B16					U2022G13
U2015B20					U2034D7
U2015F20					
U2022F5					
U2034D1					

Samples listed are a composite of all the IDPs analyzed via LM/MS technique and have been previously reported by Hartmetz et al.[2&3] or Bustin et al.[5]

VOLATILES WITHIN 10 TO 20 MICRON SIZE PIECES OF CI AND CM METEORITES

To give some idea about how representative a 10-20 micron sized particle is compared to its parent body, 10-20 micron-size particles of Murchison (CM) and Orgueil (CI) were analyzed (Figs 4-7). There was considerable diversity in the volatiles released from individual 10 to 20 micron particles analyzed from the carbonaceous chondrites. Analysis of two separate 10 micron size Murchison particles (Figs. 4 and 5) showed the widely different variations in volatile contents for the CM chondrite. Murchison Particle 7 (Fig. 4) was typical of the groundmass of the parent meteorite and contained ten times more volatiles than Murchison Particle 8 which contained a carbonate phase along with traces of a hydrocarbon. Orgueil Particle 5 (Fig. 6) contained an abundance of aliphatic hydrocarbons, polycyclic aromatic hydrocarbons, carbonates and sulfur components, where Orgueil Particle 7 (Fig. 7) contained only trace amounts of hydroxyl and carbon monoxide resulting from the oxidation of a carbon phase such as graphite. Orgueil Particle 5 contained approximately 70 times more volatiles than Orgueil Particle 7. The diversity is consistent with the heterogeneity observed by SEM/EDX. Some particles appeared to be only mineral grains (i.e. released only CO and CO_2 from carbonates, or SO_2 from sulfates, or CS_2 and/or COS from sulfides, or OH or H_2O from hydrated minerals); some particles did not release volatiles typical of meteoritic phases, and others were fairly representative of the meteorite matrix as observed by Hartmetz et al.[11] In most cases, the larger particles (greater than 20 micron-sized particles) gave spectra most similar to the parent meteorite. For both Orgueil and Murchison, a composite of all meteorite particles analyzed yielded a spectrum similar to that of the parent meteorite (Figs. 8 and 9). Sulfur-bearing species, aliphatic hydrocarbon groups, polycyclic aromatic hydrocarbons, carbonates, and water were present in each of these.

Based on the total ion chromatograms, the IDPs had the least amount of total volatiles; small Murchison particles had only slightly more volatiles (1.15 times); and small Orgueil particles had about 2.3 times as much volatile material as the IDPs. Direct analysis of the groundmass in an individual piece of Murchison with the LM/MS technique produced only 1.4 times more volatiles than the IDPs, but the analysis of 40 to 50 micron-sized particles of Orgueil evolved 20 times as much volatile material as the IDPs. These results are in keeping with previous studies which noted volatile differences.[11]

COMPARISON OF CI AND CM VOLATILES WITH "COMPOSITE" IDP VOLATILES

In order to envision what a parent body represented by the 14 IDPs selected for our study might resemble, a "composite" spectrum was prepared (Fig. 10). Backgrounds were subtracted, and the "composite" spectrum was obtained by combining all of the spectral peaks for the 14 IDPs. This can be considered to give a representative spectrum of a possible "parent body". The IDP "composite" spectrum (Fig. 10) is very much like the spectrum obtained from the CM carbonaceous chondrite groundmass (Fig. 8). The most obvious difference is the decreased intensity of the sulfur-related peaks. Although clearly present, the SO and SO_2 peaks are much smaller for the IDP composite spectrum as compared to either the CM or CI spectrum (Figs. 8 and 9). The composite IDP spectrum (Fig. 10) most resembles the Murchison spectrum and may be another indicator that the IDPs parent body is related to the CM carbonaceous chondrite-type parent bodies.

CONCLUSIONS

IDPs show a variety of volatile abundances. Particles range from essentially depleted of volatiles to concentrations approaching those found in the most primitive carbonaceous chondrite. A comparison of volatiles found in 10 to 20 micron size regions of primitive meteorites with those present in 14 randomly selected IDPs suggests that volatiles within these IDPs are similar to those observed for the groundmass of CM chondrites.

ACKNOWLEDGMENTS

This work was supported in part by NASA's Planetary Biology Program Office. R.B. wishes to acknowledge support from NASA's Summer Faculty Research Program.

REFERENCES

1. L.J. Allamandola et al., *Science* **237**, 56-59 (1987).
2. S.J. Clemett et al., *Lunar PlanetaryScience XXIV*, 309-310 (1993).
3. C.P. Hartmetz et al., *Proc. 20th LunarPlanetary Sci. Conf.* 343-355 (1990).
4. C.P. Hartmetz et al., *Proc. 21st Lunar Planetary Sci. Conf.* 557-567 (1991).
5. R. Bustin et al., *Lunar and Planetary Science XXIV* 239-240 (1993).
6. A.O. Nier and D.J. Schlutter, *Lunar Planetary Science XXIV* 1075-1076 (1993).
7. I.D.R. Mackinnon and F.J.M. Rietmeijer, *Rev. Geophys.* **25**, 1527-1553 (1987).
8. E.K. Gibson, Jr., *J. Geophys. Res.-Planets* **E3**, 3865-3875 (1992).
9. J.P. Bradley et al., *Science* **226**, 1432-1434 (1984).
10. G.J. Flynn et al., *Lunar Planetary Science XXIV*, 497-498 (1993).
11. C.P. Hartmetz et al., *Proc. 21st Lunar Planetary Sci. Conf.* 527-539 (1991).
12. E.K. Gibson, Jr. and R.H. Carr, *U.S. Geol. Survey Bull. 1890*, 35-49 (1989).
13. M.E. Zolensky and R.A. Barrett, Olivine and pyroxene compositions of chondritic Interplanetary Dust Particles. This volume (1994).
14. K. Thomas et al., *Geochim. Cosmochim Acta* **57**, 1551-1566 (1993).
15. K. Thomas, L. Keller and D.S. McKay, Quantitative analyses of carbon in anhydrous and hydrated Interplanetary Dust Particles. This volume (1994).

FIGURES

In Figures 1-10 the tops of the stippled bar is the mean while the tops of the filled and unfilled bars represent ± 1 sigma, respectively. The average background present in the vacuum system prior to the laser ablation and the average signal from four laser shots into the same piece of gold are subtracted from the IDP and meteorite signal. The data processing procedures used are described fully in Hartmetz et al.[4]

180   Volatiles in Interplanetary Dust

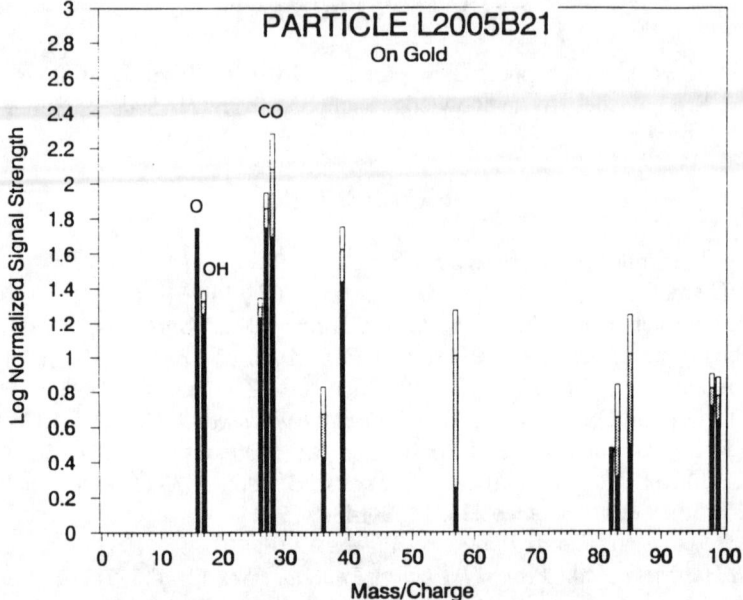

Figure 1. Log-normalized signal strength vs. m/z for Particle L2005B21 containing very few volatiles.

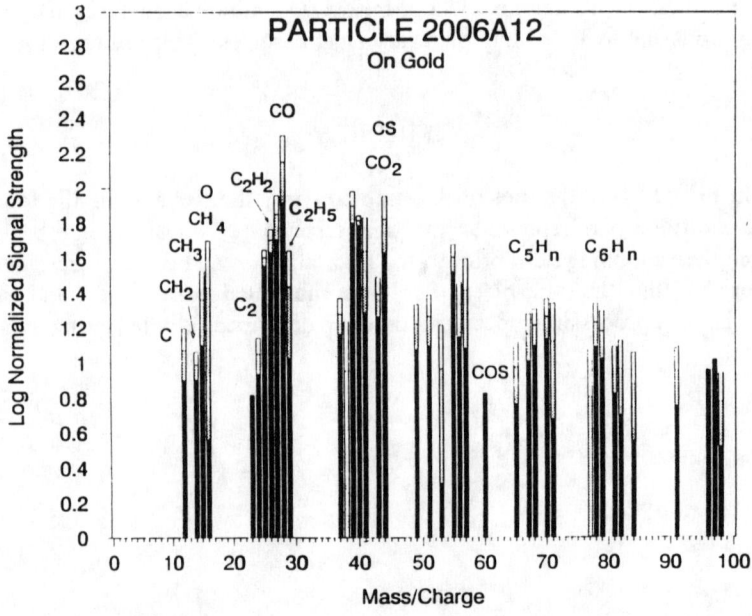

Figure 2. Mass spectrum of a volatile-rich IDP L2006A12.

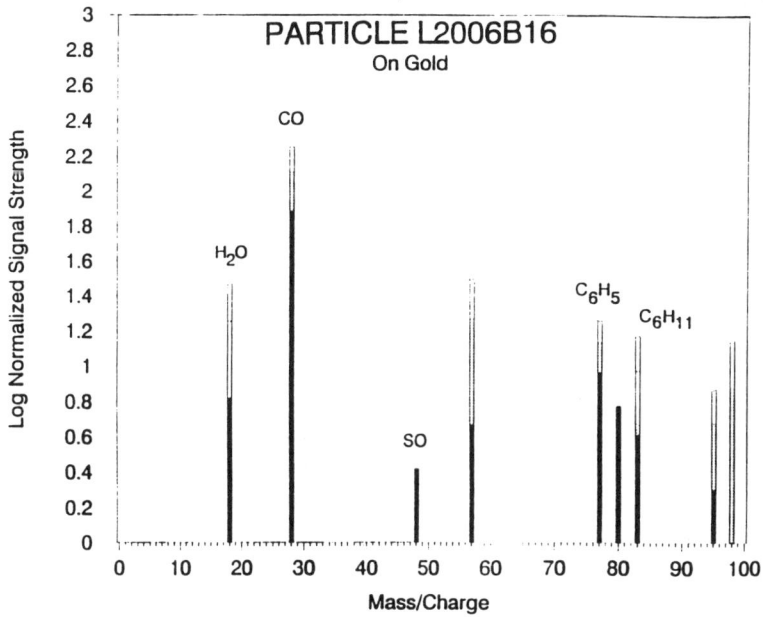

Figure 3. Mass spectrum of an IDP containing water or hydrated components.

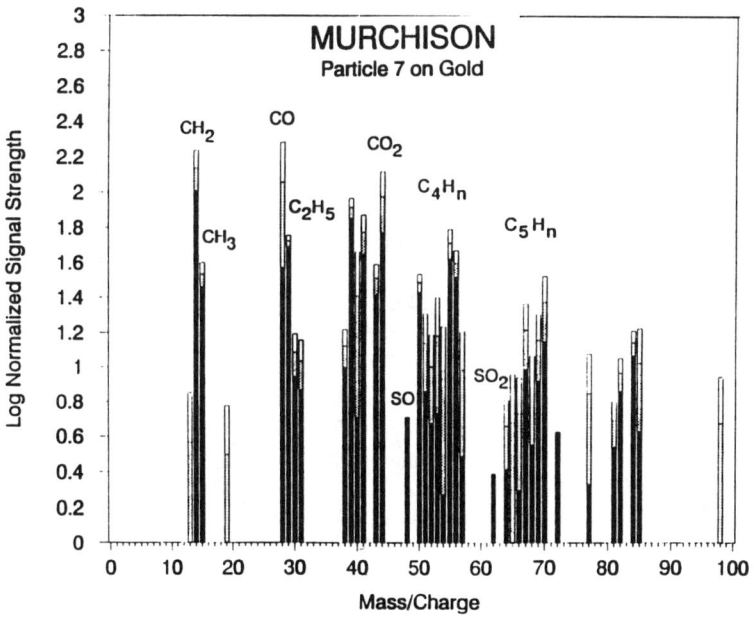

Figure 4. Mass spectrum of a volatile-rich Murchison particle.

182 Volatiles in Interplanetary Dust

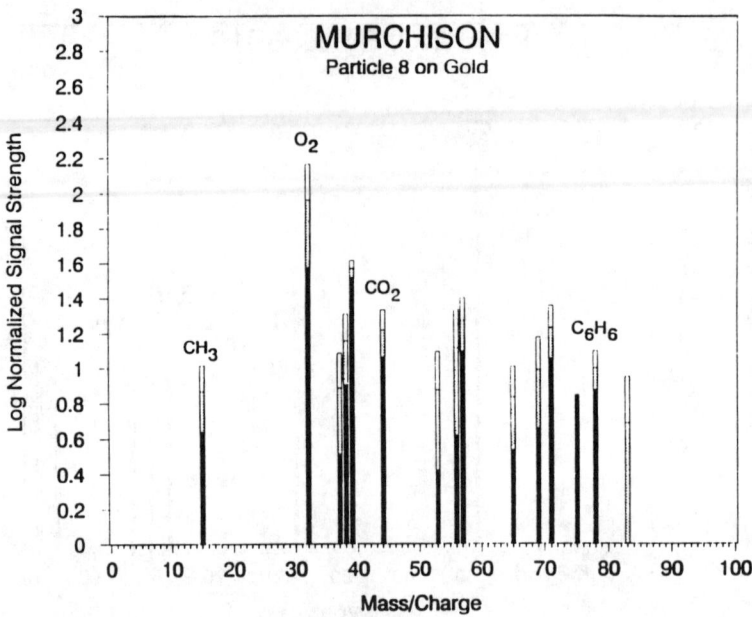

Figure 5. Mass spectrum of a Murchison particle containing few volatile-bearing components.

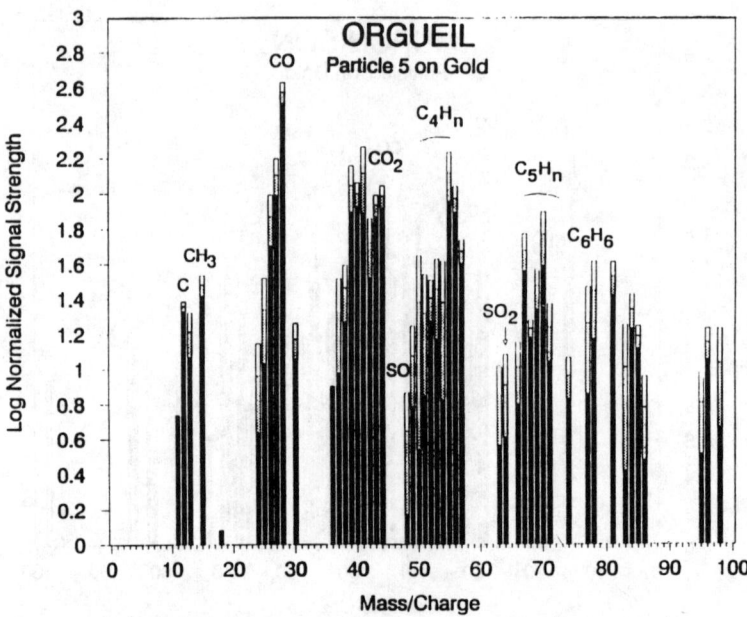

Figure 6. Mass spectrum of a volatile-rich Orgueil particle.

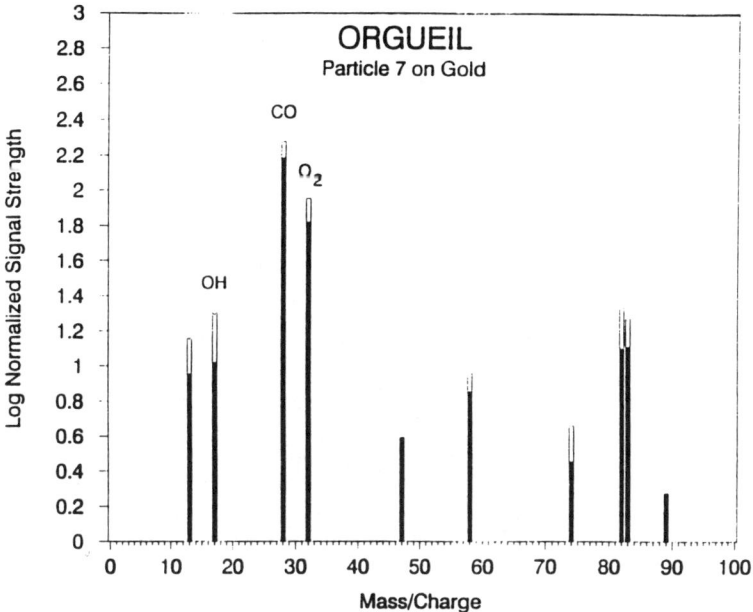

Figure 7. Mass spectrum of an Orgueil particle with few volatile-bearing components.

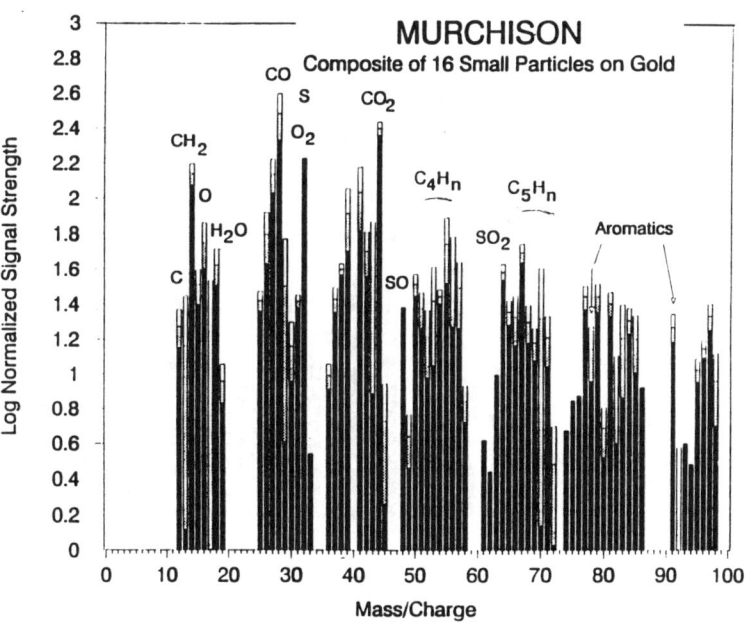

Figure 8. Composite mass spectrum of 16 particles from Murchison carbonaceous chondrite.

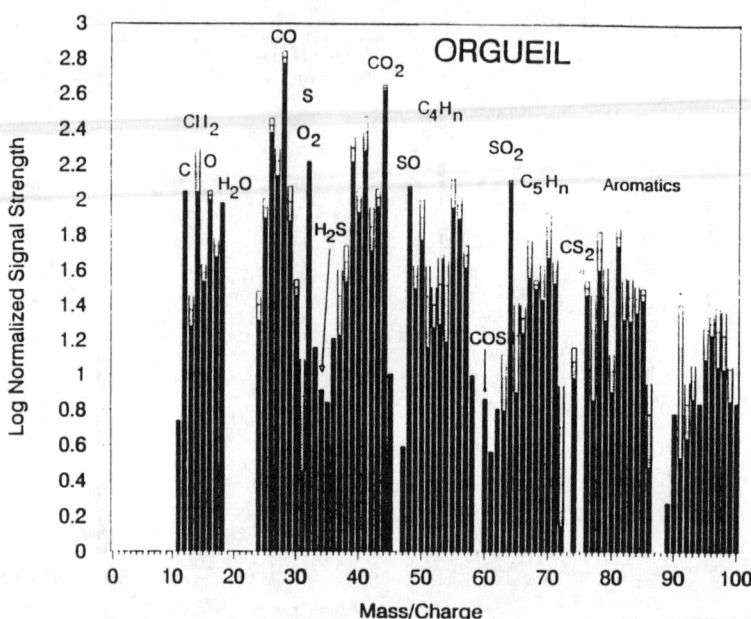

Figure 9. Composite mass spectrum of 12 particles from Orgueil carbonaceous chondrite.

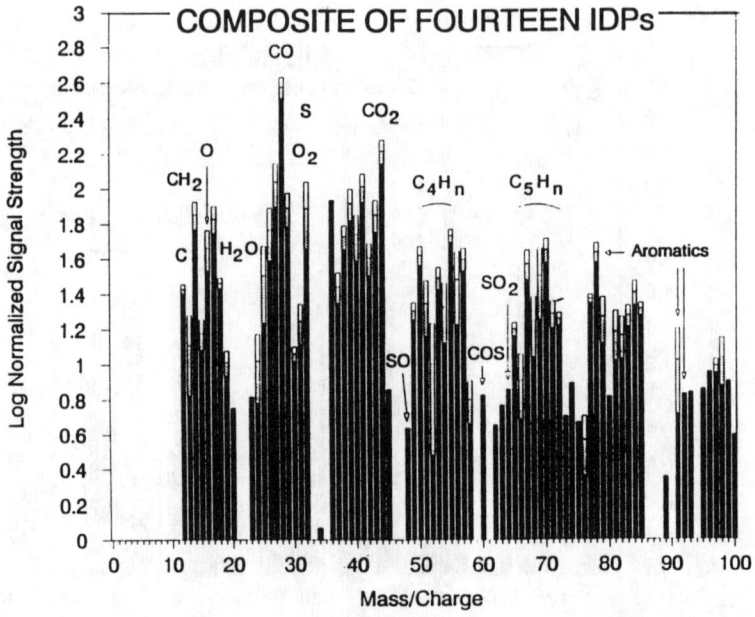

Figure 10. Composite mass spectrum of 14 IDPs examined in this study. Note the similarity with the spectrum shown in Figure 8.

ORIGIN OF THE HYDROCARBON COMPONENT OF INTERPLANETARY DUST PARTICLES

Thomas J. Wdowiak and Wei Lee
Department of Physics, University of Alabama at Birmingham
Birmingham, AL 35294 USA

ABSTRACT

Using experiments as a basis, we have developed a scenario for the origin of the hydrocarbon material of carbonaceous chondrites. This scenario can also serve as an explanation for the origin of the hydrocarbon component of interplanetary dust particles (IDPs). The formation of polycyclic aromatic hydrocarbon (PAH) molecules in the atmospheres of carbon stars undergoing a late stage of stellar evolution is indicated by the observed unidentified infrared (UIR) emission bands. Those molecules are then transported through interstellar space where they become enriched with deuterium through ion molecule reactions when passing through cold dark clouds. Many of those PAH molecules are subsequently hydrogenated and cracked in a hydrogen-dominated plasma such as that which would have occurred in the solar nebula. The resulting mixture of alkanes and residual deuterium-rich PAH molecules was then incorporated into the mineral fraction of the parent bodies of carbonaceous chondrites and IDPs.

INTRODUCTION

It is highly likely that there exists a relationship between the hydrocarbon material of chondritic IDPs and that found as a component of carbonaceous chondrites. Earlier research on the origin of such hydrocarbons focused largely on the Fischer-Tropsch process where CO and H_2 were converted at high temperatures with the aid of a catalyst to meteoritic hydrocarbons.[1&2] Some very recent laboratory experiments have suggested a pathway that is different from those previously considered. In this new model polycyclic aromatic hydrocarbons (PAHs) first formed in the atmospheres of carbon stars are transported through the interstellar medium. Such molecules are capable of surviving the harsh interstellar environment where they are subjected to ultraviolet radiation and shocks. While in cold dark clouds they become enriched in deuterium through ion molecule reactions. Finally much of the aromatic material is converted to the alkane form found to be the principal hydrocarbon component of carbonaceous chondrites.[3] The last process appears to have probably taken place in the plasma environment of the solar nebula prior to incorporation of the hydrocarbon material into the parent bodies of carbonaceous chondrites and IDPs. The T-Tauri stage of the Sun was probably an important factor in the production of the necessary plasma conditions. The hydrogenation will yield an alkane product having a lower deuterium content than the aromatic precursor by virtue of simple dilution. This is consistent with what is observed for meteoritic hydrocarbons.

The presence of a hydrocarbon component in chondritic IDPs has been demonstrated by Wopenka[4] where a laser-induced luminescence was observed when 514-nm light was directed on an IDP sample. She found the emission spectrum peaked at about 600 nm for what she called a type 3 particle. Another chondritic IDP known as Viburnum exhibited a more extreme emission peaking at about 720 nm. This emission was interpreted as being due to a mixture of large polycyclic aromatic hydrocarbons in the microcrystalline form as had been proposed earlier by Wdowiak.[5] Following that initial report a more comprehensive paper argued on the basis of Raman and laser induced luminescence for a connection between interstellar PAH molecules and the carbonaceous component of IDPs.[6] Recently, mass spectroscopy of the chondritic IDPs Aurelian and Florianus has shown the presence of PAH molecules with those having a mass around 250 amu being the most abundant. A second group around 370 amu was also found to be fairly abundant. The very low abundance of PAHs in the 78 - 192 amu range was attributed to their being heated during entry into the terrestrial atmosphere.[7] Calculation of the peak atmospheric entry temperatures of micrometeorites indicates that it is highly probable that many IDPs will be flash-heated to temperatures in excess of 600°C.[8&9] Experiments involving the heating of milligram amounts of the acid insoluble residue of the Orgueil (CI) carbonaceous chondrite to specific temperatures in a vacuum suggest that thermal alteration of the hydrocarbon component will take place during atmosphere entry.[9&10] While not an attempt to simulate IDP entry, these experiments are a good indicator of the chemical change that occurs when chondritic hydrocarbons are heated.

EXPERIMENTAL TECHNIQUES

We have carried out experiments[3] with the simple PAH naphthalene ($C_{10}H_8$) which was chosen for the ease in which it can be placed into the gas phase. The naphthalene was converted through hydrogenation in a plasma, to a material having a mid-infrared (4000 - 400 cm^{-1}) spectrum that is remarkably similar to that obtained by Cronin and Pizzarello of the hydrocarbon substance extracted with a benzene-methanol mixture (9:1) from the Murchison CM2 carbonaceous chondrite.[11] A high-molecular-weight residue was formed when the gaseous naphthalene and hydrogen mixture was subjected to a 9400-VAC electrical discharge in a specially constructed tube. It was deposited at two places on the water jacket-cooled internal glass surface of the discharge tube which have direct exposure to the electrodes placed in side arms. Approximately 300 milligrams of naphthalene was placed inside the sapphire tube that was then inserted into the discharge tube to a position halfway between the two electrodes. Additional deposits were formed on the internal walls of the sidearms in close vicinity to the electrodes, and on the inner wall of a sapphire tube through which the discharge passed. The hydrogen gas was introduced through a small side inlet. A mechanical pump held the pressure at 0.5 torr as measured with a McLeod gauge. Based upon the pumpdown pressure of 0.05 torr prior to the introduction of

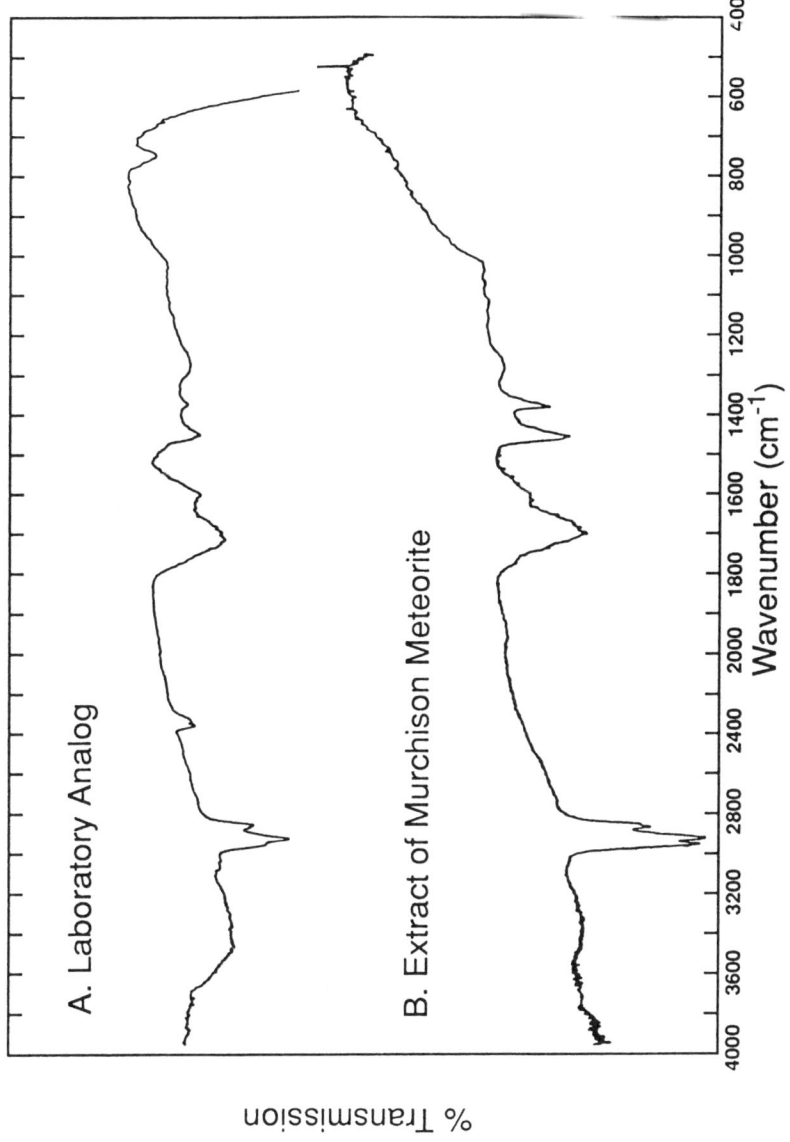

Figure 1. Comparison of the spectrum of a portion of the laboratory analog (A) and the spectrum obtained by Cronin and Pizzarello[11] of the benzene-methanol (9:1) extract of the Murchison meteorite (B). Spectrum A also shows artifacts due to H_2O (around 3450 cm^{-1}) and incomplete cancellation of atmospheric CO_2 (around 2350 cm^{-1}).

188 Origin of the Hydrocarbon Component

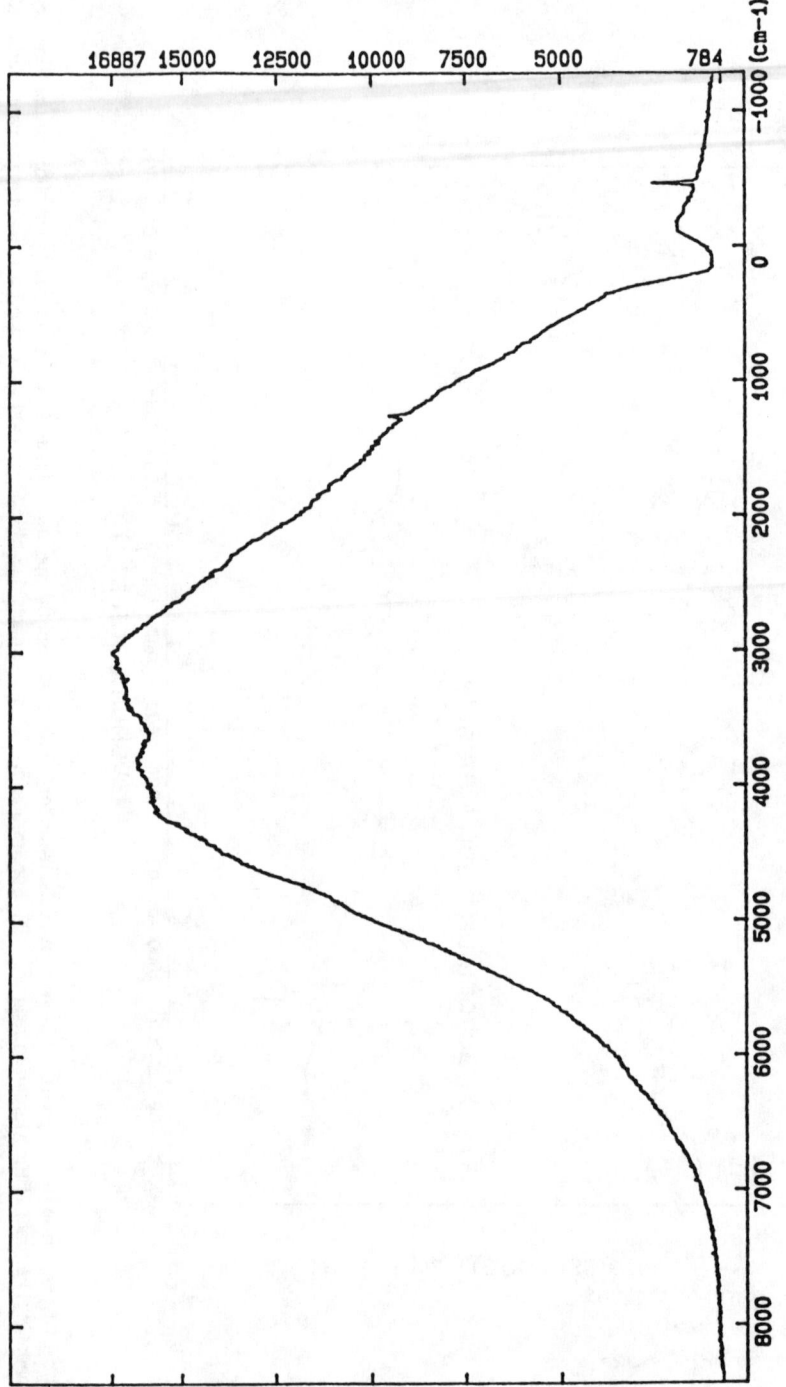

Figure 2. Laser induced luminescence spectrum of a piece of film of the laboratory analog. The excitation wavelength is 514 nm.

the hydrogen, the composition of the gas mixture is estimated to be 90% H_2, 8% N_2, and 2% O_2. The nitrogen and oxygen components are due to backstreaming of atmospheric gases from the pump. The presence of oxygen appeared to play an important role in the character of the deposited material as deduced from its IR spectrum as discussed later. The amount of recovered material is approximately 15 milligrams and residues are not present in the absence of naphthalene. A nitrogen-cooled trap is incorporated in the line between the discharge tube and the vacuum pump, and was used in some of the experimental runs.

The samples were prepared for spectroscopy by removing the film-like material from various positions of the discharge tube, washing it in benzene and methanol, drying it under vacuum, and then mounting the material on a KBr or NaCl disk for IR measurements. The IR spectra were obtained with a Mattson Polaris FT-IR spectrometer capable of a resolution of 4 cm^{-1}. The laser-induced luminescence measurements were performed using a modified EG&G Dilor XY Raman system where both Raman and luminescence spectra can be recorded from the same micron-sized area of the sample. The light signal was directed into the entrance slit of the spectrometer which employed a focal plane CCD detector. The luminescence spectrum of the deposit sandwiched between a cover glass and a glass slide was taken with a 150 grooves / mm grating (resolution of 17 cm^{-1}) during a data acquisition time of 5 s. The exciting Ar^+ laser was selected to have an output of 10 mW at the excitation wavelength of 514.53 nm.

RESULTS AND DISCUSSION

Infrared Spectroscopy

Spectrum A in Figure 1 shows the mid-infrared spectrum of the film-like deposit removed from the water-cooled surface facing one of the electrodes. It was mounted on a NaCl disk. The spectrum was measured against air. Spectrum B reported by Cronin and Pizzarello[11] is of the hydrocarbon component extracted from the Murchison meteorite and mounted on a KBr substrate. The similarity between the two spectra is remarkable; all the more because the meteorite sample which is the product of a rigorous attempt to obtain an uncontaminated sample, is the major organic constituent of the meteorite. As such, it may be representative of the complex hydrocarbon material present throughout the early solar system. The differences that exist between the spectra can be accounted for easily. There are features in the laboratory analog spectrum that are due to a small amount of residual aromatic material that has been minimally altered or has escaped alterations altogether. Those at the aromatic C-H stretch position of 3000 cm^{-1} are very weak. There is slightly more absorption around 1600 cm^{-1} which is due to the aromatic C-C deformation mode. Finally there is a definite band at 750 - 765 cm^{-1} again probably due to an aromatic vibration. This band most likely arises from the C-H out-of-plane wag for the situation of 3 or 4 H-atoms attached per external aromatic ring. This last band is normally one of the strongest infrared bands of an aromatic molecule. None of these bands is as strong as the alkane bands discussed in the following paragraph. There is also an artifact due to atmospheric carbon dioxide at 2350 cm^{-1} due to incomplete cancellation between the sample and the reference spectra.

Very relevant to the question of our deposit being a good laboratory analog of the meteoritic hydrocarbon component are the relative strengths of the alkane C-H stretching and deformation bands due to -CH_3 and -CH_2- groups in the 2900 and 1400 cm^{-1} regions. Comparison of those bands due to -CH_3 group at 2960 and 2870 cm^{-1} and -CH_2- group at 2925 and 2860 cm^{-1} in both spectra suggests the laboratory deposit has a slightly lower fraction of -CH_3 than exists in the meteorite extract. The -CH_2-/-CH_3 ratio deduced from the meteorite material spectrum is about 2, whereas in our residue it is 3 to 4. The observation that the meteorite extract has a greater fraction of material with -CH_3 terminations suggests it may contain a greater fraction of low molecular weight material. However the -CH_3 content of our sample, as indicated by the absorption bands, can be considered to be high. Substituting a PAH molecule with more rings than 2-ringed naphthalene, which was chosen for its volatility, would probably result in a high proportion of -CH_3 terminations when the ring system is cracked. Experiments to investigate this are being initiated. Hydrogenation without total cracking can result in cycloalkanes, which have been reported to be present in the Murchison CM2 meteorite.[11] There is also a very strong and broad absorption in both spectra at around 1700 cm^{-1} which is quite probably due to the C=O stretch in the carboxyl group. We consider it likely that this functional group in the laboratory material results from the presence of atmospheric oxygen in the gas mixture as discussed earlier. Apart from these strong agreements, the spectral detail exhibited by our laboratory analog is also very consistent with that of the meteoritic component.

Laboratory-Induced Luminescence Spectroscopy

The laser-induced luminescence of a piece of film of the laboratory analog is displayed in Figure 2. The 514-nm excited luminescence spectrum has a similar shape as those exhibited by the chondritic IDPs Lea and Calrissian.[6] The peak of emission between 3000 and 4000 cm^{-1} relative to the excitation, corresponds to 600 - 650 nm in wavelength. The match of this spectrum of the laboratory analog along its infrared spectrum, with both the luminescence spectra of two chondritic IDPs and the infrared spectrum of the hydrocarbon component of a CM2 carbonaceous chondrite indicates it is probably not only an analog for meteoritic material, but IDP hydrocarbons as well.

CONCLUSIONS

Our ability to produce a hydrocarbon material exhibiting an infrared absorption spectrum matching very closely that of the hydrocarbon extract of the Murchison CM2 carbonaceous chondrite, and having a laser-induced luminescence spectrum similar to carbonaceous IDPs, strongly suggests that the processes involved in the experiment simulate cosmic conditions. From the standpoint of astrophysics the utilization of a mixture of hydrogen and the PAH naphthalene in an energetic experimental environment is a valid combination. This is because PAH molecules remain as the best hypothesized source of the cosmic unidentified infrared (UIR) emission bands first observed in 1973[12] which are ubiquitous in the Milky Way and other galaxies.

In summary our model is that PAH species are formed in stellar atmospheres and then ejected (R CrB, planetary nebulae, etc.) into the interstellar medium. These molecules probably become deuterated in cold dark clouds through exchange with interstellar deuterium. Much of the material is then hydrogenated into alkanes in a plasma perhaps at the time of the formation of the Sun prior to incorporation into the parent bodies of carbonaceous chondrites and IDPs. Importantly this has the effect of diluting the deuterium content of those molecules that have been hydrogenated. Remaining deuterium rich aromatics can be considered to be unaltered interstellar molecules.

ACKNOWLEDGMENTS

We are grateful for the assistance of Thomas McCauley and Yogesh Vohra in obtaining the laser-excited luminescence spectrum. This work was supported by NASA grant NAGW-749.

REFERENCES

1. M. H. Studier, R. Hayatsu, and E. Anders, *Science* **149**, 1455 (1965).
2. E. Anders, R. Hayatsu, and M. H. Studier, *Science* **182**, 781 (1973).
3. W. Lee and T. J. Wdowiak, *Astrophys. J.* **417**, L49 (1993).
4. B. Wopenka, *Earth Planet. Sci. Lett.* **88**, 221 (1988).
5. T. J. Wdowiak, in *NASA Conf. Proc. 2403*, A-41 (1986).
6. L. J. Allamandola, S. A. Sandford, and B. Wopenka, *Science* **237**, 56 (1987).
7. S. J. Clemett, C. R. Maechling, R. N. Zare, P. D. Swan, and R. M. Walker, *Science* **262**, 721 (1993).
8. S. G. Love and D. E. Brownlee, *Icarus* **89**, 26 (1991).
9. Flynn, G.J., *Icarus* **77**, 287 (1989).
10. T. J. Wdowiak, G. C. Flickinger, and J. R. Cronin, *Astrophys. J.* **328**, L75 (1988).
11. J. R. Cronin and S. Pizzarello, *Geochim. Cosmochim. Acta* **54**, 2859 (1990).
12. F. C. Gillett, W. J. Forrest, and K. M. Merrill, *Astrophys. J.* **183**, 87 (1973).

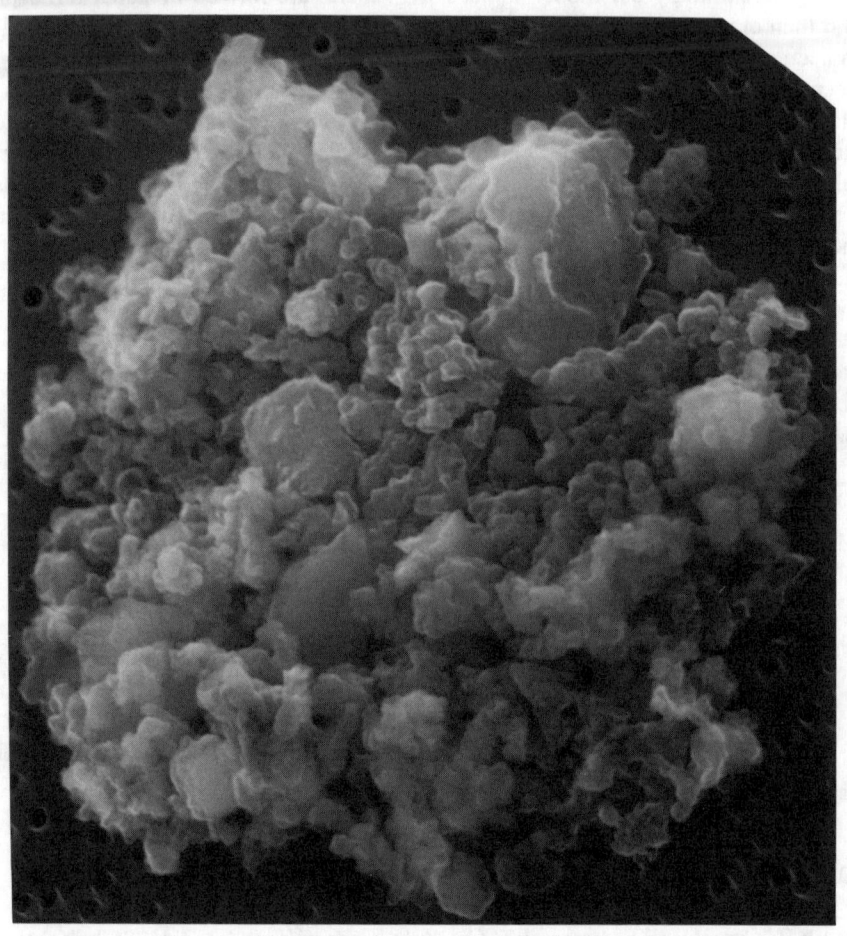

A secondary electron image of chondritic IDP W7029 B13. This IDP is clearly an aggregate of mineral grains of widely varying sizes. The overall particle measures 14 μm in maximum dimension. (NASA photo S82-27575)

CARBON IN COMET HALLEY DUST PARTICLES

Marina Fomenkova
California Space Institute, University of California San Diego
La Jolla, CA 92093-0216 USA

Sherwood Chang
NASA Ames Research Center, MS 204
Moffett Field, CA 94035 USA

INTRODUCTION

Comets are small bodies of the Solar System containing primarily a mixture of frozen gases, carbonaceous and mineral grains. They are likely to preserve volatile material from cold regions of the protosolar nebula and remnants of interstellar dust and gas. Various organic and inorganic carbon-containing components were discovered in comet Halley dust particles.[1&2] Their nature and abundance provide information about possible astrophysical sources of carbon, constrain models of interstellar grains and their alteration during solar system formation.

More than 2500 mass spectra of cometary grains with masses in the range of 5×10^{-17} - 5×10^{-12} g were measured *in situ* by PUMA-1 and PUMA-2 dust-impact mass-spectrometers on-board the VEGA spacecraft during flyby missions to Comet Halley.[3] Dust particles from the comet Halley dust envelope collided with the silver target of the instrument with a relative velocity of about 80 km/s and were ionized by hypervelocity impact. On passing through a time-of-flight analyzer, ions were separated according to their mass-to-charge ratio. The resulting spectrum was recorded by a secondary electron multiplier. Most of the analyzed ions were singly charged atomic ions. Quantization errors introduced in the amplitude by converting from analog to digital spectra were estimated as the range of numbers of ions corresponding to one digital unit; they are 8-17% for PUMA-1 and 5-15% for PUMA-2, and decrease with an increase of the amplitude.

Corrections for ionization yields and registration efficiencies are necessary for calculating elemental abundances from ion abundances. Quantitative laboratory calibrations of PUMA instruments were impossible because there are no facilities to accelerate appropriate dust grains to 70 km/s. Crude model calculations of atom-to-ion yields have been made[4] based on extrapolation of laboratory experiments at velocities ≤ 10 km/s and yields of secondary ion emission processes. Empirical coefficients for the spectra correction have been derived.[5] In all those cases, the suggested corrections were checked by comparison of resulting atomic abundances to chondritic ones. However, it was shown[6] that, without correction factors, the bulk ion abundances of rock-forming elements in comet Halley dust are close to solar if they are calculated with the mass of dust particles taken into account. Thus, the approach we used, was that the numbers of ions in spectra reflect the number of atoms.[1&2]

Ion yields may differ significantly between biogenic elements (H, C, N, O) of low atomic mass and rock-forming elements (Na and above) of high atomic mass due to large differences in their electronegativity. Also, PUMA instruments had two modes of operation, wide and narrow energy windows, with differences in instrument transmission which contributed to different ion yields for low-mass and high-mass elements.[7] However, the classification scheme[2] is based on relative abundances of low-mass elements only and we do not attempt to compare the abundance of the biogenic (low-mass) and rock-forming (high-mass) elements.

Another source of instrumental error is the absence of direct proportionality between the dust particle mass and the number of ions in its mass-spectrum.[8] That is accounted for by considering relative abundances of elements instead of absolute values of the ion amplitude.

The total mass of measured dust particles is about 0.5 ng[1], which is comparable to the mass of only one interplanetary dust particle (IDP) collected from the stratosphere. The incompatibility of spatial scales should be kept in mind when comparing the Halley solid material and other extraterrestrial samples, such as meteorites or IDPs, as it may be of principal importance for the correct choice of data for the comparison.

In this paper we discuss the relative abundances and the mass distribution of carbon-containing components in comet Halley dust, and the distribution of the total bulk carbon among these components.

CARBON PARTICLES ([C]-PHASES)

Of the grains analyzed, 97 in PUMA-1 data and three in PUMA-2 data are ~99 atom% carbon.[2] These particles, called "[C]", comprise about 6% of the bulk carbon in the observed comet Halley dust. All [C]-grains contain a minor quantity of rock-forming elements. Specific [C]-phases known in meteorites and IDPs - diamonds, graphite, or amorphous carbon - cannot be identified in cometary particles because the mass-spectra do not provide information about the grain structure. In the Murchison CM meteorite, however, characterized [C]-phases represent only a small fraction of the total, ~2% non-carbonate carbon: graphite, ~10^{-4}, and diamond, ~10^{-3}.[9]

Individual IDPs are more carbon-rich than meteorites and may have a very high carbon content: C/Si up to 4.[9] Various C-phases have been observed in chondritic porous (CP) IDPs [reference 11 and references therein]: amorphous carbon, poorly graphitised carbon and microcrystals of graphite. The distribution of carbon among these phases has not yet been determined.

In the coma of comet Halley, the [C]-particles were most abundant at the outbound segment of the VEGA-1 trajectory and uniformly present otherwise (Fig.1a). Comparison of the distribution of the intervals between occurrences of the [C]-grains with a Poisson law (Fig. 1b) shows that the [C]-grains were encountered as clusters along the trajectory and suggests there was a mechanism for disaggregation of clusters of grains or disintegration of larger grains in the coma. Chemical transformations or evaporation seem to be less probable than physical

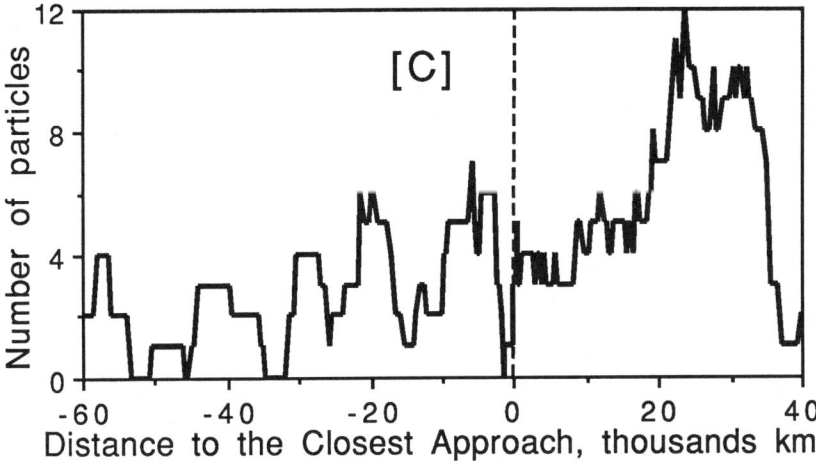

Figure 1a. Spatial distribution of [C]-grains in the coma (PUMA-1 data). The number of encountered [C]-grains was averaged with a 60 sec moving window. At the closest approach the distance to the comet nucleus was ~8900 km.

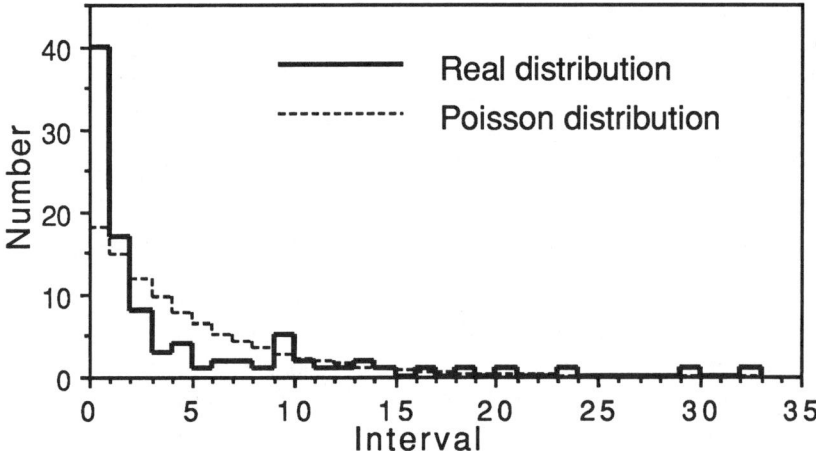

Figure 1b. Distribution of intervals between occurrences of [C] particles. A high peak at the interval equal to one indicates that, frequently, several grains followed closely upon each other. The long tail of the distribution indicates that such clusters were separated by relatively large distances. A Poisson law would approximate random occurrence.

destruction (e.g. electrostatic disruption[12]) because the known [C]-phases are chemically stable and non volatile.

The [C]-particles prevail among small grains[13] (Table 1). This is consistent with observations of submicron fluffy grains of poorly graphitised carbon in IDPs[14] and with estimates of the sizes of circumstellar and interstellar carbon particles.[15] Mass loss from carbon stars may have contributed up to 50% of all matter injected into the interstellar medium, and interstellar grains may represent a major fraction of cometary dust.[16] In terms of their size and spatial distribution the [C]-particles differ from other classes of comet Halley dust particles, and they probably represent a special population of preserved interstellar carbon grains.

Table 1

Mass distribution of carbon-containing components of comet Halley dust. PUMA-1 and PUMA-2 data are combined together. N - number of spectra. Percentages are to the total number of spectra in each mass range.

mass range	$<10^{-15}$ g		$10^{-15} - 10^{-13}$ g		$>10^{-13}$ g	
	N	%	N	%	N	%
[C] grains	51	9	33	3	16	2
Hydrocarbons	17	2-3	18	2	13	2-3
Complex organic	20	3-4	84	7-8	67	11
[C, N], [C, O] polymers	8	1	10	1	4	<1
Mixed particles	193	33	601	52	358	57
Mg-carbonates	20	3-4	44	4	46	7-8
Rock particles	237	40	291	25	63	10
[H] and [O] grains	44	7-8	71	6	59	9
Totals	590	100	1152	100	626	100

Hydrocarbons

In PUMA-1 and PUMA-2 data 42 and six particles respectively contain primarily C and H. Of these only four from PUMA-1 are "pure [H, C]-particles" lacking any other elements. The class of [H, C]-grains implies the presence of C-H bonds[2], which exist in hydrogenated amorphous carbons and polycyclic aromatic and highly branched aliphatic hydrocarbons which are thought to occur in comets and interstellar clouds[17] and have been observed in carbonaceous chondrites[18] and IDPs.[19&20] Anomalously high D/H ratios in meteoritic and IDP hydrocarbons signify an interstellar origin[19&21] (see Walker, in this volume[22]). Based on the H/C ratio, some subgroups of the [H, C]-class of comet Halley dust particles were characterized as mixtures of C-phases, polycyclic aromatic and highly branched aliphatic hydrocarbons. About 10% of the total carbon detected is contained in these grains which are uniformly distributed in all particle mass ranges.[13]

Complex Organic Matter

Complex organic material occurs in 145 particles of the PUMA-1 data (of which 25 are "pure" CHON) and 26 particles of the PUMA-2 data. Classes of CHON particles such as [H, C, N], [H, C, O] and [H, C, N, O] are thought[2] to be mixtures of various carbon and organic components rather than a single component species. These particles are more abundant among large grains (Table 1) which also suggests they may be composed of 'building blocks' of smaller size observed in other mass groups.

The bulk composition of these complex organic grains is ~ $C_{100}H_{115}N_9O_{65}S$ (~ $C_{100}H_{90}N_8O_{45}S$ if [C]-grains and hydrocarbons were included) and they contain about 25% of the total carbon content. This compares in meteoritic samples with the 20-30% of C extractable as relatively nonvolatile identifiable organic compounds and the 70-80% in the form of insoluble kerogen[18] Abundances of H and O relative to C are higher in these cometary grains than in kerogen observed in meteorites, ~$C_{100}H_{71-48}N_{1-3}O_{12}S_2$ [18], or in pregraphitic carbons observed in IDPs[11], where the C/(H+N+O) ratio is ~1:1 and relative proportions of H, N, O are unknown. Some of the cometary grains may contain water ice mixed with organic matter or, alternatively, some of the H and O could have been chemically bonded to C and lost along with water as a result of depletion of volatiles from meteorite parent bodies. Bulk meteoritic kerogen samples contain isotopically anomalous H and N and are actually made up primarily of minute particles 0.1-0.01 mm[23] (about the size of comet Halley CHON grains). Large D/H fractionation is also associated with the carbonaceous matrix of IDPs.[22&24] Thus, these carbon-rich materials may have originated from the processing of mixtures of cometary CHON grains of interstellar origin.

There are also 22 grains in PUMA-1 data, representing ~1% of the total carbon, which contain [C, O] and [C, N] polymers.[2] These may be related to compounds identified as cyanopolyynes and multi-carbon monoxides in the interstellar media.

MIXED PARTICLES

In PUMA-1 and in PUMA-2 data, respectively, 897 and 255 particles were classified as "Mixed". By definition[1], in these grains the ratio of carbon to rock-forming elements is in the range 0.1-10 and they comprise about 40% of the total carbon abundance. The Mixed particles were assumed to consist of various Mg/Fe silicates, sulfides and organic matter (which is consistent with an increase of their proportion among larger grains - Table 1). From mass-spectra only, we cannot exclude the possibility that a fraction of carbon in these grains is bonded in minerals with rock-forming elements. More detailed analysis and classification of the organic part of the Mixed particles has not yet been done.

MG-CARBONATES

The presence of Mg-rich grains in comet Halley was previously noted.[6&25] Mg-carbonates were tentatively identified[1] in 77 particles of the PUMA-1 data and 33 particles of the PUMA-2 data. The presence of carbonates was also inferred from the weak 6.8 mm emission feature in the infrared spectrum of comet Halley.[26] The fraction of the total carbon in carbonates is estimated as 4%. In some C-chondrites carbonates contain ~2-10% of carbon. They are thought to be products of aqueous alteration which occurred in a parent body.[27] Carbonates are present in hydrated IDPs[28] and are minor, but persistent phases in anhydrous chondritic porous IDPs.[29] The former are generally supposed to have an asteroidal origin, while some fraction of the latter probably originated from short-period comets.

Carbonates in comet Halley dust could have formed *in situ* by hydrocryogenic alteration. Such activity in nuclei of active short-period comets has been hypothesized[30] to take place in interstitial water layers at dust/ice interfaces below the melting point of water-ice. On the other hand, the 6.8 mm feature was observed in protostellar spectra and the interstellar occurrence of carbonates has been suggested.[31]

CARBIDES

So far, SiC grains have not been identified in comet Halley dust. Grains containing C, Si, and no Mg were considered as possible candidates, but all of them contain at least H and O, and sometimes also minor quantities of Na, Al, S, Cl, and Fe, thus prohibiting unambiguous identification. The absence of SiC in comet Halley data can be explained by a statistical lack of data. In primitive meteorites the proportion of SiC grains is ~6-8 ppm.[32] If the proportion of SiC in cometary solids were the same then we would expect to discover one spectrum out of 10^5. Only 2500 spectra were measured by the PUMA-1 and PUMA-2 instruments, and 3000 spectra were measured by the PIA instrument. The probability of one of them being SiC is therefore ~0.05. This consideration also provides an upper limit on the concentration of SiC in cometary dust: <150 ppm.

The presence of iron carbide is plausible in 2-3 spectra of the Halley dust particles. Iron carbide has been observed in several chondritic IDPs[14] in association with carbon filaments which may have formed by heterogeneous catalysis of carbon compounds.[33] However, in iron-rich grains discovered in comet Halley, grains also containing sulfur have more carbon present than metal grains.[34] This contradicts the model of a carbonaceous coating formed on the grain because sulfur is known to poison organic synthesis reactions, and rather suggests that at least a part of cometary organics is a separate independently formed component.

ROCK PARTICLES

The analysis[1] revealed 430 grains in the PUMA-1 and 161 grains in the PUMA-2 data in which the ratio of C to rock-forming elements is <0.1. They contain <1% of the total carbon, which is below the level of measurement uncertainty. The proportion of "Rock" particles decreases with an increase of the particle mass (Table 1). Based on indirect evidence, cometary grains were suggested[2] to be built of small silicate grains embedded in a carbonaceous matrix. How much do they resemble "granular units" (GU) of IDPs[14] which are about 0.3-2.0 mm in size and consist of carbonaceous material mixed with ultrafine-grained silicates, sulfides and oxides?

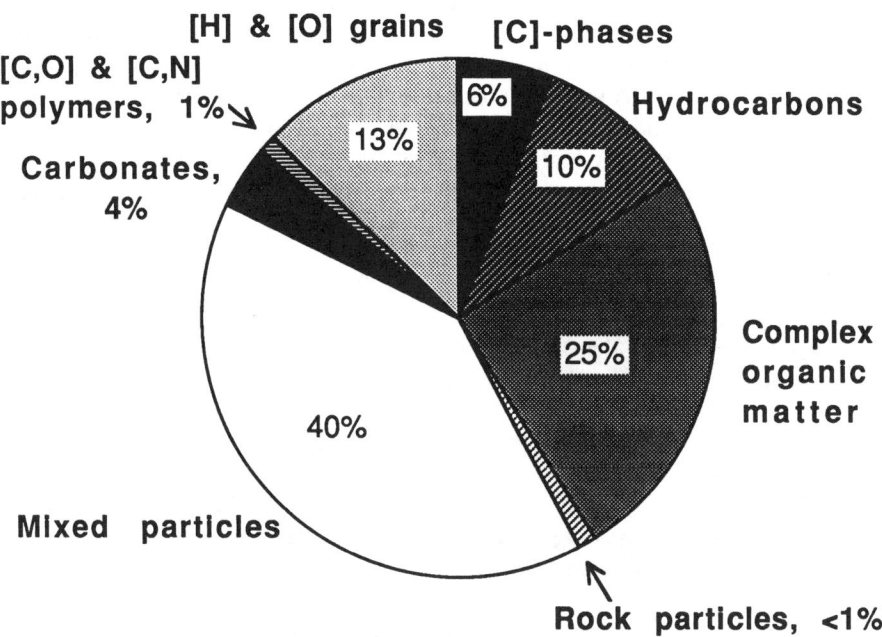

Figure 2 A summary of the data on carbon distribution among various carbon-containing components in the comet Halley dust.

This question can not be answered without a direct analysis of a returned cometary sample. See the papers in this volume by Bradley[35], Rietmeijer[36] and Klock and Stadermann[37] for a complete description of GUs.

[H] AND [O] -RICH GRAINS

These particles (158 ones in the PUMA-1 and 16 ones in the PUMA-2 data) contain ~13% of the total carbon. The nature of these grains remains a mystery because the abundance of H and O is too high for a chemically reasonable interpretation. Excess of H may be explained by the release of hydrogen dissolved in the target material[38] and the abundance of O in spectra may be strongly influenced by hypervelocity impact ionization processes.[39] On the other hand, the overabundance of H and O in some spectra may indicate the presence of water ice in these particles.

CONCLUSIONS

Figure 2 represents the summary of the data on the carbon distribution among various carbon-containing components in comet Halley dust.

ACKNOWLEDGMENTS

MNF acknowledges the support of California Space Institute. We thank D. Summers for a critical reading of the manuscript. We acknowledge constructive comments and suggestions made by reviewers.

REFERENCES

1. Fomenkova M., Kerridge J., Marti K., McFadden L. (1992) *Science* **258**, 266-269.
2. Fomenkova M., Chang S., Mukhin L. (1993) submitted to CGA.
3. Kissel J. *et al.* (1986) *Nature* **321**, 280-282, 336-338; Sagdeev R. Z. *et al.* (1986) *Proc. 20th ESLAB Symp. on the Exploration of Halley's Comet*, **ESA SP-250**, 349-352.
4. Kissel J. and Krueger F. R. (1987) *Appl. Phys. A* **42**, 69-85.
5. Lawler M.E., Brownlee D.E., Temple S., Wheelock M.M. (1989) *Icarus* **80**, 225-242.
6. Mukhin L.M. *et al.* (1991) *Nature* **350**, 480-481.
7. Sagdeev R.Z. *et al.* (1987) *Astr. Astrophys.* **187**, 179-182.
8. Fomenkova M.N., Evlanov E.N., Mukhin L. M., Prilutsky O. F. (1991) *Lunar and Planet. Sci. Conf.* **22**, 397-398.
9. Zinner E. *et al.* (1990) *Lunar and Planet. Sci. Conf.* **21**, 1379-1380; Lewis R. S. *et al.* (1987) *Nature* **326**, 160-161; Blake D. F. *et al.* (1988) *Nature* **332**, 611-613.

10. Thomas K.L., Keller L.P., Blanford G., Klock W., McKay D.S. (1992) *Lunar Planet Sci. Conf.* **23**, 1425-1426.
11. Rietmeijer F.J. (1991) *Geochim. Cosmochim. Acta* **56**, 1665-1671.
12. Mendis D.A. (1991) *Astrophys and Space Sci* **176**, 163-171.
13. Fomenkova M.N. and Chang S. (1993) *Lunar and Planet. Sci. Conf.* **24**, 501-502.
14. Rietmeijer F.J. (1992) *Trends in Mineral.* **1**, 23-41.
15. Jura M. (1987) In *Carbon in the Galaxy*, NASA CP3061, 39-45.
16. Greenberg J.M. and Hage J., *Astrophys. J.* **361**, 260-266 (1990).
17. Encrenaz T. and Knacke R. (1991) in *Comets in the Post Hall-ey Era*, R. L. Newburn et al., Eds. (Kluwer Academic, Netherlands), 107-130; Colangeli L. et al. (1990) *Astrophys. J.* **348**, 718-726; Allamandola L.J., Tielens A.G.G. M., Barker J.R. (1989) *Astrophys. J. Supplement series* **71**, 733-781.
18. Cronin J.R., Pizzarello S., Cruikshank D.P. (1988) in *Meteorites and the early solar system*, J.F. Kerridge and M.S. Matthews, Eds. (The University of Arizona Press, Tucson), 819-857.
19. Allamandola L. J., Sandford S. A., and Wopenka B. (1987) *Science* **237**, 56-59.
20. Gibson E.K. and Bustin R. (1992) *Lunar and Planet. Sci. Conf.* **23**, 501-502.
21. Kerridge J.F. and Chang S. (1985) in *Protostars and Planets II*, D. C. Black and M. S. Matthews, Eds. (The University of Arizona Press, Tucson), 735-754.
22. Walker R.M. (1994) Isotopic Constraints on Interstellar Material in Chondritic IDPs. This volume.
23. Reynolds J.H. et al. (1978) *Geochim. Cosmochim. Acta* **42**, 1775-1789.
24. McKeegan K.D., Walker R.M., Zinner E. (1985) *Geochim. Cosmochim. Acta* **49**, 1971-1987.
25. Clark B.C., Mason L.W., Kissel J. (1987) *Astron. Astrophys.* **187**, 779-784; Halliday I. (1987) *ibid.* , 921-924.
26. Bregman J.D. et al. (1987) *Astron. Astrophys.* **187**, 617-619.
27. Zolensky M. and McSween H.Y. (1988) in *Meteorites and the early solar system*, J. Kerridge and M. Matthews, Eds. (Univer-sity of Arizona Press, Tucson), 114-143; Bunch T.E. and Chang S. (1980) *Geochim. Cosmochim. Acta* **44**, 1543-1577.
28. Schramm L.S., Brownlee D.E., Wheelock M.M. (1989) *Meteoritics* **24**, 99-112.
29. Rietmeijer F.J. (1990) *Meteoritics* **25**, 209-213.
30. Rietmeijer F.J. (1991) *Earth and Pl. Sci. Let.* **102**, 148-159.
31. Sandford S.A. and Walker R.M. (1985) *Astrophys. J.* **291**, 838-844.
32. Ming T., Anders E., Hoppe P., Zinner E., (1989) *Nature* **339**, 351-354.
33. Bradley J.P., Brownlee D.E., Fraundorf P. (1984) *Science* **223**, 56-58.
34. Fomenkova M., Kerridge J., Marti K., McFadden L. (1992) *Lunar and Planet. Sci. Conf.* **23**, 381-382.
35. Bradley J.B. (1994) Mechanisms of Grain Formation, Post-accretional Alteration, and Likely Parent Body Environments of Interplanetary Dust Particles. This volume.
36. Rietmeijer F.J.M. (1994) On the Possibility of Petrological Classification of Carbonaceous Chondritic Micrometeorites. This volume.

37. Klöck W. and Stadermann F.J. (1994) Mineralogical and Chemical Relationships of Interplanetary Dust Particles, Micrometeorites and Meteorites. This volume.
38. Jessberger E.K., Cristoforidis A., Kissel J. (1988) *Nature* **332**, 691-695.
39. Evlanov E.N. et al. (1991) in *Chemistry in space*, Greenberg J. M. and Pirronello V., Eds. (Kluwer Academic Publishers, Netherlands), 383-397.

ISOTOPIC CONSTRAINTS ON INTERSTELLAR MATERIAL IN CHONDRITIC IDPs

Robert M. Walker
McDonnell Center for the Space Sciences and Physics Department,
Campus Box 1105, Washington University,
One Brookings Drive, St. Louis, MO 63130, USA

INTRODUCTION

Many discussions of the origin of comet dust, and by extrapolation at least some IDPs, focus on the similarity of the physical and chemical properties of the dust to those inferred, either directly or indirectly, for interstellar material.[1] Unfortunately there is nothing unique about the physical and chemical properties considered that require an interstellar origin; thus the inferences drawn from such discussions are at best plausible, and at worse misleading. In contrast, certain isotopic signatures can be satisfactorily explained only by invoking interstellar material. This is certainly true of the individual grains of SiC and graphite found in acid residues of primitive meteorites; the isotopic structures of both major and minor elements differ dramatically from average solar system values and show that the grains formed around stars whose nucleosynthetic signatures differed from that of the Sun.[2] In this paper we summarize data on the isotopic compositions of IDPs of different IR spectral classes and compare them with complementary data from primitive meteorites. The purpose is to set limits on the interstellar content of IDPs and, more generally, to see how the isotopic properties of IDPs may contribute to the understanding of the history of the solar system or its precursor material.

ISOTOPIC DATA IN CHONDRITIC IDPS

The data given here is contained in the (unpublished) Ph.D. theses of K. McKeegan[5-9] and F. Stadermann[4], both of whom performed extensive ion microprobe analyses of IDPs. Much of the data has previously appeared in scattered form in abstracts or occasional articles.[5-9]

McKeegan[3] made ion probe measurements of the isotopic abundances of H, C, O, Mg, and/or Si in 31 particles which had chondritic major element compositions.* An additional 15 chondritic particles were measured by Stadermann[4]

---

*By "chondritic" we mean particles whose EDS spectra show major peaks at Si, Mg, Fe, and Al with either Si, Mg or Fe as the dominant element and with Al < Mg, and with variable amounts of Cr, Ca, S, and/or Ni also present. Particles with Mg/Si < 0.5 or with large proportions of K, Ti or other elements not normally seen in the spectra of an Allende powdered standard are not considered chondritic. As discussed by Stadermann[4], this definition gives a natural separation between extraterrestrial and terrestrial samples. Some nonchondritic particles also show isotopic anomalies, but a discussion of these data is beyond the scope of the present article.

who made isotopic measurements of H, C and O in almost all of these particles. In addition, he made the first N isotopic determinations in IDPs, including some of those previously studied by McKeegan. Infrared absorption measurements were also made for 34 of the particles. Ten of the particles were classified as pyroxene type, sixteen as layer lattice silicates, and eight as olivine IR types.[10]

No isotopic anomalies clearly outside of terrestrial values were found in C, O, Mg, or Si in any of the chondritic particles. In contrast, significant D-enrichments were found in 18 particles, as well as apparent D-depletions in 4 particles. In addition ^{15}N-enrichments were found in 12 of the particles, one of the largest values being $\delta^{15}N = 411‰$ in a pyroxene-type particle (Florianus). Many, but not all, of the ^{15}N-enriched particles show δD-enrichments but the converse was not true.

Enrichments of D and ^{15}N were found in both the anhydrous pyroxene and layer-lattice classes of IDPs. Specifically, D-enrichments were found in 6 of the 10 identified pyroxene type particles and in 5 of the 16 layer lattice types. In contrast, no D or N anomalies were found in the 8 olivine type particles measured.

In summary, bulk enrichments of D and/or ^{15}N are common in chondritic IDPs occurring in ~ 40% of all particles studied. The anomalies have been found in both anhydrous pyroxene type particles and in hydrated layer lattice type particles, but not, as yet, in olivine type particles.

COMPARISON OF ISOTOPIC ANOMALIES IN IDPS WITH THOSE IN BULK METEORITES

The isotopic anomalies in hydrogen and nitrogen found in IDPs are also found in certain primitive meteorites. Specifically, the maximum D-enrichment yet seen in IDPs is $\delta D \approx 2,500$ in two different particles.[7] Similar or higher δD values have been reported for bulk samples of the CO meteorite Ornans[11] and for the unequilibrated ordinary chondrites, Semarkona and Bishunpur.[12] Similarly, the maximum $\delta^{15}N$ value of 442‰ seen in the chondritic IDP Santa Fe is only about twice that seen in bulk samples of the carbonaceous meteorite Renazzo[11] and less than the bulk value of $\delta^{15}N = +600‰$ recently measured in the unusual chondrite Acfer 182.[13]

Based on existing measurements, it thus appears that chondritic IDPs are no more primitive in the isotopic sense than the matrices of some primitive meteorites.

The matrices, in turn, probably consist primarily of nebular not interstellar materials. It follows that if some of the chondritic IDPs examined in the ion probe derive directly from cometary materials, then comet dust is not primarily interstellar. However, it must be admitted that individual phases in IDPs have very small grain sizes. Each grain could have an unusual isotopic composition, but the ensemble of many grains might have an average composition close to the solar value. Testing of this possibility will require the development of more sensitive methods for measuring the isotopic compositions of small grains.

INTERSTELLAR GRAINS IN IDPS

An obvious question is whether IDPs contain similar (or larger) concentrations of distinct interstellar grains such as those that have been found in meteorites. Unfortunately, as we now show, such grains can probably be found in IDPs only if their concentrations are much higher than those in primitive meteorites. We take the results of Amari et al.[14] on the isolation of SiC, graphite, and diamonds in acid residues of the Murchison meteorite as our starting point. We first show that the probability for finding an interstellar SiC grain in a given IDP is very low. Since SiC is far more abundant than interstellar graphite, it follows that graphite interstellar grains would be even harder to find.** We leave the discussion of the more abundant diamond fraction to last.

The size distributions and concentrations given by Amari et al.[14] show that the largest number density per unit volume of SiC grains is in their KJB fraction with measured sizes from 0.32 μm to 0.70 μm and a nominal average size of 0.49 μm. Given their measured weight concentration of 1.91 ppm for this size fraction, we estimate that only one in a hundred IDPs, 10 μm in diameter, will contain even a single SiC grain. The probability of finding either smaller or larger SiC grains in a given IDP is even less. Of course, the larger the IDP, the larger the probability that it will contain at least one SiC grain.

The above calculation may be too pessimistic in that very tiny grains of SiC (< 0.05 μm) might be very abundant, but would have been lost and not counted in the separation technique used by Amari et al.[14] (S. Amari, private communication). In favorable cases it is possible to find small grains of a distinctive phase at very low concentrations. For example, Bernatowicz et al.[15] were able to measure 200Å size TiC grains present at a level of only 20 ppm in interstellar graphite grains.

Indeed, Bradley and Brownlee reported the presence of SiC microcrystals (< 250Å in diameter) imbedded in an amorphous carbon matrix in an early TEM study of IDPs.[16] However, they pointed out the possibility that the grains could be terrestrial contaminants. This raises an important issue that puts further, rigorous constraints on the problem. The only way to be sure that a graphite or SiC grain, once found, is indeed interstellar and not a contaminant is to measure the isotopes of one or more of the major elements. With existing ion probe capabilities this means locating grains that are at least 1 μm in size, and preferably somewhat larger.

This is the same problem that we previously faced in locating and measuring interstellar SiC grains in situ in meteorites[17] using an X-ray mapping technique. Based on our extensive mapping experience in Murchison, we estimate that we would have to map an impossibly large number (> 10^4) of 10 μm diameter IDPs in order to locate a single SiC grain that could be measured in the ion microprobe.

In collaboration with G. Kurat, we are engaged in such a mapping search for SiC grains in Antarctic micrometeorites samples, which are much larger than typical IDPs. We estimate that mapping of ~ 10^3 micrometeorites will yield one or two SiC

** However, unlike SiC, interstellar graphite nodules have distinct spherical morphologies that would make them easier to identify if they were present in sufficient abundance.

grains – provided that the concentration of SiC is similar to that in Murchison. Negative results have so far been obtained demonstrating that the concentration of interstellar SiC is not dramatically greater (x10) in these particles than in Murchison.

Diamonds were actually the first identified interstellar species.[18] However, it is not currently possible to measure the isotopic properties of individual diamonds which have typical sizes of only 15Å to 20Å. Moreover, the carbon isotopic ratios of the ensemble of small diamonds ($\delta^{13}C$ -32 to -39‰) falls within the range of terrestrial values and the interstellar identification is based on the unusual isotopic composition of noble gases associated with the diamond-rich material, notably Xe-HL. Although the concentration of diamond in some meteorites is very high (~400 ppm) compared to either SiC or graphite, the noble gas concentration is quite low. So low, in fact, that an IDP 30 μm in diameter would have to consist of pure interstellar diamonds in order to get a measurable signal for Xe-HL using the highly sensitive noble mass spectrometers built by C. Hohenberg at Washington University (R. Nichols, private communication). Thus, although finding diamonds might be easier because of their relatively high concentration, proving that they were interstellar might be very difficult. If it were possible to find diamond-rich regions in IDPs, the measurement of the isotopic composition of associated N (typically depleted with $\delta^{15}N$ of -250‰ to -360‰)[19&20] might provide the best means of establishing the probable interstellar nature of the material.

IS THERE ANY INTERSTELLAR MATERIAL IN IDPS?

The answer is probably yes, but it is difficult to prove at this time. For one thing it would be quite surprising if IDPs, which otherwise appear chemically and texturally primitive compared to meteorites, did not contain preserved pre-solar grains of SiC, graphite, diamonds, and Al_2O_3[21] (as well as other interstellar phases yet to be discovered). However, as we have seen in the preceding section, the small sizes of IDPs, coupled with the low concentrations of pre-solar grains so far measured in primitive materials, makes it extremely difficult to find individual pre-solar grains in IDPs.

At one point, I had naively hoped that the micrometeorites discovered by Maurette[22&23] in Arctic and Antarctic sediments would prove to be simply large IDPs, thus giving ample material in which to search for pre-solar grains. Unfortunately, this appears not to be the case[24]; micrometeorites appear to be quite different from IDPs. The pre-solar grain content of micrometeorites is an interesting problem in itself, but appears to be independent of the question of their presence in IDPs.

ARE THE D AND N ANOMALIES CAUSED BY SURVIVING INTERSTELLAR MOLECULAR CLOUD MATERIALS?

It is known that the deuterium anomalies seen in IDPs are associated with carbon and the anomalies could be due to organic materials produced in ion-molecule reactions in cold interstellar clouds.[6&25] Although it appears to be stretching things too far, similar processes give a possible explanation of the ^{15}N-enriched material[26] also seen in IDPs. However, the two types of anomalies are quite likely to be unrelated to each other.

The interstellar molecular cloud model provides at least a plausible explanation for the observed effects and the prospect of better understanding molecular cloud chemistry certainly provides an incentive for continuing detailed studies of IDPs. However, proving an interstellar cloud origin may be difficult. The same physical processes proposed for isotopic separations in molecular clouds – e.g., molecular dissociation energies that are isotope dependent, low-temperatures to produce significant population shifts even with small energy differences, and ion-molecule reactions to eliminate kinetic barriers – might equally well work in the outer fringes of the solar nebula itself. But whether they took place in the solar nebula or further removed in interstellar molecular clouds, it seems likely that continued studies of the isotopic anomalies in IDPs will yield fundamental insights into important processes that operated before and during the formation of the solar system.

FUTURE WORK

One of the many challenges for the future is to identify the specific carriers of both the D and ^{15}N-enrichments in IDPs. Different fragments of a given IDP have quite different D-enrichments, suggesting a variable (small) carrier phase. This was confirmed directly by hydrogen ion-imaging of the particle Butterfly[3] where small "hot spots" of a highly D-enriched phase were found. Butterfly remains the only IDP in which D-hot spots have actually been seen though they must be present in other IDPs as well.

Additional ion-imaging work using improved imaging systems[21] in conjunction with other, complementary experimental techniques such as transmission electron microscopy and Raman spectroscopy may allow identification of the carrier phase.

Although it is likely to be purely coincidental, it is amusing to note that PAHs (polycyclic aromatic hydrocarbons), which are believed by many astronomers to be important constituents of interstellar clouds, were found in the only two (of seven measured), IDPs that had large D anomalies.[27]

The mass spectra of the PAHs observed in the two IDPs that gave positive results are dominated by odd mass species in the intermediate molecular weight range from 200 to 300 amu.[28] This is in contrast to the results previously obtained on both bulk and acid residue samples of primitive meteorites[29-32] using the same technique. The odd mass peaks could be due to the substitution of functional species containing odd numbers of nitrogen atoms such as cyano (-CN) and amino (-NH$_2$) groups. Again, it is amusing to note that Florianus had one of the largest δ^{15}N anomalies yet seen in IDPs. Additional work is clearly needed.

ACKNOWLEDGMENTS

The ion microprobe techniques that made possible the isotopic measurements of IDPs summarized here were developed by E. Zinner. I wish to thank all of my colleagues at the McDonnell Center for the Space Sciences for fruitful discussions, particularly, for this paper, C. Alexander, S. Amari, S. Messenger, R. Nichols, F. Stadermann, and P. Swan. I am grateful for the reviews by F. Stadermann and G. Flynn that helped clarify the text. Flynn also noted that for completeness I should refer to the precise thermal ionization mass spectrometric measurements of Mg isotopic compositions of IDPs reported by Esat et al.[33] and Esat and Taylor.[34] Evidence for $\delta^{26}Mg$ anomalies at the level of 3 to 4 parts per thousand was found in several particles but this does not change the conclusions of the present paper regarding interstellar material in IDPs. Measurements of deuterium enrichments in IDPs reported by Blake and Chang[35] likewise do not change the discussion given here.

REFERENCES

1. See, for example Encrenaz T., Puget J. L. and D'Hendecourt L. (1991) *Space Sci. Rev.* **56**, 83-92.
2. Anders E. and Zinner E. (1993) *Meteoritics* **28**, 490-514.
3. McKeegan K. D. (1987) Ph. D. Thesis, Washington University, St. Louis, MO.
4. Stadermann F. J. (1991) Ph.D. Thesis, University of Heidelberg, Germany (in German).
5. Zinner E., McKeegan K. D. and Walker R. M. (1983) *Nature* **305**, 119-121.
6. McKeegan K. D., Walker R. M. and Zinner E. (1985) *Geochim. Cosmochim. Acta* **49**, 1971-1987.
7. McKeegan K. D., Swan P., Walker R. M., Wopenka B. and Zinner E. (1987) *Lunar Planet. Sci. XVIII*, 627-628.
8. Stadermann F. J., Walker R. M. and Zinner E. (1989) *Meteoritics* **24**, 327.
9. Stadermann F. J., Walker R. M. and Zinner E. (1990) *Lunar Planet. Sci. XXI*, 1190-1191.
10. Sandford S. A. and Walker R. M. (1985) *Astrophys. J.* **291**, 838-851.
11. Kerridge J. (1985) *Geochim. Cosmochim. Acta* **49**, 1707-1714.
12. McNaughton N. J., Borthwick J., Fallick A. E. and Pillinger C. T. (1981) *Nature* **294**, 639-641.
13. Grady M. M., Pillinger C. T. and Arden J. W. (1992) *Meteoritics* **27**, 226.
14. Amari S., Lewis R. S. and Anders E. (1993) *Geochim. Cosmochim. Acta* submitted.
15. Bernatowicz T. J., Amari S., Zinner E. K. and Lewis R. S. (1991) *Astrophys. J.* **373**, L73-L76.
16. Bradley J. P. and Brownlee D. E. (1983) *Lunar Planet. Sci. XIV*, 67-68.
17. Alexander C. M. O'D., Swan P. and Walker R. M. (1990) *Nature* **348**, 715-717.
18. Lewis R. S., Tang M., Wacker J. F., Anders E. and Steel E. (1987) *Nature* **326**, 160-162.

19. Russell S. S. (1993) Ph.D. Thesis, The Open University, Milton Keynes, U.K.
20. Virag A., Zinner E., Lewis R. and Tang M. (1989) *Lunar Planet. Sci. XX,* 1158-1159.
21. Nittler L. R., Walker R. M., Zinner E., Hoppe P. and Lewis R. S. (1993) *Lunar Planet. Sci. XXIV* 1087-1088.
22. Maurette M., Hammer C., Brownlee D. E., Reech R. and Thomsen H. H. (1987) *Science* **233**, 869-872.
23. Maurette M., Pourchet M., Bonny P., de Angelis M. and Siry P. (1989) *Lunar Planet. Sci. XX* 644-645.
24. Alexander C. M. O., Maurette M., Swan P. and Walker R. M. (1992) *Lunar Planet. Sci. XXIII* 7-8.
25. Zinner E. (1988) In *Meteorites and the Early Solar System,* (eds. J. F. Kerridge and M. S. Matthews), University of Arizona Press, Tucson, 956-983.
26. Geiss J. and Bochsler P. (1982) *Geochim. Cosmochim. Acta* **46**, 529-548.
27. Clemett S. J., Maechling C. R., Zare R. N., Swan P. D. and Walker R. M. (1993) *Lunar Planet. Sci. XXIV,* 309-310.
28. Clemett S. J., Maechling C. R., Zare R. N., Swan P. D. and Walker R. M. (1993) *Science* **262**, 721-725.
29. Hahn J. H., Zenobi R., Bada J. L. and Zare R. N. (1988) *Science* **239**, 1523.
30. Zenobi R., Philippoz J.-M., Buseck P. R. and Zare R. N. (1989) *Science* **246**, 1026.
31. Zenobi R., Philippoz J.-M., Zare R. N., Wing M. R., Bada J. L. and Marti K. (1992) *Geochim. Cosmochim. Acta* **56**, 2899.
32. Clemett S. J., Maechling C. R., Zare R. N. and Alexander C. M. O'D. (1992) *Lunar Planet. Sci. XXIII,* 233-234.
33. Esat T. et al. , *Science* **206**, 190-197 (1979).
34. Esat T. and Taylor S.R., *Lunar Planet. Sci. XVIII*, 269-270 (1987).
35. Blake D. and Chang S., *Lunar Planet. Sci. XIX,*, 92-93 (1988).

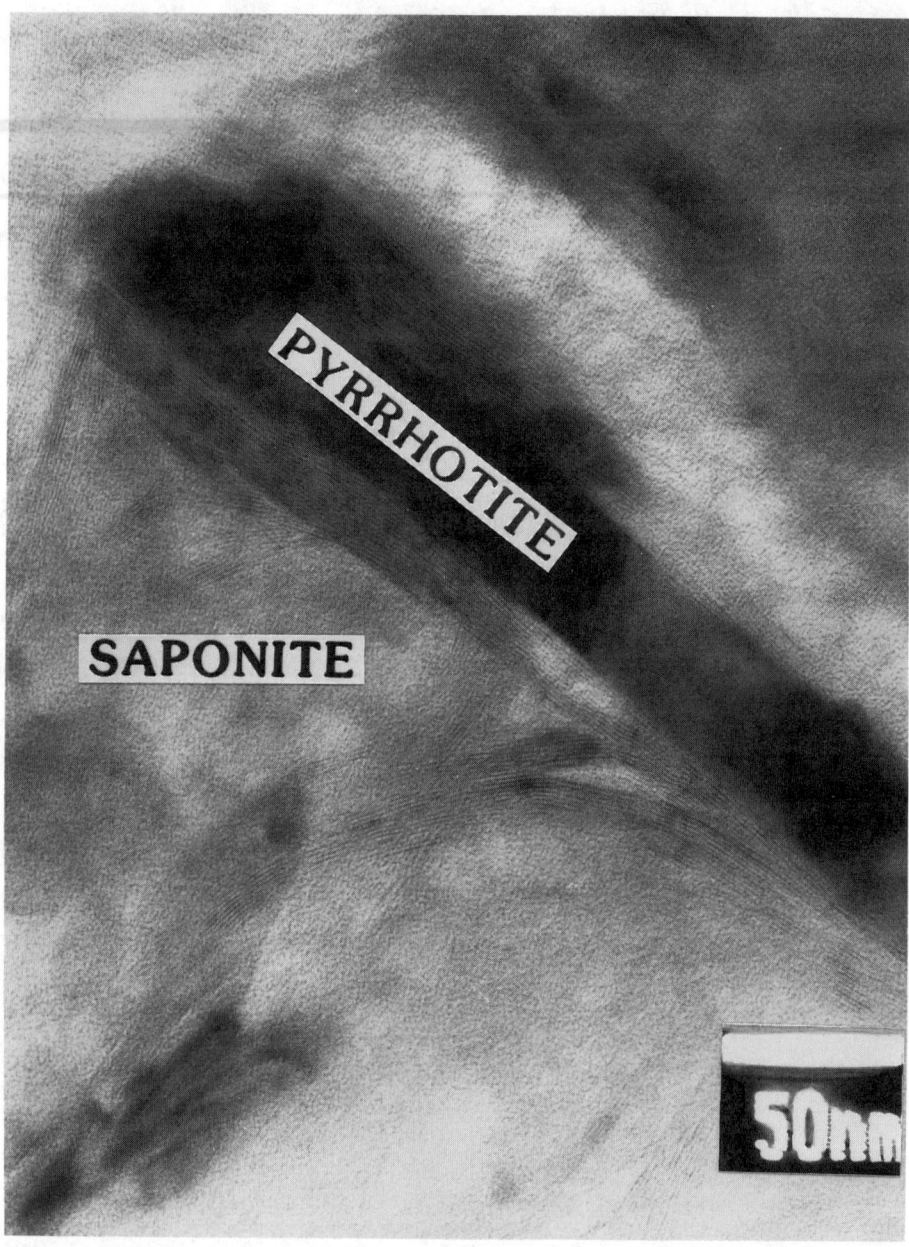

A TEM image of an ultramicrotomed slice of hydrous chondritic IDP L2007 15. This smectite-type IDP contains numerous plates of pyrrhotite, which have apparently served as nucleation surfaces for smectite. In this image 1 nm lattice fringes of flaky saponite (a smectite) run parallel to (some flaking off) the (001) faces of a pyrrhotite crystal (black, seen edge on). Scale bar measures 50 nm. (Photo by M. Zolensky)

$^6Li/^7Li$, $^{10}B/^{11}B$ AND $^7Li/^{11}B/^{28}Si$ IN INDIVIDUAL INTERPLANETARY DUST PARTICLES

Xu Yin-Lin
Purple Mt. Observatory, Nanjing, PRC

Song Ling-Gen, Zhang Yong-Xia
Fudan University, Shanghai, PRC

and
C. Y. Fan
Department of Physics, University of Arizona, Tucson, Arizona, USA

ABSTRACT

We measured $^6Li/^7Li$, $^{10}B/^{11}B$ and $^7Li/^{11}B/^{28}Si$ in four interplanetary dust particles (IDPs). The average values of $^6Li/^7Li$ and $^{10}B/^{11}B$ are $(8.40 \pm 0.37) \times 10^{-2}$ and $(2.40 \pm 0.18) \times 10^{-2}$, respectively. The variation of $^7Li/^{11}B/^{28}Si$ is so large that it makes the average meaningless. The physical implication of these values and their variations related to the evolution of the solar system is discussed.

INTRODUCTION

At the very initial stage of the development of our solar system, the solar nebula was presumably composed of ^1H, ^2H, ^3He, ^4He and ^7Li which were made during the Big Bang[1], and C, N, O, ... nuclei which were products of nearby supernova explosions. ^6Li (together with about an equal amount of ^7Li), ^9Be, ^{10}B and ^{11}B were later produced by cosmic ray particles bombarding the local interstellar and interplanetary C, N, O, ... nuclei.[2] Accordingly, $^{10}B/^{11}B$ in a specimen should equal the ratio of the spallation cross sections of C, N, O, etc., for the production of ^{10}B and ^{11}B by cosmic ray bombardment, weighted by the relative elementary abundances in its parent material and integrated over the cosmic ray spectrum. On the other hand, since most of the ^7Li nuclei are of primodial origin, $^6Li/^7Li$ in a specimen depends on the ratio of the spallation cross sections as well as the exposure history of the parent material to high energy particles. Accepting the current model of the evolution of the solar system, three evolutional stages can be cited. At the very initial stage, the production of ^6Li nuclei occurred throughout the solar system but probably at a higher rate in the outer than in the inner region since the cosmic ray intensity in the inner region was likely reduced due to nebular material absorption.

As the solar nebula began to condense to form the proto-sun and planets, the solar wind commenced, carrying away the angular momentum of the solar system and modulating the cosmic ray intensity. The intensity gradient would certainly have been reflected in the production rate of ^6Li nuclei. Furthermore, in the vicinity of a planet, the production is limited only to the surface material. In a later stage, the Sun may become very active, emitting large fluxes of solar cosmic rays. ^{6}Li nuclei produced would be buried in the surface material of inner planets. All the cited cases imply that ^{6}Li/^{7}Li in a sample depends very much on the origin of the parent material. Unfortunately, the evolutional history recorded in terrestrial and lunar samples were completely scrambled by subsequent volcanic activity and meteoritic impacts. This makes the measurement of ^{6}Li/^{7}Li in IDPs valuable because they are primitive. In this paper, we shall report the measurement of ^{6}Li/^{7}Li, ^{10}B/^{11}B and ^{7}Li/^{11}B/^{28}Si in four IDPs and discuss the findings.

EXPERIMENTAL METHOD AND RESULTS

The detection of Li, Be and B isotopes in two IDPs was previously reported by us in 1991.[3] The equipment used for the measurement was a CAMECA IMS-3F Secondary Ion Microanalyzer at Fudan University. At that time, however, we did not have standards for the calibration and the apparatus was not properly tuned for the measurement. Therefore, we reported the ^{6}Li/^{7}Li and ^{10}B/^{11}B values but could not make a precise estimate of the experimental errors. By calling numerous laboratories in the country, we were fortunate to find a standard Li isotope source in the form of Li_2CO_3 (^{6}Li/^{7}Li = (8.32 ± 0.02) x 10^{-2}) from Dr. G.D. Flesch of Iowa State University.[4] In the meantime, the microanalyzer was upgraded for measuring isotopes in small samples. With a boron isotope standard purchased from the National Bureau of Standards in the form of H_2BO_3, ^{10}B/^{11}B = 0.2473 ± 0.0002), and a mixture of Li, B and SiO_2 prepared in our own laboratory, we were ready to refine our measurements.

The microanalyzer was first set to measure the standards. Taking the Li standard as an example, the profile of ^{6}Li and ^{7}Li were obtained first, as shown in Figure 1. From the profiles, we decided that a resolution of m/Δm = 2000 was sufficient for the isotopic abundance measurement.

The microanalyzer was then set to scan the ^{6}Li and ^{7}Li intensity, alternating, at a rate of about 3 seconds per scan. It took about 183 seconds to complete 65 cycles. Standard deviations were calculated every 5 cycles. Finally, the ratio of ^{6}Li to ^{7}Li was calculated. The difference between the measured and the standard value gives us the needed correction factor which is about 5%. A similar procedure was followed for ^{10}B to ^{11}B and ^{7}Li/^{11}B/^{28}Si.

The standard sources were then replaced by one of the IDPs. The intensities of ^{6}Li and ^{7}Li, ^{10}B and ^{11}B and then finally ^{7}Li/^{11}B/^{28}Si were measured by following the procedures which were used for measuring the standards. The experiment was finished in one day and we found that the microanalyzer was extremely stable by re-checking it with the standards. The plots of these scans for IDP W7074C3 are displayed in Figure 2 as an illustration. The final results of the four IDPs are listed in Table 1.

Table 1

IDP	$^6Li/^7Li \times 10^2$	$^{10}B/^{11}B \times 10$	$^7Li/^{11}B/^{28}Si$
W7074I8	8.53 ± 0.44	2.32 ± 0.26	47.1 ± 1.1 / 8.31 ± 0.83 / 10^6
W7074C3	8.17 ± 0.34	2.43 ± 0.15	204.2 ± 1.1 / 7.06 ± 0.45 / 10^6
W7074A7	8.49 ± 0.47	2.35 ± 0.12	57.0 ± 1.2 / 8.33 ± 0.25 / 10^6
W7074C15	8.47 ± 0.22	2.53 ± 0.19	103.1 ± 2.7 / 7.10 ± 0.39 / 10^6
Average	8.40 ± 0.37	2.40 ± 0.18	

CONCLUSIONS AND DISCUSSION

There are 28 values of $^6Li/^7Li$ and 30 values of $^{10}B/^{11}B$ published in the Journal of Physical and Chemical Reference Data Vol. 13.[5] These values, excluding two exceptionally small values of $^6Li/^7Li$, (which are 11.0 and 11.32), are plotted together with the values given by Anders and Grevesse[6] and the inverse of our measured quantities in Figures 3a and b. It is apparent that the $^{10}B/^{11}B$ values are almost constant, which is ~4.0, whereas $^6Li/^7Li$ have large variations with our measurements near the lower end. The large error bars of our experimental values were entirely due to the limited sample sizes (~1 ng) available for the measurements.

In order to relate the isotopic abundances with the evolutionary history of the solar system, we need the production rates of these isotopes. Meneguzzi et al.[7] studied in detail the production of Li, Be and B by cosmic ray particles in the local interstellar space. They assumed that (a) the cosmic intensity in the interstellar space is about 1 particle/(cm^2 sec ster) and (b) the elementary abundances of C, N, O, etc. were the same as that at present. By using semi-empirical values for the spallation cross sections, they calculated the rates of production of the five light isotopes, which are given in Table 2.[2]

Table 2

	$^6Li/H$	$^7Li/H$	$^9Be/H$	$^{10}B/H$	$^{11}B/H$
dn/dt × 10^{28} sec^{-1}	2.5	4.0	0.7	2.7	7.0

On the basis of this theory, the ratio of ^{11}B to ^{10}B is a constant, independent of the specimen material; it is the ratio of the spallation cross-sections of C, N, O, etc. for the production of ^{11}B and ^{10}B, weighted by the relative abundances of the heavier elements and the shape of the cosmic ray spectrum. Figure 3b seems to indicate that this is the case, except that the measured ratio is 4.0 instead of 2.6 as deduced from the production rates in Table 2. This discrepancy can be explained by the inaccuracy of the spallation cross-sections, especially in the energy range below about 1 BeV/nucleon where the energy spectrum is poorly known. The variation of the ^{7}Li/^{6}Li values depicted in Figure 3a can be explained as the result of the differences in radiation levels received by the experimental samples.

We shall assume that the parent materials of the four chondritic IDPs we analyzed were in the outer region of the solar system, and that they had little shielding against cosmic ray bombardment. Let $(^7$Li/H$)_o$ and $(^7$Li/H$)_c$ be the abundance of ^{7}Li relative to hydrogen by Big Bang synthesis and from spallation respectively, and $(^6$Li/H$)_c$, the abundance of ^{6}Li relative to hydrogen due to cosmic ray production, (denoted respectively by R_o, R_c and r_c). Since R_c/r_c is about 1.6 (according to Table 2). Then, the measured ^{7}Li/^{6}Li, in IDPs, taken to be 11.9, is:

$$11.9 = (R_o + R_c)/r_c = R_o/r_c + 1.6$$

or,

$$r_c/R_o = 0.097 \qquad (1)$$

To calculate the value of R_o, we use our measured value of ^{7}Li/^{28}Si and ^{28}Si/^{1}H compiled by Anders and Grevesse[6], which is $9.22 \times 10^5 / 2.79 \times 10^{10}$. From the four measured values, we obtain R_o as 1.6, 6.7, 1.9 and 3.4×10^{-9}. The two low values fall within the range of R_o from the consideration of the hot Big Bang model.[1] For the estimate of the age of our solar system we shall use 1.6×10^{-9} as R_o, which corresponds to ^{4}He mass fraction 0.26, ^{2}H/^{1}H = 10^{-5} and a nucleon density 10^{-30} g/cm^{-3}.[1] Let T be the age of the solar system. From Eq. (1) and Table 2, we have:

$$(dn/dt)\,T = R_o = 1.5 \times 10^{-10}$$

$$T = 6.0 \times 10^{-17} \text{ sec} = 1.9 \times 10^{10} \text{ years}$$

This is the age of our solar system counting from the time when the system did not contain any ^{6}Li. It should be emphasized that this is an order of magnitude calculation. In order to make a better estimate, we must refine the measurement of ^{6}Li/^{7}Li and have more accurate spallation cross-sections.

ACKNOWLEDGMENTS

This work is partially supported by NASA under the Grant NAGW 2249 and by the Chinese Academy of Science. The authors are grateful to the Curatorial Facility of NASA Johnson space center for supplying the IDPs, and to Dr. Flesch for supplying the Li_2CO_3 standard.

REFERENCES

1. Yang J., Turner M.S., Steigman G., Schramm D.N. and Olive K.A. (1984) *Ap. J.* **281**, 493; Boesgaard A. M. and Steigman G. (1985) *Ann. Rev. Astron. Astrophys.* **23**, 319.
2. Reeves H. (1974) *Ann. Rev. Astron. Astrophys.* **12**, 437.
3. Xu Y. L., Xie P., Fan C.Y. and Cao Y.M. (1991) *Science in China* **34** no.2, 209.
4. Flesch G.D., Anderson A.R.and Svec H.J. (1973) *Int. J. Mass Spectrom. Ion Phys.* **12**, 265.
5. De Biévre P., Gallet M., Holden N. E. and Barnes I.L. (1985) *J. Phys. Chem. Ref. Data* **13**, 814.
6. Anders A. and Grevesse N. (1989) *Geochim. Cosmochim. Acta* **53**, 197.
7. Meneguzzi M., Andouze J. and Reeves H. (1971) *Astron. Ap.* **15**, 337.

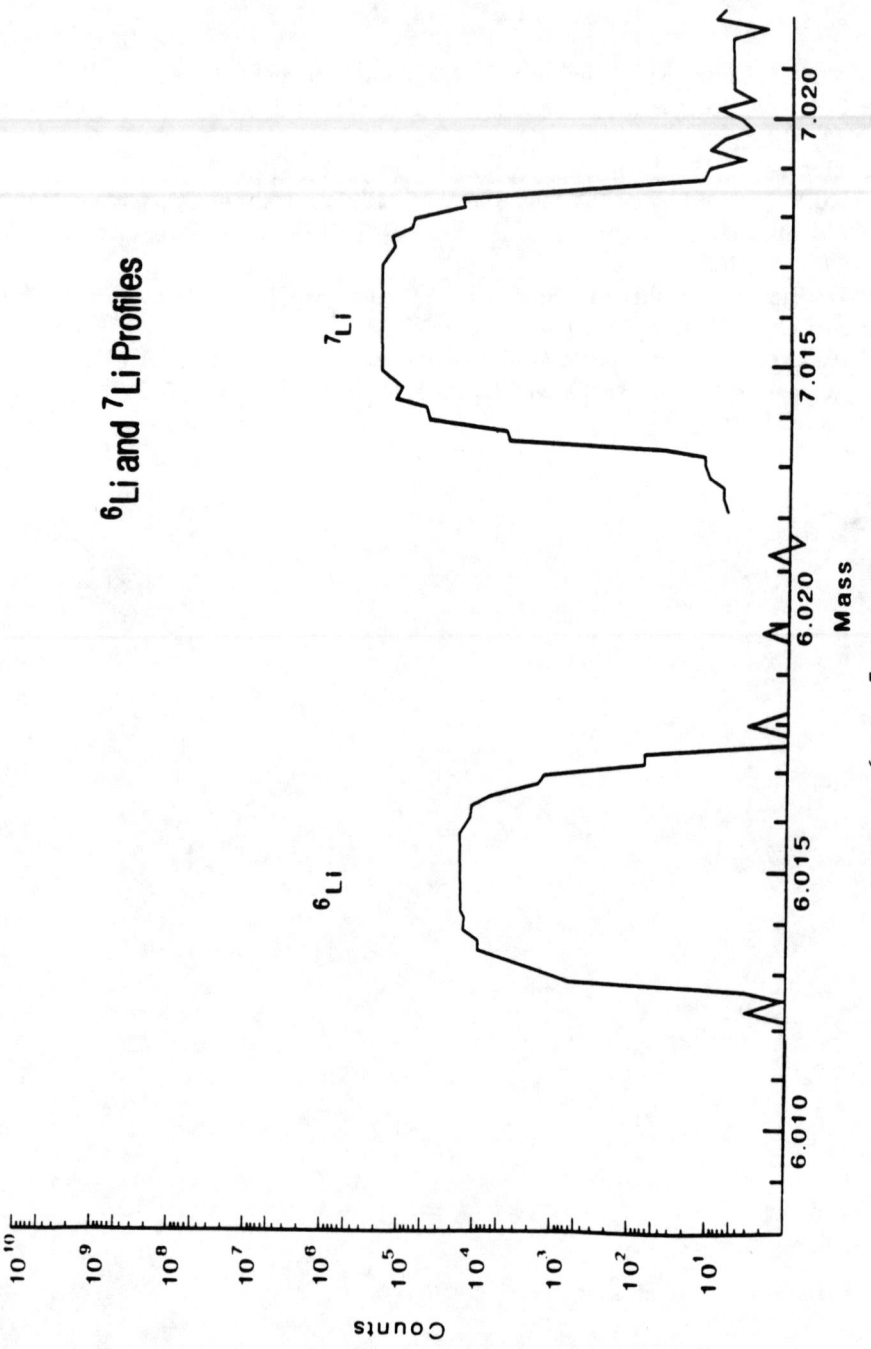

Figure 1. ^{6}Li and ^{7}Li intensity profiles.

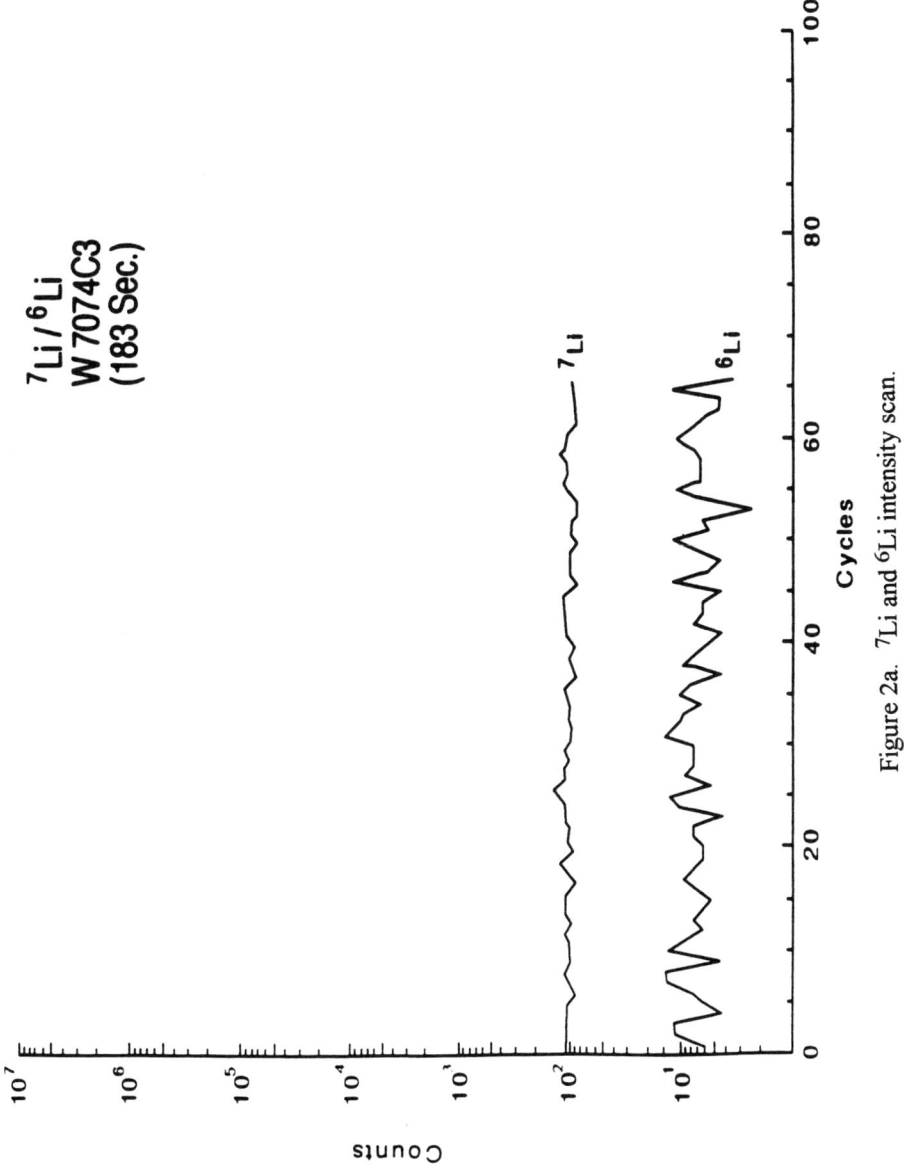

Figure 2a. ^{7}Li and ^{6}Li intensity scan.

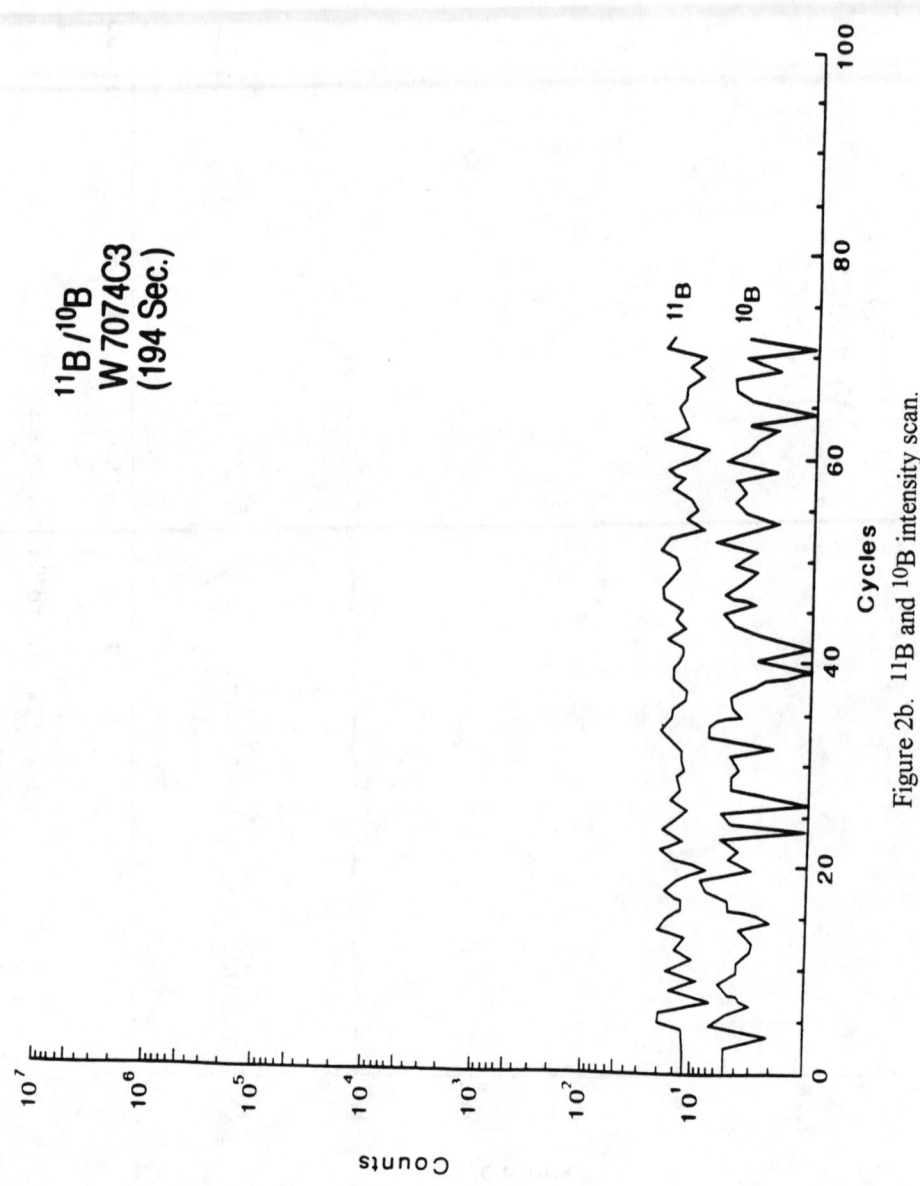

Figure 2b. ^{11}B and ^{10}B intensity scan.

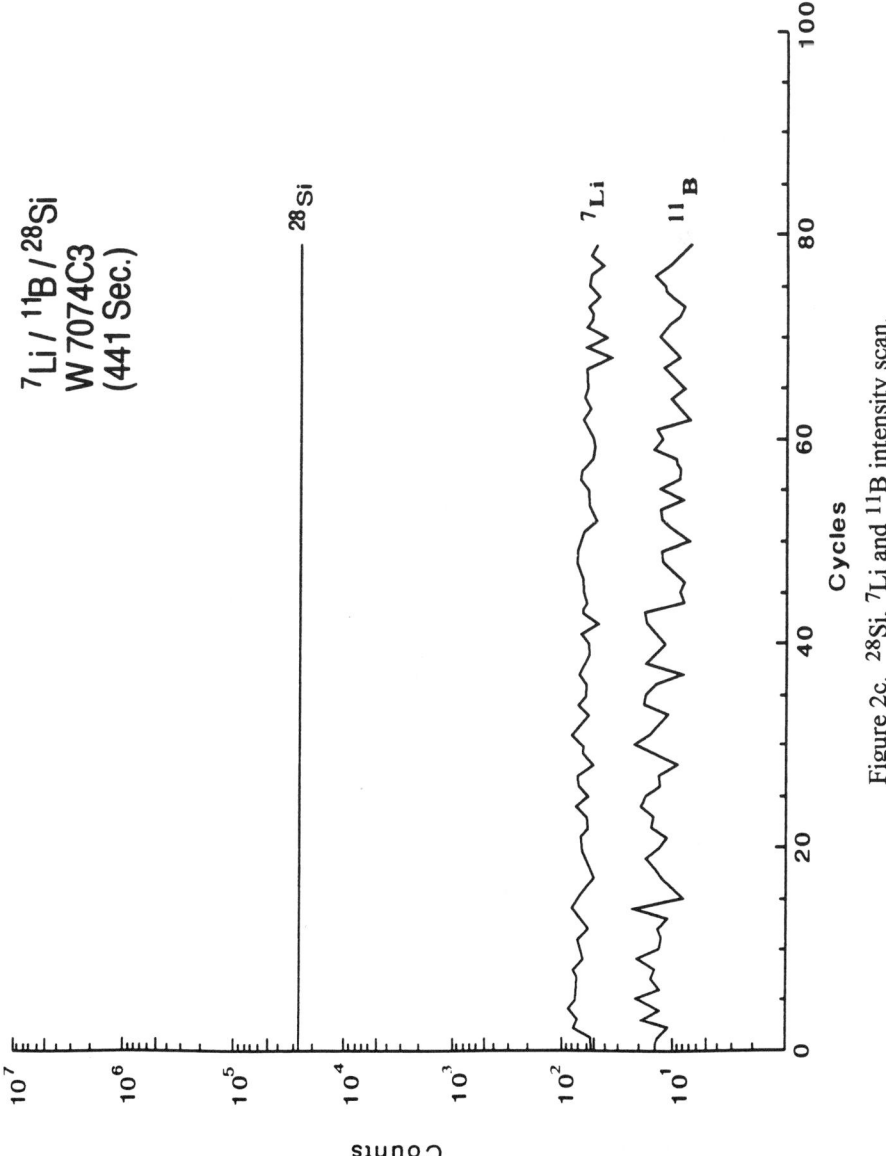

Figure 2c. ^{28}Si, ^{7}Li and ^{11}B intensity scan.

220 $^6Li/^7Li$, $^{10}B/^{11}B$ and $^7Li/^{11}B/^{28}Si$

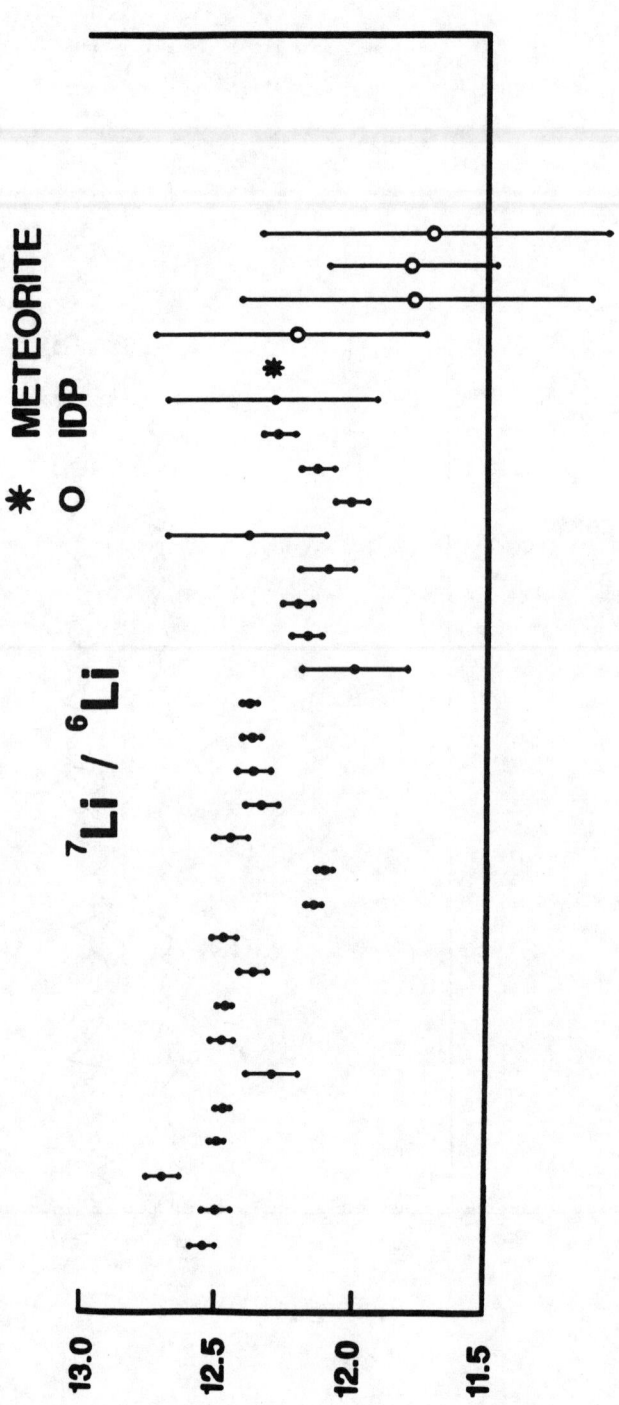

Figure 3a. $^7Li/^6Li$ in various samples.

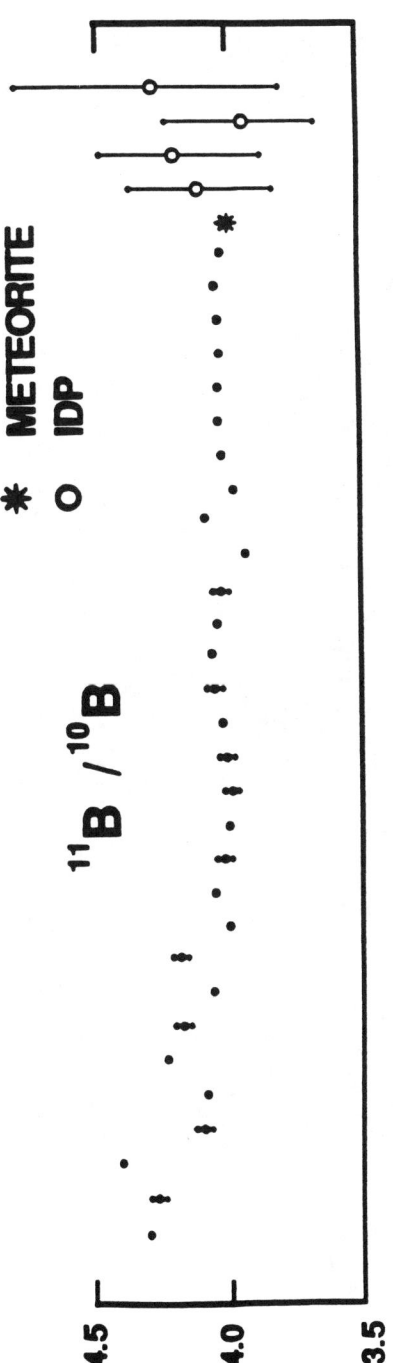

Figure 3b. $^{11}B/^{10}B$ in various samples.

222 $^6Li/^7Li$, $^{10}B/^{11}B$ and $^7Li/^{11}B/^{28}Si$

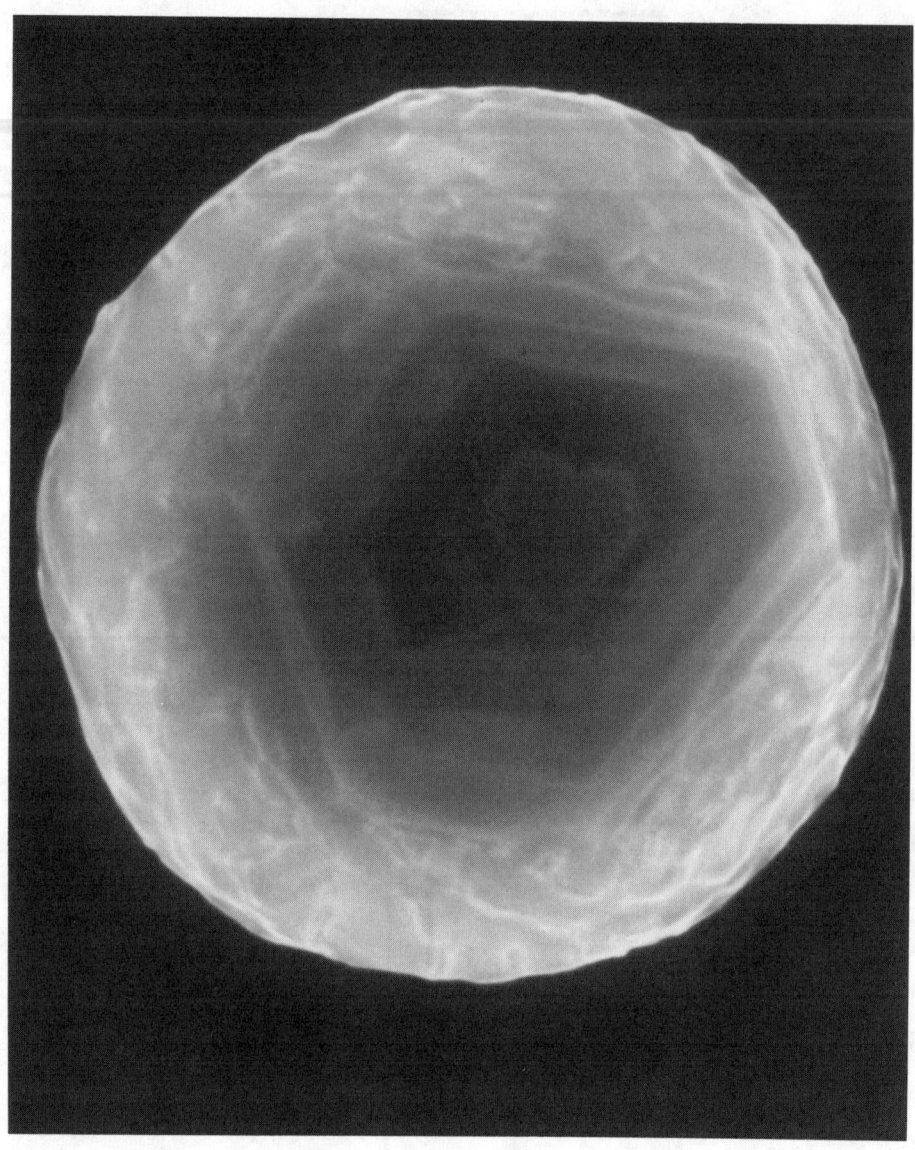

Stratospheric particle W7027 B15, which consists entirely of Fe-Ni sulfides (one large crystal is visible in the center). The composition and spherical shape suggests that this is a sulfide-dominated IDP which was melted upon atmospheric entry. The particle measures 12 μm across. (NASA photo S83-26574)

COMETARY DUST: A THERMAL CRITERION TO IDENTIFY COMETARY SAMPLES AMONG THE INTERPLANETARY DUST COLLECTED FROM THE STRATOSPHERE

G. J. Flynn
Dept. of Physics, SUNY-Plattsburgh,
Plattsburgh, NY 12901 USA

ABSTRACT

The interplanetary dust particles collected from the Earth's stratosphere are believed to sample both the asteroids and the comets. Near-earth gravitational segregation favors collection of the lowest velocity interplanetary dust, from the main-belt asteroids, enhancing its proportion in near-earth collections over its proportion in space. Nonetheless, some cometary dust must be present on the stratospheric collectors. A thermal criterion has been developed to separate the interplanetary dust collected from the stratosphere into two groups: one heated more extremely than is possible for main-belt asteroidal particles, and a second heated less extremely. The first group should consist of cometary particles, while the second group should include both main-belt asteroidal particles and cometary particles which encountered the Earth under favorable entry conditions.

INTRODUCTION

Both comets and main-belt asteroids are known to contribute to the interplanetary dust population, since dust associated with both of these sources was detected by the Infrared Astronomical Satellite.[1&2] The relative proportions of the cometary and the asteroidal contributions to the Zodiacal Cloud and to the interplanetary dust collected from the Earth's stratosphere have not yet been definitively established. These proportions may vary with time as a result of major catastrophic disruptions in the main-belt[3] or the appearance of fresh, active comets.[4] Dermott et al.[5] and Reach[6] suggest the debris from catastrophic collisions in the main-belt can account for much, perhaps most, of the Zodiacal Cloud particles. Earth collection of these asteroidal particles is strongly favored by near-Earth gravitational enhancement, which increases the effective Earth-capture cross section for interplanetary dust particles (IDPs) having low geocentric velocities.[7] However, comets are observed to produce interplanetary dust, and entry heating calculations indicate that some of these particles should survive Earth atmospheric entry without melting.[8] The identification of these unmelted cometary IDPs would allow inferences of the compositions, mineralogies, and physical properties of their cometary parent bodies.

SEPARATION BY VELOCITY

Flynn[8] has shown that IDPs evolving to Earth intersecting orbits under the influences of solar gravity and Poynting-Robertson radiation drag from main-belt asteroids, from comets with perihelia >1.5 AU, and from comets with perihelia ≤1.5 AU separate into three distinct velocity groups, with the main-belt asteroidal dust having geocentric velocities ≤5 km/sec, while the cometary dust has higher geocentric velocities.

These geocentric velocities, calculated for nodal crossing at 1 AU in the absence of Earth's gravity, can be transformed into atmospheric entry velocities using energy conservation:

$$v_{entry} = [(11.1 \text{ km/s})^2 + (v_{nodal\ crossing})^2]^{1/2} \quad .$$

Main-belt asteroidal particles, which encounter the Earth from nearly circular orbits, have Earth entry velocities ranging from escape velocity (11.1 km/s) up to about 12.4 km/sec (nodal crossing velocity = 5.5 km/sec) in the model developed by Flynn.[8]

Jackson and Zook[9] performed a more complete simulation of the orbital evolution of interplanetary dust including the effects of planetary gravitational perturbations. They have modeled the evolution of dust grains released from 15 main-belt asteroids, 15 short period comets with perihelia >1 AU, and 5 long period comets with perihelia <1 AU. Their results have confirmed the existence of the three velocity regimes, although there is some overlap, resulting from planetary gravitation perturbations particularly among the two cometary groups. Almost all 10 μm radius dust from main-belt asteroids had nodal crossing velocities at 1 AU of ≤7.5 km/s. This corresponds to atmospheric entry velocities ≤13.4 km/s, only slightly higher than the ≤12.4 km/s atmospheric entry velocity limit for dust from main-belt asteroids previously suggested.[8]

A few main-belt asteroids, such as Hungaria and Phocaea, have unusually high orbital inclinations, approximately 23° in each case. These two asteroids, and other members of their families, produce IDPs indistinguishable from cometary dust based on their entry velocities. However, the dust contribution from these two main-belt asteroid families is likely to be small. The major dust bands detected by the IRAS satellite are associated with the Eos, Themis, and Koronis families of asteroids, all of which have much lower orbital inclinations and produce IDPs which are easily distinguished from cometary IDPs by the entry velocity criterion.

In the Jackson and Zook[9] calculations, the lowest geocentric velocity for a cometary IDP was 5.3 km/sec, and the comets which they selected for inclusion were those most likely to overlap in their Earth encounter velocities with the main-belt asteroidal particles (Zook, Pers. Comm. 1993).

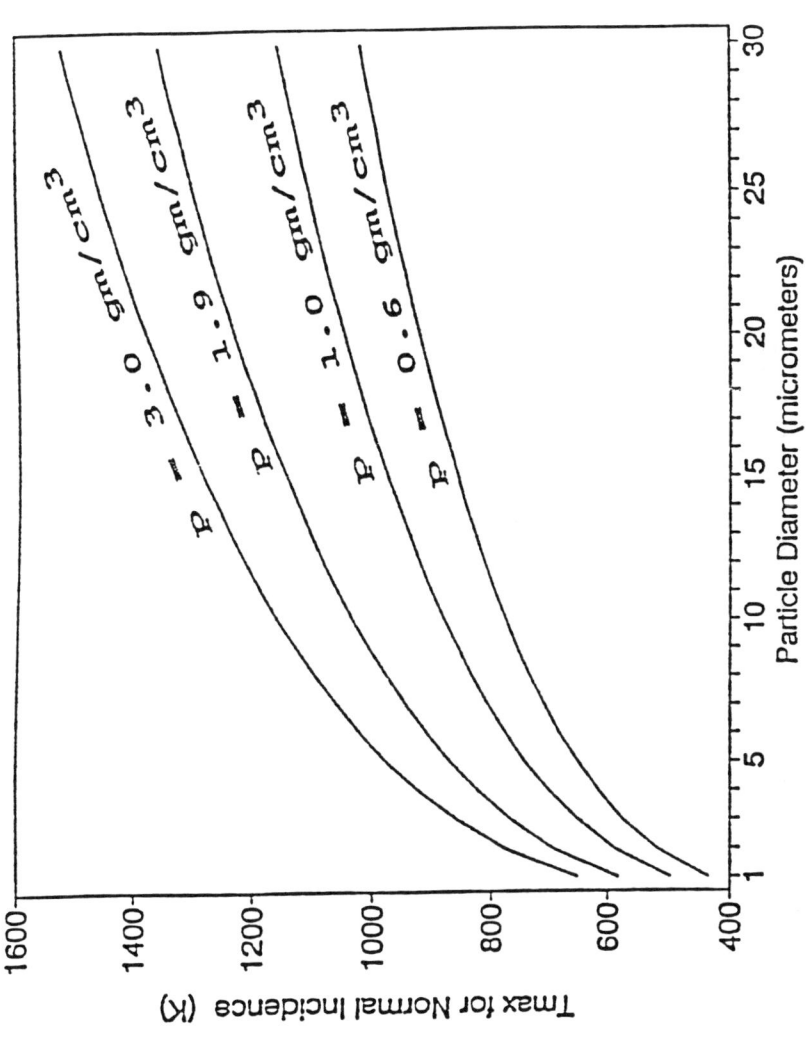

Figure 1. Peak temperature (T_{max}) reached by main-belt asteroidal IDPs (v_{entry} = 13.4 km/s, normal incidence) during atmospheric entry versus particle diameter and particle density (p).

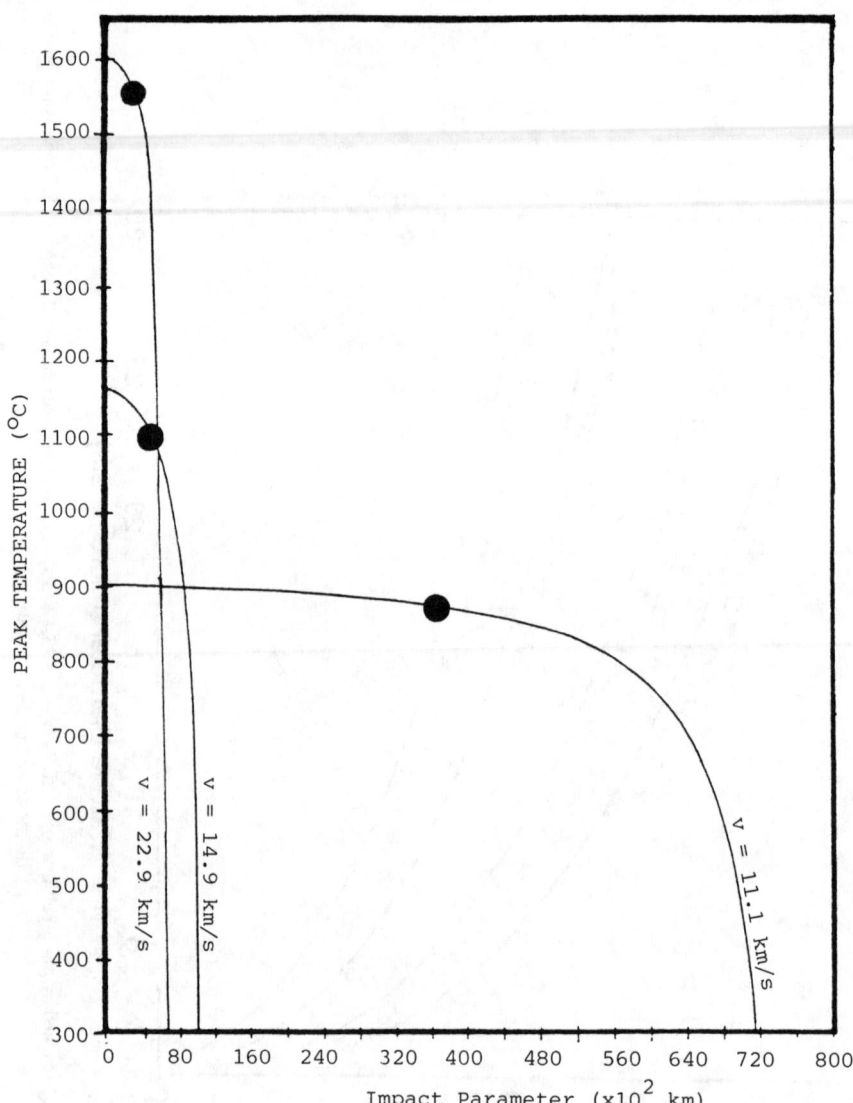

Figure 2. Peak temperature versus impact parameter for 20 μm diameter, density 1 gm/cm^3 interplanetary dust particles having atmospheric entry velocities of 11.1, 14.9, and 22.9 km/sec. The black circle indicates the peak temperature for the 45° entry angle while the value for a 90° entry angle corresponds to an impact parameter of 0.

Table 1
Minimum Particle Diameter for Solar System Retention
for Dust Particles Emitted at Perihelion

Source Comet	D_{min}*
Schwassman-Wachmann 1	2 μm
Temple 1	5 μm
Temple 2	5 μm
Kopff	5 μm
Giacobini-Zinner	8 μm
d'Arrest	6 μm
Encke	15 μm
Swift-Tuttle	58 μm
Halley	71 μm

* Assumes density = 1 gm/cm^3, spherical particle

SEPARATION BY PEAK ENTRY TEMPERATURE

The peak temperatures reached by IDPs on atmospheric entry can be used to indicate the distribution of IDP velocities.[8] Each chondritic IDPs contains many internal thermometers: minerals which are altered above certain temperatures (e.g., the oxidation of Fe-bearing olivine to laihunite, formation of magnetite, and Fe-Mg reequilibration of olivines and pyroxenes), volatile elements which are lost sequentially with increasing temperature, solar flare tracks which anneal at different temperatures in different minerals, and solar implanted noble gases which outgas progressively with temperature[8] (see also Nier[10] in this volume). Thus, in principle, limits on the peak temperature reached by each IDP on Earth atmospheric entry can be established.

On an individual particle basis, the separation of IDPs into cometary and main-belt asteroidal groups is complicated because the peak temperature reached on entry depends on the entry angle, with normal incidence resulting in the most extreme heating.[8] Thus a cometary IDP encountering the Earth at grazing incidence is heated less extremely than a main-belt asteroidal IDP with the same physical properties (size, shape, density, and emissivity) encountering the Earth at normal incidence.[8]

In principle, it is possible to distinguish some cometary IDPs from their main-belt asteroidal counterparts since the normal incidence case results in the most extreme entry heating for any given particle size and density. Thus IDPs can be separated into two groups: (1) particles heated above the maximum temperature possible for main-belt asteroidal particles of that size and density, and, (2) particles which are less severely heated.

The first group contains only cometary particles (and a minor contribution from asteroids in highly inclined or highly elliptical orbits, such as the Earth-crossing asteroids). The second group includes main-belt asteroidal particles and cometary particles having small incidence angles.

Using the atmospheric entry interaction model developed by Whipple[11] and extended by Fraundorf[12] the peak temperature reached during atmospheric entry was calculated for normal incidence at a velocity of 13.4 km/sec for IDPs with diameters ranging from 1 to 50 μm and densities of 0.6, 1.0, 1.9, and 3.0 gm/cm^3 (see Figure 1). The 13.4 km/s entry velocity (nodal crossing velocity = 7.5 km/s) is the upper limit for main-belt asteroidal IDPs, except rare IDPs from asteroids in high inclination orbits.

Once the size and density of an individual IDP are determined, the location of the particle can be plotted on Figure 1. If any internal thermal indicator gives evidence that this IDP was heated above the maximum temperature for a main-belt asteroidal particle of that size and density, then the particle is a likely candidate to be from a cometary parent body.

Rietmeijer[13] points out that volatile element loss and mineral transformation, two of the indicators used to determine the peak temperature reached on atmospheric entry, are likely to be sensitive to diffusion effects, and thus to the size of the mineral grains. However, this can only reduce the inferred peak temperature, moving some high velocity particles into the low temperature group. Since the low temperature group already consists of a mixture of low velocity particles and high velocity particles with favorable entry angles, this effect introduces no further ambiguity. The identification of the high temperature group with a high velocity component (i.e., likely cometary particles) remains unaffected.

Brownlee et al.[14] have proposed a similar thermal criterion to distinguish main-belt asteroidal IDPs from cometary IDPs. In their approach the peak temperature reached by the average IDP (one arriving at a 45° entry angle) is taken to distinguish main-belt asteroidal from cometary particles. Figure 2 shows the peak temperatures reached by IDPs having a density of 1 gm/cm^3 and a diameter of 20 μm for entry velocities of 11.1 km/sec, 14.9 km/sec, and 22.9 km/sec. Differences between the peak temperatures using this model and the Brownlee et al.[14] technique are typically about 50° C to 100° C, with this model imposing a more restrictive criterion. Since we propose to identify as cometary those particles exceeding the peak temperature for main-belt asteroidal IDPs, it seems appropriate to use the most rigid criterion to establish that temperature.

The use of a thermal indicator to distinguish the cometary subset of the IDPs naturally results in all cometary particles identified by this technique exhibiting some degree of thermal alteration. An understanding of the process of thermal alteration, by the pulse heating of stratospheric IDPs and analog materials, is required if the pre-atmospheric properties of these cometary particles are to be inferred.

OPTIMUM PARTICLE SIZE

IDPs ranging from less than 5 to over 60 µm in size have been recovered from the stratospheric collection surfaces. However, the search for cometary IDPs can focus on a more narrow size range. Most of the large cometary particles (>20 or 30 µm) melt on atmospheric entry[8], thus the search for cometary IDPs should concentrate on particles below 20 µm in size.

These peak temperature calculations assume blackbody emissivity at temperatures up to the peak experienced by each IDP. Rizk et al.[15] performed entry heating calculations using a theoretical expression for the emissivity appropriate for spheres of forsteritic olivine. Their peak temperatures agree to within 100 K with predictions assuming a blackbody emissivity for particles ≥10 µm in diameter.[15] However, they calculated that particles significantly smaller than 10 µm are heated to peak temperatures several hundred degrees higher than predicted using the blackbody assumption. Thus the peak temperatures in Figure 1 are likely to be valid for particles ≥10 µm in diameter, but the validity of these calculations for smaller particles (particularly those ≤5 µm in diameter) cannot be assessed because the actual emissivity of small aggregate particles consisting of silicate grains in a high carbon matrix has not been determined.

The collection of small cometary particles at Earth is hampered because solar radiation pressure imposes a lower limit on the size of cometary particles which remain bound by solar gravity.[16] For example, dust ≤8 µm in diameter emitted by comet Giacobini-Zinner near perihelion is expelled from the Solar System. This minimum diameter decreases for particles emitted at larger distances from the Sun, however the dust production by comets peaks near perihelion (as shown in Table 1).

Because of the uncertainty in the peak temperature calculations for particles ≤10 µm in diameter, and the likelihood that most of the small particles emitted by the major dust producing comets are ejected from the Solar System, cometary IDPs can best be distinguished from main-belt asteroidal IDPs near 10 µm in diameter.

CONCLUSIONS

The degree of heating of stratospheric IDPs can provide important information on their sources. A program of study of heated IDPs, including sequential measurements of major and trace element contents, particle densities, mineralogies, solar flare tracks, and He^4 content, has the potential of identifying a cometary subset of the IDPs. The identification of this cometary subset should allow the chemical, physical, mineralogical, and isotopic properties of their cometary parent bodies to be established.

ACKNOWLEDGMENTS

This paper benefited from critical reviews by F. J. M. Rietmeijer and K. L. Thomas. This work was supported by NASA Grant # NGT3587.

REFERENCES

1. Low, F. J. et al. (1984) *Astrophys. J.* **278**, L19.
2. Sykes, M.V. Lebofsky, L.A., Hunten, D.M., and Low, F (1986) *Science* **232**, 1115-1117.
3. Sykes, M. V. and Greenberg, R. (1986) *Icarus* **65**, 51-69.
4. Whipple, F. L. (1967) in *The Zodiacal Light and the Interplanetary Medium, NASA SP-150*, 409-426.
5. Dermott, S. F. et al. (1992) in *Asteroids, Comets, and Meteors 1991*, Lunar and Planetary Institute, Houston, pp. 153-156.
6. Reach, W. T. (1991) in *Origin and Evolution of Interplanetary Dust*, 211-214.
7. Flynn, G.J. (1990) *Proc. 20th Lunar Planet. Sci. Conf*, 363-371.
8. Flynn, G. J. (1989) *Icarus* **77**, 287-310.
9. Jackson, A. A. and Zook, H. A. (1992) *Icarus* **97**, 70-84.
10. Nier, A.O. (1994) Helium and Neon in Interplanetary Dust Particles. This volume.
11. Whipple, F. L. (1950) *Proc. Nat. Acad. Sci.* **36**, 687-695.
12. Fraundorf, P. (1980) *Geophys. Rev. Lett.* **10**, 765-768.
13. Rietmeijer, F. J. M. (1993) *Lunar Planet. Sci. XXIV*, 1261-1262.
14. Brownlee, D.E. (1993) *Lunar and Planetary Science XXIV*, 205-206.
15. Rizk, B. et al. (1991) *J. Geophys. Res.* **96**, A2, 1303-1314.
16. Burns, J. A. (1979) *Icarus* **40**, 1-48.

A PROPOSAL FOR A PETROLOGICAL CLASSIFICATION SCHEME OF CARBONACEOUS CHONDRITIC MICROMETEORITES

Frans J.M. Rietmeijer
Department of Earth and Planetary Sciences,
University of New Mexico, Albuquerque, NM 87131, USA.

ABSTRACT

I explore a classification scheme for carbonaceous chondritic interplanetary dust particles that are mixtures of at least two principal components, *viz.* granular and polyphase units. The proposed classes are based on the carbon and silicate mineralogy in the principal components. This classification scheme monitors the chaotic non equilibrium conditions that characterize the onset of mineralogical activity in solar system protoplanets. Small-scale mineralogical, but not chemical, heterogeneity will be the most conspicuous property of CC IDPs. A lack of chemical fractionation but pervasive non equilibrium mineralogy will be diagnostic for the least altered solar system materials.

INTRODUCTION

Properties of 2-50 micron-sized interplanetary dust particles (IDPs) that can be determined without prejudice by individual investigators include chemical composition, morphology, shape, color, transparency, opacity, and translucence.[1-4] These properties can ultimately allow unambiguous IDP classification. For example, energy dispersive spectroscopic analysis using scanning and analytical electron microscopes will place an IDP in either a chondritic[5] or nonchondritic[6] group. Some caution is necessary. That is, some low-nickel particles with CI-normalized nickel abundances have trace element distributions that are similar to those of terrestrial basalts[7], but this property does not uniquely exclude an extraterrestrial origin.[7&8] Refinements of chondritic IDP classification use the dominant mineral species in a particle, *e.g.* olivine, pyroxene, or phyllosilicates which is determined by FTIR[9], and manual and automated chemical point-count analysis.[10]

Further petrographic analysis allocates a phyllosilicate-rich IDP to either the smectite- or serpentine-rich subtypes [Table 1]. Chondritic IDPs are also classified based on their morphology into aggregate and non-aggregate particles. The former are dominated by olivines and pyroxenes but aggregate IDPs are not strictly "anhydrous" as they may contain small amounts of layer silicates and salts.[7&8] Non-aggregate IDPs are dominated by phyllosilicates. Aggregate particles occur as chondritic porous (CP) and chondritic filled (CF) IDPs that mainly differ in their degree of porosity. The CP IDPs are fluffy, porous aggregates that contain rare silicate whiskers and euhedral grains[13&14], *e.g.* CP IDP W7010*A[13] and those shown in Figures 11.1.2.a & 11.1.3.a of ref. 14. The CF IDPs do not have the fluffy (i.e. porous) morphology that typifies CP IDPs. A typical phyllosilicate-rich, non-aggregate chondritic IDP has a smooth surface. They are known as chondritic smooth (CS) IDPs such as particle W7017B12[15] and those shown in Figures 11.1.2.b & 11.1.3.b of ref. 14.

More than a decade of interplanetary dust studies have produced a much-used classification scheme[9,14-16] that is summarized in Table 1. With the aid of Table 1, an individual IDP is quickly placed among those already described in the literature. I want to explore the possibility of a petrologic classification scheme for chondritic IDPs using mineralogical properties that will be a refinement of the general scheme presented in Table 1.

TABLE 1: INTERPLANETARY DUST CLASSIFICATION

(1) Particles with a chondritic bulk composition

   (1.1) particles dominated by amorphous materials plus carbon materials
   (1.2) particles dominated by neso- and ino- (anhydrous) silicates plus carbon materials
        (1.2.1) olivine-rich particles
        (1.2.2) pyroxene-rich particles
   (1.3) particles dominated by phyllosilicates (i.e. hydrous silicates) plus carbon materials
        (1.3.1) smectite-rich particles
        (1.3.2) serpentine-rich particles
   (1.4) particles containing silicates only

(2) Particles with a nonchondritic bulk composition

   (2.1) single mineral grains, such as iron-sulfides, olivine
   (2.2) aggregates of refractory minerals, mostly fine-grained hibonite, melilite and perovskite

The mineralogy and textures of chondritic aggregate IDPs, and most CS IDPs, define a unique group of ultra fine-grained extraterrestrial materials that differ significantly in form and texture from components of carbonaceous chondrites. They contain mineral assemblages that do not occur in any meteorite class[15] and commonly contain compositionally-variable amorphous materials [cf. ref. 16]. I note that the presence of amorphous materials hampers classification that relies on automated, point-count chemical analysis. The elemental distribution pattern of IDPs containing amorphous materials will be indistinguishable from typical elemental distribution pattern of phyllosilicate-rich IDPs. Hence, this is an argument in favor of non-automated point-count analyses.

CARBON ABUNDANCES: CARBONACEOUS CHONDRITIC IDPS

The carbon materials (Table 1) include hydrocarbons and amorphous carbons[17], turbostratic pre-graphitic carbons[18], poorly-graphitised carbons, and lonsdaleite[16]. The high carbon content of some CP IDPs was already evident from petrologic point-count analyses[19]. Subsequent energy dispersive analyses of chondritic IDPs[20,21] established bulk carbon contents of many chondritic IDPs that are higher than those in carbonaceous chondrite meteorites, *viz.* C/Si = 2.39 x CI in CP IDPs, and 1.32 x CI in CS IDPs.[20,21] The ranges of CI-normalized carbon abundances are 0.7-4.9 in anhydrous aggregate IDPs[21] and 3.1-6.3 in hydrated CS IDPs.[22]

The carbon abundances in CS IDPs can be as high as those in CP and CF IDPs which means that (1) distinction between aggregate and CS IDPs is not possible based on carbon content alone, and (2) many chondritic IDPs are a chemically unique class of extraterrestrial materials. The distinctive value between chondritic IDPs and conventional meteorites is provisionally set at 2 x CI, or ~5 wt% carbon[21], or more conservatively at 3 x CI.[23] Chondritic IDPs with less carbon typically include particles with an affinity to the type CI[24] and CM[25&26] carbonaceous chondrite meteorites.

I suggest that the bulk carbon content of a chondritic IDP is used as a discriminating parameter in appreciation of the uniqueness of the carbon-rich subset of chondritic IDPs. Taking license from the existing IDP nomenclature, I propose the label *carbonaceous chondritic (CC) IDPs* for the carbon-rich subset of chondritic IDPs. Hence, a fraction of the currently identified CP, CF and CS IDPs will become CCP, CCF and CCS IDPs to highlight carbon contents in excess of 2 or 3 times the CI abundance.

PETROLOGY

Principal Components

Carbonaceous chondritic IDPs consist of four major components, *i.e.* (1) granular units (GUs)[13], (2) polyphase units (PUs), (3) single-crystal platelets (~0.5-2.0 micron in size), and (4) rare silicate whiskers and euhedral crystals[13&27] (Fig.1). The GUs are ~0.2-2.0 micron in diameter and form the loosely packed matrix of CCP IDPs.[28] They consist of ultra fine grains that are ~1.5 nm thick and 2 - 1000 nm in diameter. These grains are embedded in (refractory) hydrocarbons and amorphous carbons. They are mostly Mg,Fe-olivines and Ca-poor pyroxenes, Fe,Ni-sulfides and oxides[10&13] with size distributions supporting thermal annealing of amorphous precursors[28]. The ratio of carbonaceous materials to silicates plus sulfides varies from pure carbon to carbon-rich chondritic GUs.[10&16] The PUs are ~1.0 micron in diameter and they are amorphous or holocrystalline. The latter can be either coarse or fine-grained [29]. Amorphous, as well as crystalline, PUs can have one of four bulk compositions (1) magnesiosilica, (2) ferromagnesiosilica, (wt% (Mg/(Mg+Fe) = 0.23; 0.73-0.95), (3) (Na-rich), ferromagnesio-alumino-silica (both high and low

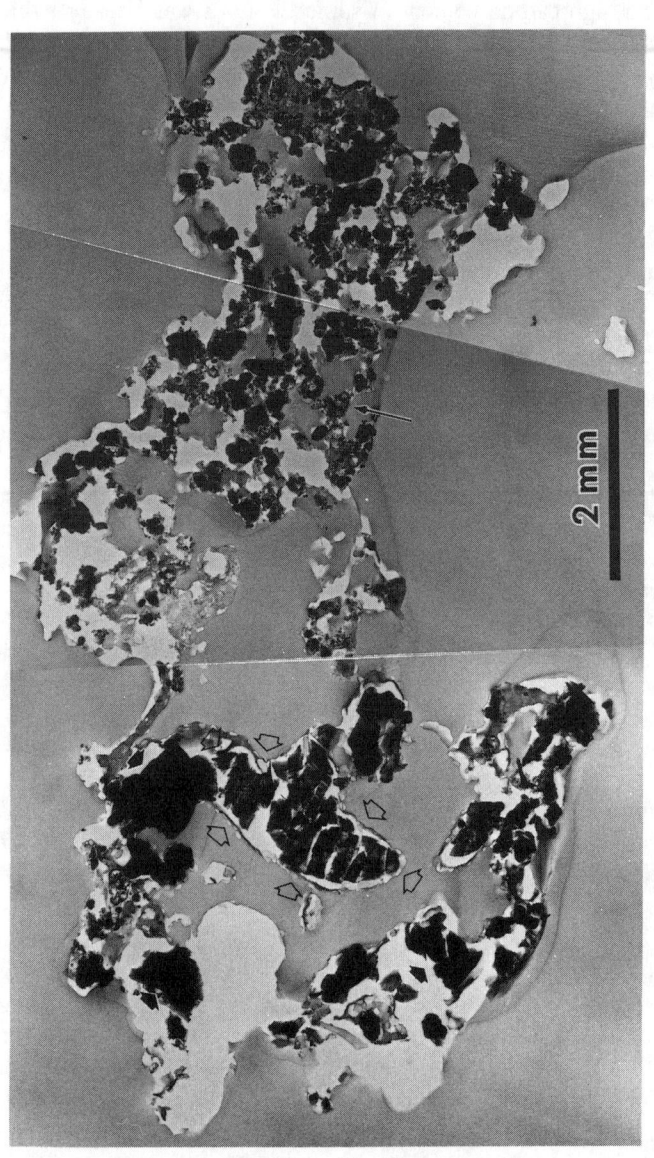

Figure 1. A transmission electron microscope photomicrograph of a ~100 nm thin section selected from among several serial ultrathin sections prepared from chondritic porous IDP L2011A9 [Rietmeijer, unpubl. data]. The gray areas are the embedding epoxy filling voids in the particle. The white areas indicate sample loss from the section and are an experimental artifact. The stubby black arrow identifies a large FeS grain. These grains in this section tend to be electron-opaque to the transmitted beam. The long arrow in the right-hand part of the IDP points to a granular unit. GUs are common in this IDP. The stubby open arrows outline an unusually large polyphase unit. The shattered appearance of the PU approximately coincides with compositionally different areas that include, generally crystalline, Ca-rich and Ca-poor clinopyroxenes, Ca-poor pyroxene and olivine. All silicates are Mg-rich.

Al$_2$O$_3$), (4) silica-rich, or (5) (rare) approximately chondritic. The single-crystal platelets might be crystalline fragments of PUs.[10,12,16,29,30]

It is my contention that all CC IDPs are variable mixtures of the principal components GUs and PUs. That is, in CCP IDPs the fraction of GUs is much greater than PUs, whereas about equal amounts of GUs and PUs characterize CCF IDPs, and (fused) GUs outnumber the amount of PUs in CCS IDPs. Variations in the bulk carbon content within, and among CCP, CCF and CCS IDPs, might reflect the amount of carbon relative to silicates and sulfides in the GUs. It may be possible to use the petrologic properties of principal components to refine CC IDP classification.

Petrology of Principal Components

The petrologic properties of GUs and PUs contain valuable clues to the study of protoplanet alteration of CC IDPs; the data support hydrocryogenic (T < 0°C)[31] and low-temperature aqueous (T well below 300°C)[15&16] alteration, and thermal annealing[28] of CC IDPs. For example, phyllosilicates in chondritic aggregate IDPs include poorly-ordered nonstoichiometric smectites[19] and well-ordered single crystal kaolinite, talc and various smectite minerals.[12,30] A scenario that explains co-occurrence of these two different types of layer silicates in IDPs is symbolized by a reaction of nonstoichiometric smectite to well-ordered talc plus kaolinite[15]:

$$3\ Mg_5Al_2Si_7O_{20}(OH)_4 \cdot X_i \cdot n.H_2O + 5\ Si(OH)_4\ (aq.) + 5/2\ O_2 =$$
$$5\ Mg_3Si_4O_{10}(OH)_2 + 3\ Al_2Si_2O_5(OH)_4 + \{3X_i \cdot (OH)_{10} + 3n.H_2O\}$$

[X_i denotes monovalent interlayer cations in smectite; the braces symbolize the composition of the coexisting aqueous phase]

The parent phyllosilicates in the reaction could have formed under hydrocryogenic conditions when IDPs are still embedded in the water-ice of the parent body. At higher temperatures, an aqueous fluid buffered in the C-H-O-S system[32] might be conducive to the precipitation of the product phases. At these higher temperatures, well-ordered, compositionally different layer silicates might also form by hydration of feldspars[30] and olivine, or their amorphous chemical equivalents. The hydration of olivine by an aqueous fluid with silica in solution is symbolized by a reaction of the type:[32]

$$4\ Mg_2SiO_4 + X_i + 4\ SiO_2 \cdot (0.25n).H_2O + 0.5\ CH_4 + H_2S + 3.5\ CO_2 =$$
$$Mg_6Si_8O_{20}(OH)_4 \cdot X_i \cdot n.H_2O + MgSO_4 + MgCO_3 + 3\ C$$

This reaction produces magnesite and Mg-sulfate that are observed in IDP L2005K8[12], magnesite in W7010*A2[11] and lonsdaleite in CP IDP W7029*A.[33] Saponite is present in about 50% of chondritic IDPs.[12] Variations in phyllosilicate compositions might be related to the timing of aqueous alteration following dry crystallization and fractionation of amorphous PUs.[16&30] Assuming aqueous alteration occurred at thermodynamic equilibrium, the thermal regime is broadly constrained between 175-195°C, based on reactions among coexisting smectites and kaolinite.[30] Still, it remains to be shown that equilibrium conditions did indeed prevail. It is likely that hydration reactions proceeded at lower temperatures with catalytic support from carbon-rich species[33] and Ti-oxide Magneli phases[34], or metals and spinels in the IDPs. Details of these catalytic reactions in CC IDP parent bodies remain elusive but their importance might be inferred from the wide variety of carbon-rich species that support a complex petrologic history of CC IDPs.[18&35]

DISCUSSION

The silicate and carbon-rich minerals in the principal components of carbonaceous chondritic IDPs conceivably define PETROLOGIC CLASSES. The onset of mineralogical activity in solar system protoplanets was probably chaotic and non equilibrium in nature. The mineralogical reactions were sensitive to:

(1) reaction kinetics,
(2) surface free energy of the typically nanometer-sized grains,
(3) catalytic support, and
(4) spatial and temporal variations in the presence of heat sources and, thus, to local variations in the water/rock ratio, fluid transport efficiency, and salinity and pH of the fluid.[30]

The classification scheme proposed in this paper [Table 2] reflects the chaotic non equilibrium character of CC IDP metamorphosis as a function of temperature and variations in water-rock ratio, and *it is therefore dependent upon inferred parent body processes*. However, considering the putative parent bodies of CC IDPs in the outer asteroid belt (P and D infrared class asteroids) and comet nuclei, it is plausible that immediately upon accretion water-ice was a major constituent of these parent bodies. For example, *hydrocryogenic alteration* [CLASS 1] below the melting point of water-ice would occur with a limited availability of liquid water. *Aqueous alteration* at higher temperatures [CLASS 2] required higher water/rock ratios and freely-available water. If water is unavailable at these higher temperatures, CC IDP metamorphosis will be characterized by thermal metamorphism that readily affects the amorphous materials in CC IDPs. During this thermal annealing regime *dry crystallization and fractionation* [CLASS 3] of polyphase units will occur, *e.g.* the formation of plagioclase and alkali-feldspars.

The timing of thermal annealing relative to aqueous alteration in CC IDPs is not yet known but it is probably random in both time and space. The effects of diagenesis and thermal annealing may not be uniform at the scale of individual particles[16], let alone among individual fragments of so called "cluster particles"[36], or among anhydrous and hydrated IDPs.[37] I emphasize that amorphous materials and ultra fine grains in CC IDPs provide high free energy conditions for the onset of mineralogical activity in solar system protoplanets.

TABLE 2 : PETROLOGICAL CLASSES BASED ON THE MINERALOGY OF GRANULAR and POLYPHASE UNITS IN CARBONACEOUS CHONDRITIC INTERPLANETARY DUST PARTICLES

	GUs	PUs
CLASS 1	hydrocarbons; volatile-rich amorphous carbons	poorly-ordered layer silicates
CLASS 2	lonsdaleite; pregraphitic carbons; mixed-layered turbostratic carbons	well-ordered layer silicates; salts
CLASS 3	amorphous carbons; poorly graphitised carbons; graphite nanocrystals	Fe,Mg-silicates; alkali-feldspars and plagioclase; amorphous + crystalline 'silicates'

Many particles show evidence for dynamic pyrometamorphism due to flash-heating during micrometeorite deceleration in the Earth's atmosphere. The degree of dynamic pyrometamorphism seems quite variable among individual particles as can be inferred from the development of iron-oxide rims, among others. The classification scheme presented here assumes that dynamic pyrometamorphism is either negligible [18] or that its signature on constituent minerals is recognizable [38-40] and appropriate corrections can be made to constrain the pre-entry mineralogical composition.

The petrologic classes are mainly defined by the carbon species and silicate minerals in the principal components but do not include the ultra fine minerals in the granular units. The grain size distributions in the GUs support continued *in-situ* nucleation and growth. [28] The exact timing of this process cannot yet be constrained although the hydrocryogenic regime may not promote crystallization of amorphous GUs.

In the proposed classification the carbon-rich CP particle W7029C1 with lonsdaleite, poorly graphitised carbon and both poorly-ordered and well-ordered phyllosilicates [15] becomes a $CCP_{1,2,3}$ IDP. The particle L2005L2 contains amorphous material, saponite, forsterite and pyroxene [12] and would be classified as a $CCF_{0,2,3}$ IDP. In fact, this classification follows naturally from the classification schemes in Tables 1 and 2. Both predict the existence of fully amorphous CC IDPs [CLASS 0] which is partially preserved in amorphous GUs and PUs. At first glance the subscripts seem cumbersome, but they readily provide information into the chaotic onset of CC IDP parent body evolution. The proposed classification scheme relies on the petrologic properties of the principal components of the CC IDPs. The principal components represent the solar nebula dust in the CC IDP accretion regions[16] and their variable proportions constrain, to a first order, the bulk composition (i.e. high carbon content) and morphology of CC IDPs. The petrologic properties of principal components discussed in this paper can be determined without prejudice of the individual investigator. The classes presented in Table 2 are an attempt to order and correlate the petrologic properties of carbon phases and the silicate minerals in CC IDPs. This paper is an effort to begin systematic analyses of the formation and evolution of the parent bodies of CC IDPs which are a unique type of ultra fine-grained extraterrestrial material that has become recently available for laboratory studies. I have chosen for a petrologic classification scheme of the unique CC IDPs. Other classification properties that might have potential use, such as the water content, or total gas abundances, but these quantitative measurements are difficult to obtain.[41] These properties will be more sensitive to changes during dynamic pyrometamorphism than mineralogical properties[16] and will be, together with iron-oxide rims on IDPs, indicative of the extent of atmospheric entry heating.

CONCLUSIONS

The most conspicuous property of CC IDPs is their small-scale mineralogical heterogeneity. The lack of chemical fractionation but pervasive non-equilibrium mineralogy is diagnostic for these least altered solar system materials. Mineralogical heterogeneity will decrease, and chemical fractionation commence, during prolonged residency of CC IDPs in evolving protoplanetary bodies. Many chondritic IDPs have a uniquely high bulk carbon content. While the 3 times CI carbon enrichment of chondritic IDPs was for the first time firmly established by energy dispersive analyses[20-22], carbon-enriched IDPs can also be recognized by petrologic point-count analysis of transmission electron microscope photomicrographs.[19] Although the latter does not yield a reliable quantitative analysis, it nevertheless suffices to identify a particle as a CC IDP.

The classification of chondritic IDPs based on bulk chemistry and morphology alone is an excellent scheme that incorporates all major IDP types. It is readily determined at the bulk level of each particle. Here, I propose to re-name carbon-rich IDPs as *carbonaceous chondritic (CC) IDPs*. I contend that the CC IDPs are a mixture of at least two principal components, *viz.* carbon-rich to carbonaceous granular units and polyphase units. I have explored the possibility of using the petrologic properties of these principal components for classification of CC IDPs. The petrologic properties for this classification are readily, and without prejudice, determined by TEM and AEM analyses. This petrologic classification scheme is useful to monitor the onset of mineralogical activity in solar system protoplanets. I stress that at this point the classification scheme is a proposal. Its success will require considerable discussion from the community most heavily involved in the studies of interplanetary dust particles.

ACKNOWLEDGMENTS

This work was supported by grants from the National Aeronautics and Space Administration, NAG 9-160 and NAGW-3626. I thank Adrian Brearley, Rhian Jones, Horton Newsom, Steve Sutton and Jim Papike for critical comments on various drafts of this paper.

REFERENCES

1. P. Fraundorf, K.D. McKeegan, S.A. Sandford, P. Swan and R.M. Walker *Proc. 13th Lunar Planet. Sci. Conf., J. Geophys. Res.* **87** *Suppl.*, A403 (1982).
2. I.D.R. Mackinnon and D.S. McKay, *Lunar Planet. Sci.* **XVII**, 510 (1986).
3. I.D.R. Mackinnon, D.S. McKay, G. Nace and A.M. Isaacs, *Proc. 13th Lunar Planet. Sci. Conf, J. Geophys. Res.* **87** *Suppl.*, A413 (1982).
4. K.M. Kordesh, I.D.R. Mackinnon and D.S. McKay, *Lunar Planet. Sci.* **XIV**, 389 (1983).
5. D.E. Brownlee, *Ann. Rev. Earth Planet. Sci.* **13**, 147 (1985).
6. M.E. Zolensky, *Science* **237**, 1466 (1987).
7. G.J. Flynn and S.R. Sutton, *Proc. 20th Lunar Planet. Sci. Conf.*, 335 (1990).
8. F.J.M. Rietmeijer, *Proc. Lunar Planet. Sci. Conf.* **22**, 195 (1992)
9. S.A. Sandford and R.M. Walker, *Astrophys. J.* **291**, 838 (1985).
10. J.P. Bradley, *Geochim. Cosmochim. Acta* **52**, 889 (1988).
11. F.J.M. Rietmeijer, *Meteoritics* **25**, 209 (1990).
12. M.E. Zolensky and D. Lindstrom, *Proc. Lunar Planet. Sci.* **22**, 161 (1992).
13. F.J.M. Rietmeijer, *Proc. Lunar Planet. Sci. Conf.* **19**, 513 (1989).
14. J.P. Bradley, S.A. Sandford and R.M. Walker, In *Meteorites and the Early Solar System* (J.F. Kerridge and M.S. Matthews, eds), pp. 861-898 (Univ. Arizona Press, Tucson, 1988).
15. I.D.R. Mackinnon and F.J.M. Rietmeijer, *Reviews Geophys.* **25**, 1527 (1987).
16. F.J.M. Rietmeijer, *Trends Mineral.* **1**, 23 (1992).
17. B. Wopenka, *Earth Planet. Sci. Lett.* **88**, 183 (1988).
18. F.J.M. Rietmeijer, *Geochim. Cosmochim. Acta* **56**, 1665 (1992).
19. F.J.M. Rietmeijer and I.D.R. Mackinnon, *Proc. 16th Lunar Planet. Sci. Conf., J. Geophys. Res.* **90** *Suppl.*, D149 (1985).
20. L.S. Schramm, D.E. Brownlee and M.M. Wheelock, *Meteoritics* **24**, 99 (1989).
21. K.L. Thomas, G.E. Blanford, L.P. Keller, W. Klock and D.S. McKay, *Geochim. Cosmochim. Acta* **57**, 1551 (1993).
22. K.L. Thomas, L.P. Keller, G.E. Blanford, W. Klock and D.S. McKay, *Meteoritics* **27**, 296 (1992).
23. L.P. Keller, K.L. Thomas and D.S. McKay, Carbon in primitive Interplanetary Dust Particles. *This Volume* (1994).
24. L.P. Keller, K.L. Thomas and D.S. McKay, *Geochim. Cosmochim. Acta* **56**, 1409 (1992).
25. P. Bradley and D.E. Brownlee, *Science* **251**, 549 (1991).
26. F.J.M. Rietmeijer, *Lunar Planet. Sci.* **XXIII**, 1153 (1992).

27. J.P. Bradley, D.E. Brownlee and D.R. Veblen, *Nature* **301**, 473 (1983).
28. F.J.M. Rietmeijer, *Earth Planet. Sci. Lett.* **117**, 609 (1993).
29. M.S. Germani, J.P. Bradley and D.E. Brownlee, *Earth Planet. Sci. Lett.* **101**, 162 (1990).
30. F.J.M. Rietmeijer, *Earth Planet. Sci. Lett.* **102**, 148 (1991).
31. F.J.M. Rietmeijer, *Nature* **313**, 293 (1985).
32. F.J.M. Rietmeijer, *Lunar Planet. Sci.* **XVI**, 696 (1985).
33. F.J.M. Rietmeijer and I.D.R. Mackinnon, *Nature* **326**, 162 (1987).
34. F.J.M. Rietmeijer and I.D.R. Mackinnon, *Proc. Lunar Planet. Sci. Conf.* **20**, 323 (1990).
35. F.J.M. Rietmeijer, In *Asteroids, Comets, Meteors-1991* (A.W. Harris and E. Bowell, eds), pp. 513-516 (Lunar Planetary Institute, Houston, 1992).
36. K.L. Thomas, W. Klock, L.P. Keller, G.E. Blanford and D.S. McKay, *Meteoritics* **28**, 448 (1993).
37. M. Zolensky, R. Barrett and P.J. Burkett, *Meteoritics* **28**, 469 (1993).
38. F.J.M. Rietmeijer, *Lunar Planet. Sci.* **XXIII**, 1151 (1992).
39. F.J.M. Rietmeijer, *Lunar Planet. Sci.* **XXIV**, 1199 (1993).
40. L.P. Keller, K.L. Thomas and D.S. McKay, *Lunar Planet. Sci.* **XXIII**, 675 (1992).
41. E.K. Gibson, *J. Geophys. Res.* **97(E3)**, 3865 (1992).

COLLECTION OF INTERPLANETARY DUST

The Earth can be a very effective collection medium for interplanetary dust. The smallest grains are decelerated in the atmosphere and can be collected in the stratosphere in an intact state. Larger grains (principally micrometeoroids) can be recovered from polar ices, such as that to be found in Antarctica (bottom of photo). (Apollo 17 photo AS17-148-2272)

COLLECTION OF INTERPLANETARY DUST

At present, IDPs are actively being collected in the stratosphere, from polar ices, and within impact features on spacecraft. Papers by Maurette, Warren, Zolensky and coworkers cover basic aspects of these efforts. It is fair to say that all techniques being used are complementary, and that none is clearly superior in all aspects. The stratospheric collections provide the least contaminated and heated particles, but recovered particles are generally less than 70 µm in diameter. Particles from polar ices are often larger than the stratospheric particles, but are more contaminated and heated, and may actually represent a different population of objects. Polar samples can be collected from glacial ice of a predetermined age, possibly permitting the determination of temporal changes (if any) in IDP type and flux. In the present volume Maurette et al. discuss plans for dust collection at the Amundsen-Scott South Pole Station, an effort which will build on an earlier program run by Robert Witkowski and William Cassidy.

Particles collected in space can be collected with velocity and trajectory information, unlike those collected at Earth, but are generally highly shocked and/or melted at best, and at worst are vaporized. All of these collection techniques are being improved, as resources permit. In the stratosphere, collection with a medium other than silicone oil is being attempted (so far without success). Cleaner equipment is now being used in the Antarctic to collect particles. Better capture media are being developed for less disruptive collections in space. Future flight opportunities for on-orbit collection of IDPs include "reflights" of the Long Duration Exposure Facility (LDEF II), the European Recoverable Carrier (EURECA II), Mir, the Space Shuttle and the U.S. Space Station.

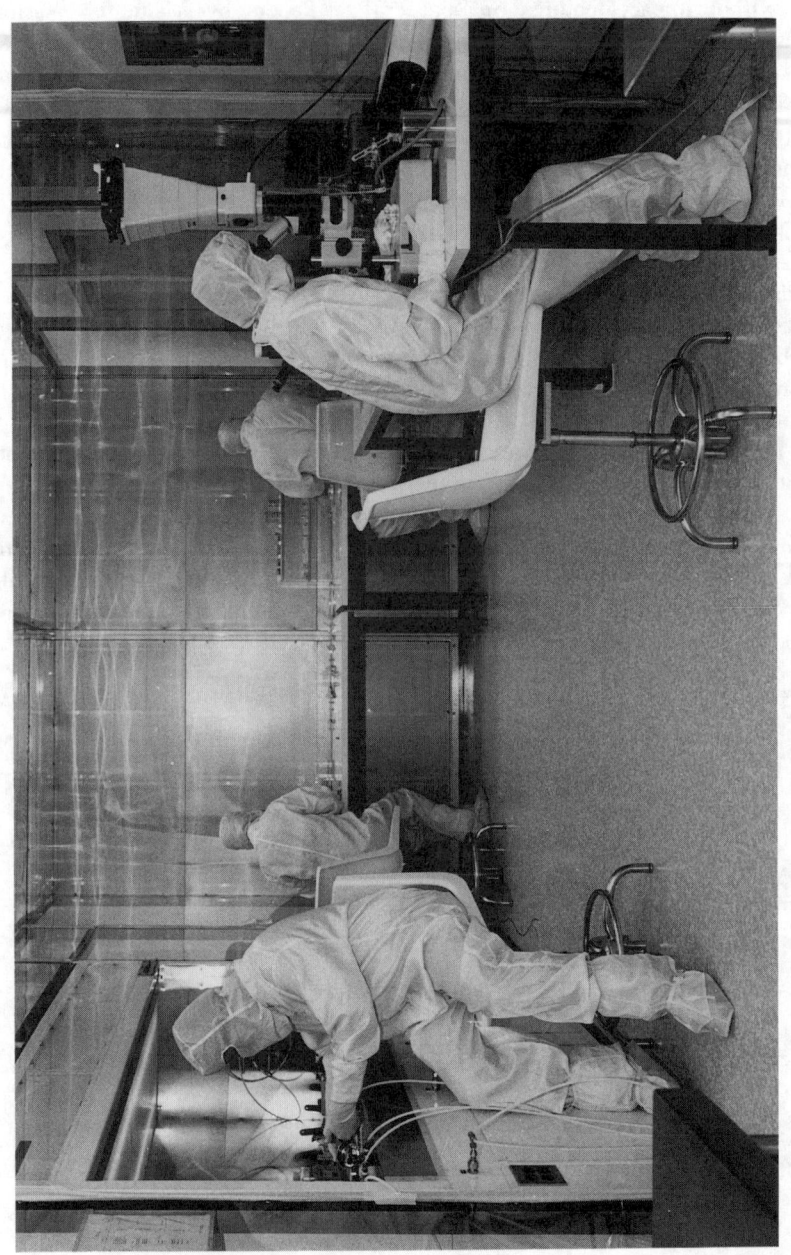

View of anonymous scientists the class 100 clean room in NASA's Interplanetary Dust Curatorial Laboratory at the Johnson Space Center. In the foreground a scientist (right) examines collection plates, another (left) works at the wet bench, and in the background two scientists observe extraterrestrial particles through binocular microscopes. In this laboratory particles collected in the stratosphere are removed from collectors for preliminary analysis and allocation to scientists worldwide. (NASA photo S85-36299)

COLLECTION AND CURATION OF INTERPLANETARY DUST PARTICLES RECOVERED FROM THE STRATOSPHERE BY NASA

Jack L. Warren
Lockheed-ESC, 2400 NASA Road 1
Houston, Texas 77058, USA

and

Michael E. Zolensky
Solar System Exploration Division, NASA, Johnson Space Center
Houston, Texas 77058, USA

ABSTRACT

Since May, 1981, the National Aeronautics and Space Administration (NASA) has used aircraft to collect interplanetary dust particles (IDPs) from Earth's stratosphere. Specially designed dust collectors are prepared for flight and processed after flight in an ultra clean (Class-100) laboratory constructed for this purpose at the Lyndon B. Johnson Space Center (JSC) in Houston, Texas. Particles are individually retrieved from the collectors, examined and cataloged, and then made available to the scientific community for research. Interplanetary dust thereby joins lunar samples and Antarctic meteorites as a critical extraterrestrial material curated at JSC.

INTRODUCTION

Of all current sampling techniques, collection in the Earth's upper atmosphere is the superior way to obtain the least altered Interplanetary Dust Particles (IDPs) for laboratory study. IDPs can, conceivably, be collected in space along with information about their associated velocities and trajectories, but at the typical cost of a partially melted or vaporized sample. (Efforts are currently underway to mitigate this effect; see papers in this volume by Borg et al.[1] and Zolensky et al.[2]) Large IDPs can be gathered from the oceans or polar glaciers (see the next paper in this volume[3]), but these are generally chemically altered, to greater or lessor extent, by residence for extended periods in ice or sea water. However, small particles (<100 μm in diameter) entering the Earth's atmosphere can be gently decelerated. The degree of heating experienced by IDPs during this deceleration depends upon initial velocity, entry angle, size and density (see George Flynn's paper in this volume[4]).

Upon atmospheric entry IDP, trajectory information is irretrievably lost, and velocity information is determinable only through the heating effects it imposes[4]. In the atmosphere, IDP settling velocities are determined mainly by Stokes Law; at altitudes of 20-30 km these velocities are on the order of cms/sec for 10 μm-sized grains[5]. Since the 1950s, particles have been collected in the atmosphere, first on filters and then, beginning in the 1970s, on inertial impaction surfaces. In order to minimize terrestrial contamination of collection surfaces, IDPs are now gathered in the lower stratosphere (~20 km altitude) by NASA stratospheric aircraft. Don Brownlee pioneered the use of impaction collection in the stratosphere[5], and many of the techniques described in this paper were developed in his laboratory.

Since 1981, NASA has conducted a program to collect a representative record of the particle load of the lower stratosphere using impaction collectors flown on stratospheric aircraft. The program is managed by the Office of the Curator, Johnson Space Center (JSC). We note that while the collection of IDPs has been the principal goal of this program, terrestrial material (including volcanic ash) and space debris particles are also collected in great numbers[6] and are likewise available for study. With the growing realization that important climatological changes can be influenced by atmospheric dust, we anticipate that the collected terrestrial material will become an increasingly important scientific resource.

COLLECTION SURFACES

The collection surfaces are flat plates of Lexan (a space-age polycarbonate plastic) and come in two convenient sizes, (1) conventional collectors and (2) Large Area Collectors, having 30 and 300 cm^2 surface areas, respectively (Figure 1). These collector surfaces are coated with high-viscosity silicone oil (dimethyl siloxane) and sealed within special airtight housings, to be opened only in the stratosphere or in the clean-lab facility. The collectors are carried into the stratosphere under the wings of NASA ER-2, WB-57F and (at one time) U-2 aircraft. These collectors are installed in specially constructed wing pylons which ensure that the necessary level of cleanliness is maintained between periods of active sampling. During successive periods of high altitude (20 km) cruise, the collectors are exposed in the stratosphere by barometric controls and then retracted into sealed storage containers prior to descent. In this manner, a total of 20-80 hours of stratospheric exposure is accumulated for each collector.

CLEAN ROOM LABORATORY

To support stratospheric dust collection and curation activities, a Class 100 clean room was established at JSC. The configuration of a horizontal flow tunnel was chosen for this task, because so much of the laboratory activity requires working over a microscope. By using horizontal flow, collectors and samples can always be positioned upstream or parallel to a worker's body, which is the greatest contamination source in the laboratory. A vertical Class 100 wet bench was installed on one side and towards the back of the tunnel, and is used to clean all items entering the tunnel. The wet bench is supplied with filtered deionized (DI) water, filtered isopropyl alcohol, and distilled/filtered freon. The freon will be replaced with ultra-pure water in 1994, in an effort to be more environmentally friendly. Thereafter, hexane will be used as the degreasing agent.

J. L. Warren and M. E. Zolensky 247

a

b

Figure 1. Views of the two types of impaction collection surfaces used to collect dust in the stratosphere. (a) Four conventional collectors are shown deployed from the wing pylon (located underneath the plane wing). In practice, barometric switches permit deployment only in the stratosphere. (b) Two deployed Large Area Collectors; they resemble pie pans and face one another when closed. The two collection plates each measure 24 cm in diameter, making them an order of magnitude larger in surface area than the conventional stratospheric dust collection surfaces.

All tools and other items introduced into the Class 100 clean room are cleaned by the following method:

A) Each item is scrubbed with DI water and soap solution. The entire surface is covered 10-15 times, then placed into a DI water soak tank. The soak prevents the soap from drying on an item while other items are being cleaned.
B) Each item is removed from the soak tank and spray-rinsed with filtered DI water at a pressure of 45 psi.
C) Spray rinse with filtered isopropyl alcohol at 45 psi, to remove the water.
D) Spray rinse with distilled/filtered freon (ultrapure water after 1994) at a pressure of 45 psi, to remove the alcohol.
E) Item are placed upstream of the cleaning area to dry in air.
F) While the items are drying, the DI water in the scrub tank and the soak tank are discarded and replaced. This time only a minimal quantity of soap is used in the scrub tank.
G) Steps A-E are repeated.

PREPARATION AND ASSEMBLY OF COLLECTION PLATES

The collection medium used on the transparent Lexan collector plates is an equally transparent, colorless, silicone oil (500,000 cs viscosity). To apply the oil to the collector surface, a solution of 20:1 of freon and silicone oil is mixed and placed into a syringe equipped with a 0.4 µm Nucleopore filter membrane. The filtered solution is then applied through a long needle to keep contamination sources (like a body) as far as possible from the collector surface. The freon is allowed to evaporate, leaving a layer of silicone oil. Then a second layer is applied to the collection surface. The final silicone layer is 10-20 µm thick. All freon must be allowed to evaporate, to prevent freon vapor pressure from building up in the collector housing after it is sealed; such pressure will cause a disturbance on opening at altitude and could cause contamination. The collector surfaces are then secured within the pylon housing boxes and bagged for transport to the aircraft for flight.

Collectors are typically exposed for 20-80 hours (total) in the stratosphere. This may necessitate months of actual flying. Following flight the collector housing boxes are returned to the JSC Curatorial Facility for processing.

PROCESSING OF PARTICLES

Particle mounts designed for the JEOL 35CF Scanning electron Microscope (SEM) and JEOL 100CX Scanning Transmission Electron Microscope (STEM) are currently the standard receptacles for dust particles in the clean room. Each mount consists of a graphite frame (size ~3x6x24 mm) onto which a Nucleopore filter (0.4 µm pore size) is attached. A conductive coat of carbon is vacuum evaporated onto the mount and then a microscopic reference pattern is "stenciled" onto the carbon-coated filter by vacuum evaporation of aluminum through an appropriately sized template.

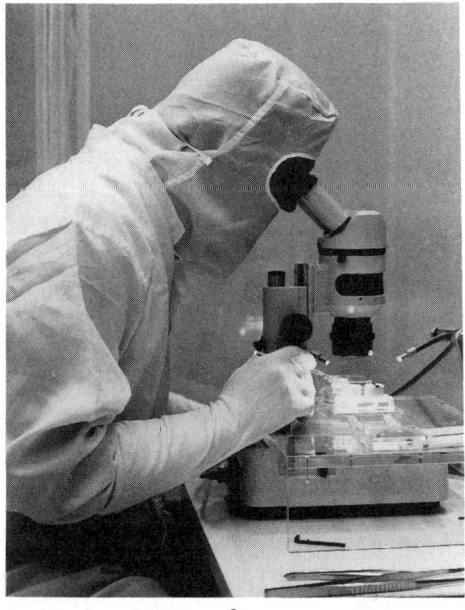

a

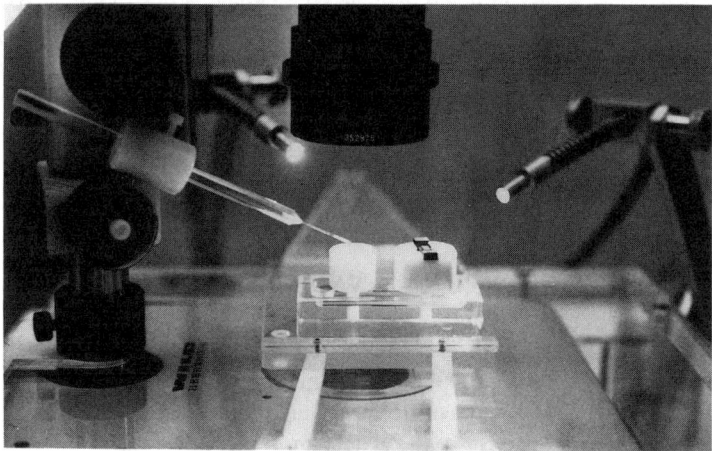

b

Figure 2. (a) In the Class 100 clean room laboratory a worker is picking a particle off of a small (conventional) stratospheric dust collection surface (the square transparent Lexan plate in the center of the image. A quartz glass fiber (coming in from the left) is used to remove each particle. The entire operation is performed under a binocular microscope. Each removed particle is then placed onto the black rectangular graphite mount at the lower right for preliminary characterization and eventual storage. All particles are available for allocation to scientists worldwide. (b) Close-up of the experimental setup.

Particles are individually removed from collectors using a glass-needle manipulator under a binocular stereo-microscope. Don Brownlee's method for manipulating particles was chosen for use in the laboratory. A Plexiglass table placed over the microscope transmitted light stand, and an attached drafting arm is used to manipulate the particle into the field of view of the microscope (Figure 2).

The needles used to transfer the particles are specially made in the laboratory Using a variable power supply, voltage is carefully applied to a thin metal strip to melt the end of a cleaned 2 mm glass pipette, which is then heated to softening and drawn away from the strip. Excess glass is cut away and the tip is annealed. The glass fiber is then placed into a needle holder with a recessed tip for use (Figure 2).

By attaching the needle holder to the microscope post, the needle is controlled within the field of view. A particle on the collection surface is moved into the field of view with the drafting arm and is picked off with the needle tip. Each particle is then positioned on an aluminum-free area of a freon-cleaned (freon 113), carbon-coated STEM mount (see above) and washed in place with hexane to remove silicone oil. Each mount is normally limited to 16 particles. The particles are rinsed using one of three methods. The usual system, which is a derivative of a method devised by Phil Fraundorf at Washington University in the 1970s, uses an ultrafine capillary to dispense small drops of hexane. The flushing-probe capillary is made by drawing a heat-softened glass tube to the point to where it seals. The tube is then cut off above the seal point and annealed until the opening at the end of the tube narrows to 10 to 20 μm in size. The probe is positioned perpendicular to the mount to be washed. Once the flushing probe is filled with hexane and touched to the mount's surface, a pool of hexane forms around the point of the probe and the particle is moved to this pool to be washed. Using the flushing-probe method, any movement of the particle can be observed and controlled. The other two, less-frequently used, flushing methods are (1) wicking the hexane up through a fritted piece of glass, with the mount setting on top; and (2) using a hexane still designed by David McKay and Uel Clanton of JSC.

PRELIMINARY EXAMINATION OF PARTICLES

Each rinsed particle is examined, before leaving the Class-100 clean room processing area, with a petrographic research microscope equipped with transmitted, reflected and oblique light illuminators. At a magnification of 500X, the size, shape, transparency, color, and luster of each particle are determined and recorded.

After optical description, each mount is examined by SEM and Energy-Dispersive X-ray Spectrometry (EDX). Secondary-electron imaging of each particle is performed with a JEOL-35CF SEM at an accelerating voltage of 20 kV. Images are therefore of relatively low contrast and resolution due to deliberate avoidance of conventionally-applied conductive coats (carbon or gold-palladium) which might interfere with later elemental analyses of particles. EDX data are collected with the same JEOL-35CF SEM equipped with a Si(Li) detector and PGT 4000T analyzer. Using an accelerating voltage of 20 kV, each particle is raster-scanned and its X-ray spectrum recorded over the 0-10 keV range by counting for 100 sec. No system (artifact) peaks of significance appear in the spectra.

Following SEM/EDS examination, each particle mount is stored in a dry nitrogen gas atmosphere in a sealed cabinet until allocation to qualified investigators.

COSMIC DUST CATALOGS

The preliminary information and images of each particle are then published by the JSC Office of the Curator in the form of the *Cosmic Dust Catalogs*. Each page in the main body of the catalog is devoted to one particle and consists of an SEM image, an EDX spectrum, and a brief summary of preliminary examination data obtained by optical microscopy (Figure 3).

Each cataloged particle receives a provisional first-order identification based on its morphology (from SEM image), elemental composition (from EDS spectrum), and optical properties. Particle types are defined for their descriptive and curatorial utility, not as scientific classifications. These tentative categorizations, which reflect judgments based on a decade of collective experience, should not be construed to be firm identifications and should not dissuade any investigator from requesting any given particle for detailed study and more complete identification. The precise identification of each particle in our inventory is beyond the scope and intent of our collection and curation program. Indeed, the reliable identification and scientific classification of cosmic dust is one of many important research tasks that we hope to stimulate.

Periodically, as warranted, newsletters are published containing late-breaking information concerning the state of collection and curatorial activities. This newsletter, called the *Dust Courier*, also includes information concerning NASA JSC's holdings of returned spacecraft parts featuring meteoroid impact features. Current holdings include surfaces from Surveyer Spacecraft and the Solar Maximum, Palapa, LDEF and EURECA Satellites.

SAMPLE REQUESTS

Scientists desiring further information concerning the allocation of interplanetary dust, or the *Cosmic Dust Catalogs* or *Dust Couriers* should contact:

Curator for Interplanetary Dust
Code SN2
NASA/Johnson Space Center
Houston, Texas 77058 U.S.A.
Telephone: (713) 483-5128
FAX: (713) 483-2911

U2034 D 7

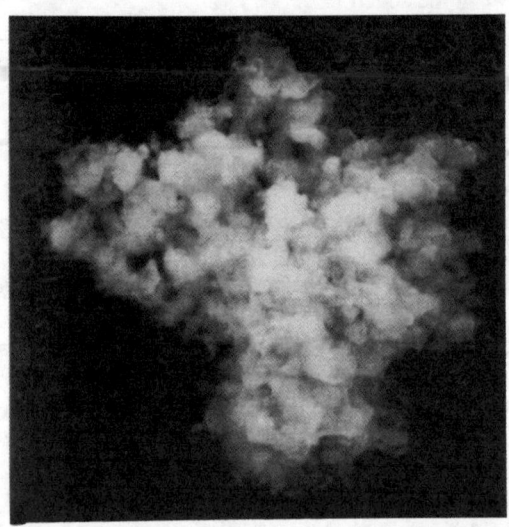

SIZE: 15
SHAPE: I
TRANS.: O/TL
COLOR: Black to yellow
LUSTER: D/SV
TYPE: C

COMMENTS:

S-87-45985

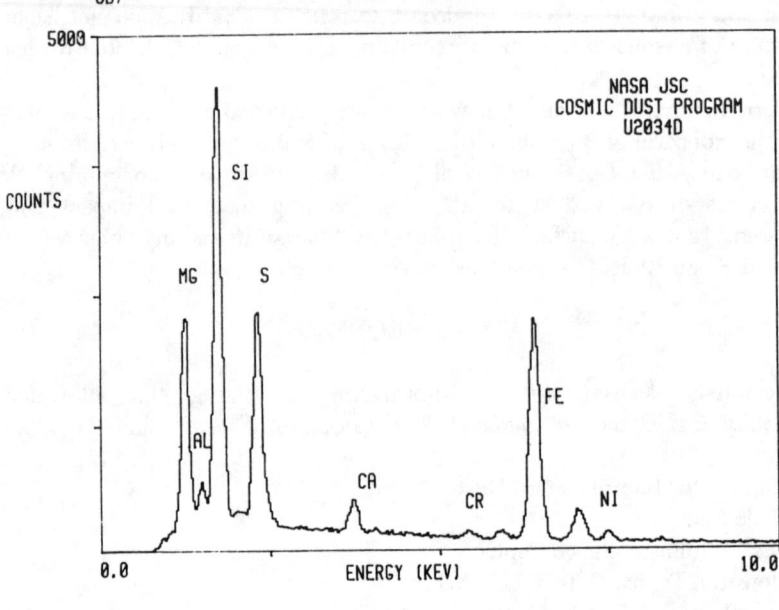

Figure 3. A page from a typical Cosmic Dust Catalog, showing an SEM image of a C (Cosmic) type particle U2034 D7, with its dimensions, optical properties and EDX spectrum.[7]

REFERENCES

1. Borg J., Bibring J-P., Maag C., Tanner W. and Alexander M. (1994) Description of the COMRADE Experiment. This volume.
2. Zolensky M.E., Hörz, F., See T., Bernhard, R.P.,Barrett R.A., Mack K., Warren J.L. and Kinard W.H. (1994) Meteoroid Investigations Using the Long Duration Exposure Facility. This volume.
3. Maurette M., Immel G., Hammer C., Harvey R., Kurat G. and Taylor S.. (1994) Collection and Curation of IDPs from the Greenland and Antarctic Ice Sheets. This volume.
4. Flynn G.F. (1994) Changes to IDP Composition and Mineralogy by Terrestrial Encounters. This volume.
5. Brownlee D.E. (1978) Microparticle studies by sampling techniques. In *Cosmic Dust* (Ed. J.A.M. McDonnell), Wiley Publ. Co., 295-336.
6. Zolensky M.E., McKay D.S. and Kaczor L.A. (1989) *J. Geophysical Research* **94**, D1, 1047-1056.
7. Zolensky, M.E., Barrett, R.A., McKay, D.S., Thomas, K.L., Warren, J.L. and Watts, L.A. (1988) *Cosmic Dust Catalog 9*. NASA/Planetary Materials Branch Publication 22744, 105p.

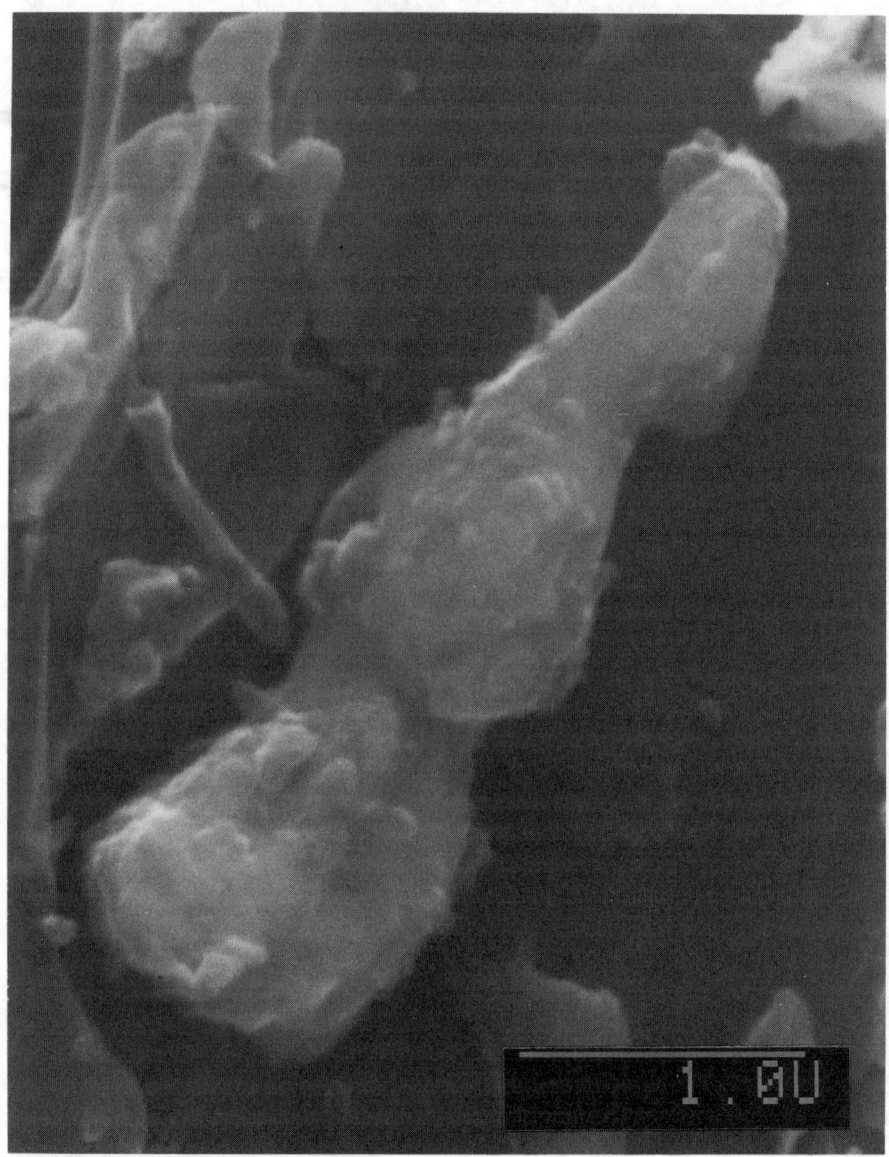

A secondary electron image of an impact residue grain from the Solar Maximum Satellite. This grain penetrated the outer blanket layer serving as thermal protection for the satellite, and is shown on the front of the blanket's second layer. The major components of the grain are silicon, iron, aluminum, calcium and oxygen, and its origin is undetermined. The scale bar measures 1 μm. (Photo by M. Zolensky)

WINDOWS OF OPPORTUNITY IN THE NASA JOHNSON SPACE CENTER COSMIC DUST COLLECTION

Frans J.M. Rietmeijer
Department of Earth and Planetary Sciences,
University of New Mexico, Albuquerque, NM 87131, USA.

and

Jack L. Warren
Lockheed Engineering and Science Company
Mail Code C23, 2400 NASA Road One, Houston, TX 77058, USA.

ABSTRACT

Non-spherical chondritic interplanetary dust particles (IDPs) have been routinely collected in the stratosphere at ~17-19 km altitude since May 1981 by the NASA Cosmic Dust Program. These IDPs show distinct morphological subtypes, viz. chondritic porous (CP) and chondritic filled (CF) aggregate particles, glazed CP and CF IDPs with 'softened' contours, and non-aggregate chondritic smooth and rough-textured (CR) IDPs. The distributions of these IDP subtypes show distinct abundance maxima as a function of collection time. Albeit conservatively at this time, these distributions support systematic intra- and inter-annual variations. Rigorous statistical treatment of the data is disabled by gaps in the collection periods and uncertainties in collection and curation. This study recommends monthly sampling of the lower stratosphere during a two year period to quantify these temporal variations in the non-spherical chondritic IDP subtypes. With frequent and systematic collection we will be able to combine temporal variations in IDP subtype distributions with their petrologic and chemical properties. It will then also be possible to correlate the temporal variations sampled in the lower stratosphere with events that deliver chondritic IDPs to the Earth's atmosphere. Hence, the study of micrometeorites collected in the stratosphere might substitute for *in-situ* sampling of asteroids and comet nuclei.

INTRODUCTION

This paper explores the existence of windows of opportunity for the collection of interplanetary dust particles (IDPs) due to temporal variations in the fluxes of particle types that were heretofore neither anticipated nor appreciated. Our study is a first pioneering attempt that may be incomplete and contain errors; with hindsight some of its interpretations may have been naive. We would like to show that the information on temporal variations in the types of IDPs in the lower stratosphere is potentially within the grasp of the NASA Johnson Space Center (JSC) Cosmic Dust Program. Our paper will identify several parameters that are currently ill-constrained and identify uncertainties in others. These parameters need to be investigated in order to obtain a significant database for quantification of temporal variations in IDP subtypes in the lower stratosphere and to derive variations in the flux of these subtypes as they survive atmospheric entry heating. There will be (long-term) *intra*-annual and (2) *inter*-annual variations. From the beginning of the NASA/JSC Cosmic Dust Collection Program it was appreciated that the collectors sampling the stratosphere at different times displayed marked differences in collected particle types. This phenomenon is still evident in the latest Cosmic Dust Catalog (CDC). For the purpose of this paper, we will emphasize possible temporal variations among non-spherical chondritic IDPs in the NASA/JSC cosmic dust collection.

All non-spherical chondritic IDPs labeled "cosmic" were selected from the NASA/JSC CDCs, volumes 1-13, compiled by the Cosmic Dust Preliminary Examination Team.[1] The elemental distribution pattern of Mg, Al, Si, Ca, Ti, Fe and Ni in the energy dispersive spectra (EDX) of these IDPs is approximately chondritic. All selected IDPs contain sulfur, albeit the EDX sulfur peak in some cases only denotes a trace amount of this element.[2] Non-spherical chondritic IDPs have two basic morphologies[3&4]:

(1) aggregate particles that include (a) fluffy, chondritic porous (CP) IDPs and
    (b) chondritic filled (CF) IDPs that lack the distinctive porosity of CP IDPs,
    and
(2) compact, non-aggregate particles that include (a) chondritic smooth (CS)
    IDPs and (b) rough-textured (CR) IDPs.

We stress that the current Cosmic Dust Program remains one of the best efforts to obtain samples of this unique type of ultra fine extraterrestrial materials that complements the 'traditional' meteorites. The main concern throughout our paper is that our effort is currently a game of small-number statistics. Yet, we would like this paper to be a call to the National Aeronautics and Space Administration for a dedicated effort to look for temporal variations in IDP subtypes.

DUST COLLECTIONS

General

Particles are collected in the lower stratosphere using flat-plate collectors of different surface area (A), viz. (1) small area collectors (SACs) with $A = 30$ cm^2 and (2) large area collectors (LACs) with $A = 300$ cm^2. During the past 10 years, 58% of the collection periods took place from February through August. The collections from August through December show an eight-year gap wherein no IDPs were collected, that is, excluding SAC W7074 (Table 1).

The period of December to March remains essentially unsampled although SAC W7074 accumulated 14 hours of dust collection from 8/20/87 through 3/22/88. It is unknown when exactly these hours were accumulated during the lifetime of the collector that sampled the lower stratosphere for 32 hours. In general, the periods of stratospheric exposure vary considerably among the SACs. An extended collection period (U2034; W7074) is conceivably a time-integrated sampling that might conceal several dust-producing events. For the purpose of this paper, *we will assume that there is no change in the flux of IDP subtypes sampled at the collection altitude in the lower stratosphere during the collection period of each collector.*

Table 1
Number of non-spherical chondritic IDPs used in the analysis of subtype distributions, excluding the rough-textured non-aggregate and spherical, chondritic filled particles (see below), and the collection periods and times in the lower stratosphere.

Collector	Collection period	Collection time (hours)	Number of non-spherical chondritic IDPs
W7017	07/07-09/15/81	45	15
W7029	09/15-12/15/81	35	35
W7013	05/22-07/06/81	65	29
W7027	09/15-12/15/81	35	32
U2001	03/13-04/08/82	31.4	15
U2011	03/15-04/01/83	35.1	5
U2015	06/22-08/18/83	39.5	31
U2022	04/09-06/26/84	41.8	38
U2034	04/27-08/28/85	30.7	13
W7074	Aug, 1987-07/01/88	32	11
L2005	10/03-10/13/89	35-40	63
L2006	10/03-10/13/89	35-40	46
L2011	June and July, 1991	35.9	37

Collector Statistics

Number of Particles

Using the complete dataset (N=13, cf. Table 1), the normal distribution of the number of particles on the collectors has a one-sigma (mean +/- 1 stand. dev.) range of 12.2 - 44.7 µm that includes 64% of the dataset. *Ibid*, but excluding SAC L2011, the population (N=12) mean is 30.4 µm (1 stand. dev. = 15.4 µm) and the one-sigma range = 15.1 - 45.7 µm. This range includes 83% of the dataset, but it excludes two of three LACs. We separate the collector types:

(1) SACs (N=9): mean = 24.3 µm (1 stand. dev. = 10.6 µm) and the one-sigma range = 13.7 - 35.0 µm (89% of the dataset),
(2) LACs (N=3)(!), mean = 48.7 µm (1 stand. dev. = 13.2 µm) and the one sigma range = 35.5 - 61.9 µm.

The use of both SACs and LACs begs the question of sampling bias with regard to particle size. We present the particle dimensions the root-mean-square (rms) size, viz. rms = $\{a^2 + b^2\}^{1/2}$, where a and b are two orthogonal dimensions across the particle. The SACs and LACs show similar rms-size ranges between 4-58 μm for all non-spherical chondritic IDPs. About 90% of these IDPs on the SACs have a Gaussian distribution with the mean rms = 17 μm and range (2 stand. dev.) = 5.5-18.5 μm (Fig. 1). On the LACs these IDPs show a skewed distribution with mean rms = 21.5 μm and rms_{mode} = 11 μm (Fig. 1). However, the different shapes of the size distributions are probably an experimental artifact due to curatorial bias whereby large particles are preferentially picked off the LACs.

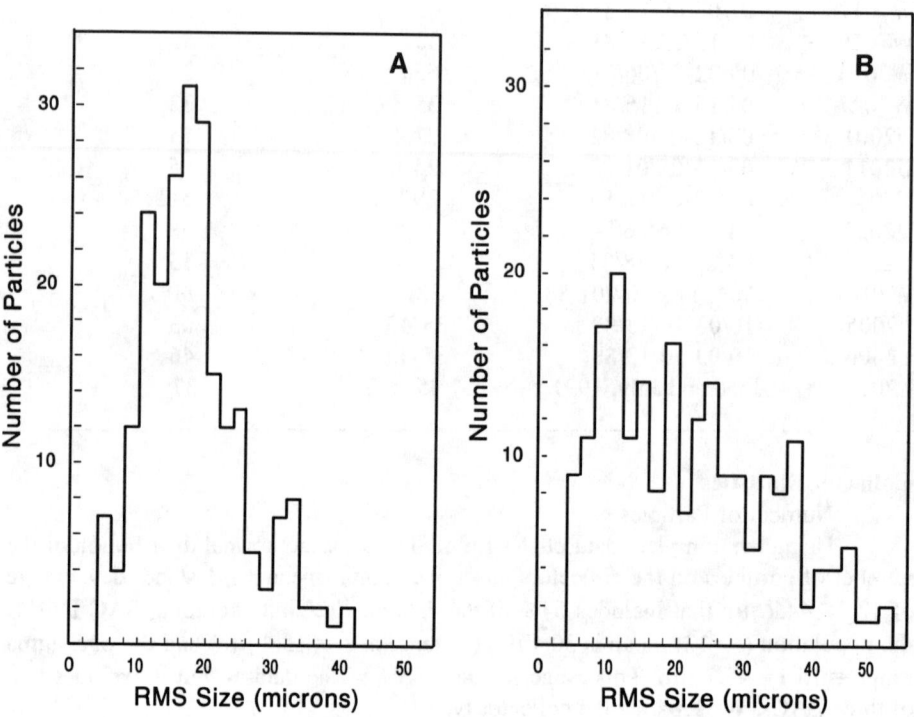

Figure 1. Histogram of rms size distributions for non-spherical chondritic IDPs on the small area collectors (2a) and large area collectors (2b). Two particles, rms size = 54 and 57 μm, were omitted from (2a). Three particles, rms size = 53, 54 and 57 μm, were omitted from (2b).

At this point it is fair to admit that despite the very best efforts, it is hard to pick off particles from a collector surface each time in exactly the same way, although it is believed that any bias will be repeated on the other collectors. Still, to some extent, the variations among IDP subtypes using different collector surfaces will include some degree of bias in the curation procedure. It is believed that this bias for aggregate particles will be negligible as generally all aggregate IDPs are removed from a collector surface. Please note that collector surfaces reported in the CDCs may still contain an unspecified amount of non-spherical chondritic IDPs. We will assume that the remaining particles are a random population with subtype distributions identical to those IDPs published in the CDCs.

Hours of Collection

The actual hours of collection for the L2005 and L2006 are not known exactly but a reasonable value of 37.5 hours will be used in our calculations. Using the complete dataset (Table 1), the population (N=13) mean is 38.6 hours and 1 stand. dev. = 8.9 hours and the one-sigma range of 29.7 - 47.5 hours includes 92% of the dataset. This range excludes SAC W7013 (65 hours).

These statistics highlight an important phenomenon that is already recognized by the Cosmic Dust Collection workers. That is, the collector's position with respect to the pylon that houses these surfaces during periods of non-collection can have a significant effect on the number of particles collected. An offset of 2-3° from perfect alignment will dramatically decrease the number of particles collected onto the surface. Differences in the position of collectors underneath the airplane wings might also affect the amount of particles collected. For example, SAC W7013 contains about the average amount of particles (that is, assuming no curatorial bias) but it has almost twice the average exposure time of dust collection. And, whilst the exposure time for SAC U2011 is similar to the average value, its surface contains only 21% of the average amount of particles. We can not be sure since there is no record of the collector orientation during the sampling flights, but both examples seem to highlight an experimental artifact. However, there is no obvious manner in which this effect could be selective with regard to IDP subtypes. *Whenever a collector shows the presence of different IDP subtypes, the absence of a particular subtype (N=0) is accepted here as a real measure of its lower stratospheric abundance.*

Lengths of Collection Periods

The length of the collection period is known for every collector except L2011, for which we will assume 10 hours similar to the other LACs (Table 1). The collection period for SAC W7074 is exceptionally long compared to the other collector periods. With the exclusion of this collector, the remaining population (N=12) has a normal distribution with mean = 56.6 days (1 stand. dev. = 36.2 days). The one-sigma range, 20.4 - 96.8 days, includes all but the SACs L2011 and U2034. A combination of the actual hours and periods of collection alone shows that all collectors can be used for comparative purposes with the exception of SAC L2011. We will use the ratio of the actual collection hours and the length of the collection period (Table 2) as the coverage factor for each collector.

Table 2
Hours of actual dust collection, collection period and coverage factor (C).

Collector	Collection time (hours)	Collection period (hours)	C (x100)
W7017	45	1,680	2.7
W7029	35	2,184	1.6
W7013	65	1,080	6.0
W7027	35	2,184	1.6
U2001	31.4	624	5.0
U2011	35.1	408	8.6
U2015	39.5	1,368	2.9
U2022	41.8	1,872	2.2
U2034	30.7	2,952	1.0
W7074	32	7,608	0.4
L2005 (*)	35-40	240	15.6
L2006 (*)	35-40	240	15.6
L2011	35.9	240 (**)	14.9

(*): 37.5 hours; (**) 10 days (for both see text).

The coverage factor for the LACs is probably the best attainable goal. The best covered SAC (U2011) is slightly better than 50% of the LAC coverage, whereas for four SACs the coverage is only ~11% of the LAC value. We submit that *the exploration of temporal variations in non-spherical chondritic IDP subtype distributions should take into account the percentage of coverage and collector area.*

For the following subtype distributions, it should be remembered that the data include *individual* IDPs published in the Cosmic Dust Catalogs. That is, when individual IDPs are identified as paired particles, they are counted separately as paired IDPs may show subtle differences in morphology. Not all paired IDPs on the collectors are listed in the catalogs.

DATABASE

Definitions and Data

Each particle in the CDCs is identified by a scanning electron microscope (SEM) photomicrograph and an EDX spectrum. Projected particle dimensions are provided along with its light-optical properties. This paper presents particle dimensions as the root-mean-square (rms) size (cf. section on Number of Particles). For 23 IDPs three-dimensional data are available[5] which show that most IDPs are flattened spheroids with a shape factor, $F = 0.875$ [$F = (b+c)/2a$, where c is a third orthogonal dimension across the particle]. One of us (FJMR) assigned all non-spherical chondritic IDPs (that include both aggregate and non-aggregate particles) in the CDCs to six subtypes, *viz.* CP IDPs, glazed CP IDPs, CF IDPs, glazed CF IDPs, CS IDPs, and CR IDPs. The exterior of many aggregate IDPs in the catalogs are softened, almost glazed. We refer to this clearly identifiable morphology as glazed CP and glazed CF IDPs. Whether the glazed appearance shows melting of a low melting mineral (e.g. Fe-Ni-sulfides) during particle deceleration in the Earth's atmosphere, or whether it expresses compositional effects (e.g. high carbon content, formation of magnetite rims, Zn and He depletion) is yet unknown and requires future study (Don Brownlee, pers. comm., 1993).

Examples of the subtypes are shown in Figs 2a-f. During a one year period, three attempts to fit these IDPs into the subtypes were reproducible, even including 11% of particles that do not readily fit a subtype. About 81% of nonclassifiable IDPs occur on LACs L2005, L2006 and L2011. In the following analysis only non-spherical chondritic IDPs other than CR IDPs will be considered because these particles occur uniquely on the LACs L2005 (N=23), L2006 (N=18) and L2011 (N=13), and SAC U2022 (N=2). The number of particles of each subtype on a collector (n) (Table 3) is normalized to the total amount of non-spherical chondritic IDPs on the collector (N) (Table 1). For comparison purpose the particle percentage, $100n/N$ (Table 3), is normalized to the collector area, A (%. cm^{-2}). Referring to the caveats stated earlier, comparisons of subtype abundances on the collectors also require a parameter to express the collection time, i.e. the coverage factor, C (Table 2) (%. C. cm^{-2}). Thus, the subtype proportions listed in Table 4 will be a rough measure of their concentration in the lower stratosphere during their collection period. The (rounded-off) data from Table 4 are used in Figure 3 to display subtype distributions in the lower stratosphere as a function of collection period during a ten-year effort. This figure provides an easy way to evaluate these subtypes distributions and it emphasizes the uniqueness of CR IDPs (Fig. 3f). It is noted that the subtype distributions are not effected qualitatively should the CR IDPs be included in the calculations.

Simultaneously-Flown Collectors

The SACs W7027 and W7029, as well as LACs L2005 and L2006 were flown simultaneously. In some crude manner the subtype distributions among these collectors might constrain the accuracy of the present endeavor. We note that all four collectors are normal with respect to number of particles on their surface and the exposure factor whilst emphasizing the inherent difference between SACs and LACs. From the available data there is no reason to suspect that these collectors behaved abnormally during dust collection. We conclude that there is therefore no obvious reason why *the relative proportions of IDP subtypes* listed in Table 3 should not accurately represent the dust load of the stratospheric airmass that was sampled by each individual collector (cf. section on Hours of collection). This conclusion presumes there is no curatorial bias (cf. section on Number of particles). We will accept each collector as an individual data point for our analysis of subtype distributions.

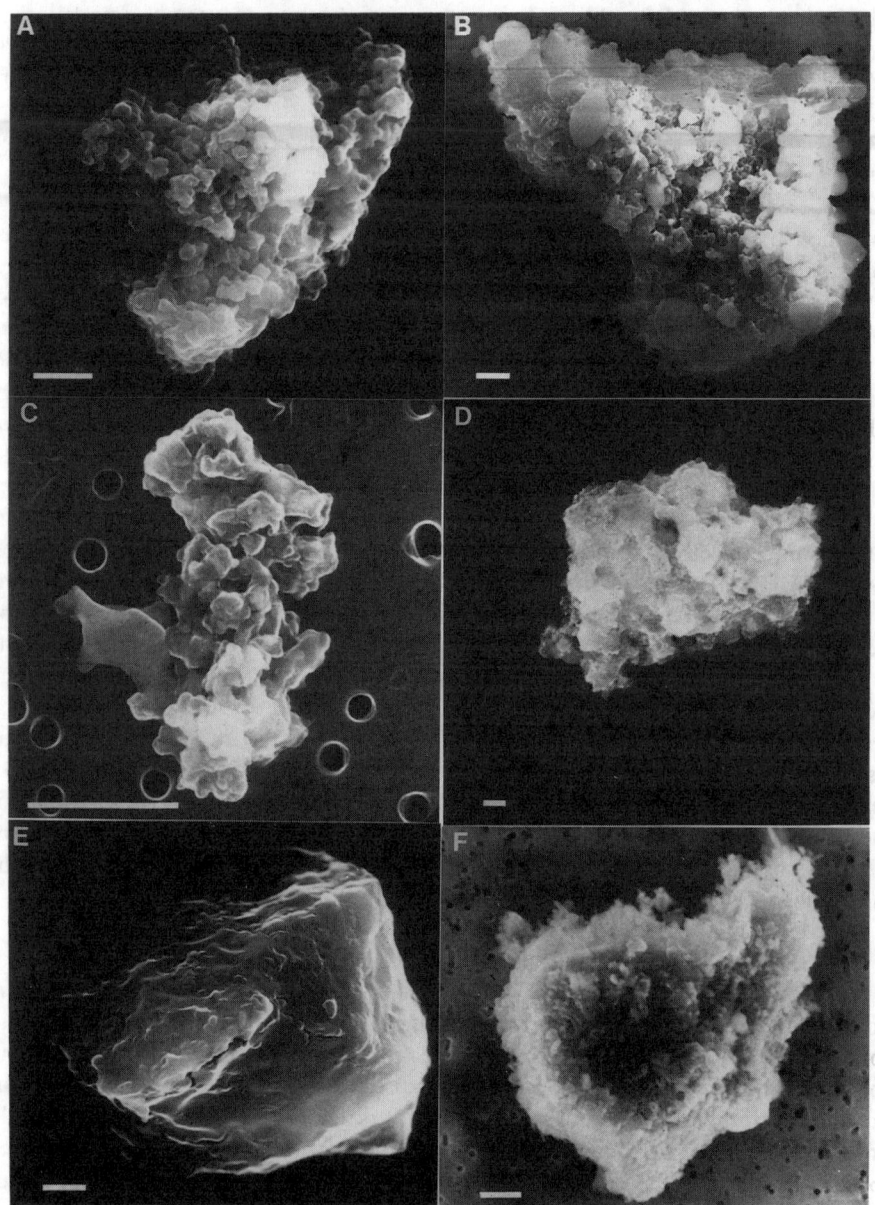

Figure 2: Examples of morphological subtypes of non-spherical chondritic interplanetary dust particles from the NASA Johnson Space Center Cosmic Dust Collection used in this paper: (a) CP IDP (W7017B2), (b) glazed-CP IDP (L2006N4), (c) CF IDP (U2015C17), (d) glazed-CF IDP (U2015F14), (e) CS IDP (W7017B12), and (f) CR IDP (L2011P11). The scale bars measure 2 μm.

Table 3. Number (n) and percentages (*), $100n/N$ (%), of morphological subtypes for non-spherical chondritic Interplanetary Dust Particles

Collector	CP IDPs N	%	glazed CP IDPs N	%	CF IDPs N	%	glazed CF IDPs N	%	CS IDPs N	%
W7017	3	20.0	0	0	4	26.7	1	6.7	7	46.7
W7029	7	20.0	9	25.7	6	17.1	9	25.7	4	11.4
W7013	2	6.9	8	27.6	2	6.9	11	37.9	6	20.7
W7027	2	6.3	5	15.6	1	3.1	13	40.6	11	34.4
U2001	2	13.3	5	33.3	3	20.0	2	13.3	3	20.0
U2011	-	-	5	100.0	-	-	-	-	-	-
U2015	3	9.7	9	29.0	0	0	10	32.3	9	29.0
U2022	2	5.3	12	31.6	4	10.5	11	28.9	9	23.7
U2034	1	7.7	4	30.8	0	0	5	38.5	3	23.1
W7074	1	9.1	5	45.5	0	0	4	36.4	1	9.1
L2005	19	30.2	11	17.5	10	15.9	17	27.0	6	9.5
L2006	22	47.8	11	23.9	5	10.9	6	13.0	2	4.3
L2011	8	21.6	13	35.1	3	8.1	7	18.9	6	16.2

(*) values for the LACs and SAC U2022 are reported exclusive of CR IDPs. The SAC U2011 is an unusual data point (cf. text).

Table 4
Percentage of morphological subtypes for non-spherical chondritic Interplanetary Dust Particles normalized to collector area (A) multiplied by the coverage factor (C):
$[C.100n/N]/A$

Collector	CP IDPs	glazed CP IDPs	CF IDPs	glazed CF IDPs	CS IDPs
W7017	1.81	0.00	2.40	0.59	4.21
W7029	1.07	1.38	0.91	1.38	0.61
W7013	1.38	5.52	1.38	7.56	4.14
W7027	0.34	0.83	0.16	2.16	1.84
U2001	2.21	5.55	3.35	2.20	3.35
U2011	-	28.6	-	-	-
U2015	0.93	2.81	0.00	3.13	2.81
U2022	0.40	2.31	0.77	2.11	1.74
U2034	0.26	1.03	0.00	1.28	0.77
W7074	0.12	0.61	0.00	0.48	0.12
L2005	1.56	0.94	0.78	1.40	0.47
L2006	2.49	1.24	0.62	0.62	0.16
L2011 (*)	1.09	1.87	0.47	0.94	0.78

(*) : see text for coverage factor

Still, there remains the question why individual collectors at different locations underneath the aircraft wings should sometimes show significant variations in IDP subtypes. Are these variations an experimental artifact related to the geometry of the aircraft/collector system and airspeed? Or do they reflect true inhomogeneities in IDP subtype abundances in the lower stratosphere as a function of specific subtype properties and stratospheric processes, such as exchange between the tropo- and stratosphere, stratospheric turbulence and the timing of collections with respect to the spring and fall stratospheric turnaround times? There is obviously a host of parameters that should be investigated before it will be possible to quantify variations in non-spherical chondritic IDP subtypes in the lower stratosphere. In the meantime, it might be recommended to study these variations using average values for as many as possible simultaneously-flown collectors (Table 5). We note that using the data in Table 5 will not make a substantial qualitative difference, which we accept as an indication that our effort may proceed albeit with an understanding of the limitations discussed so far.

TABLE 5. $[C.100n/N]/A$ values for morphological subtypes of non-spherical chondritic IDPs on simultaneously-flown collectors.

Collectors	CP IDPs	glazed CP IDPs	CF IDPs	glazed CF IDPs	CS IDPs
W7027/29	0.71	1.11	0.56	1.75	1.19
L2005/06	1.96	1.05	0.72	1.10	0.38

PARTICLE SUBTYPE DISTRIBUTIONS

Aggregate Particles
CP IDPs

Fluffy aggregates, or CP IDPs[3,6,7] are rare in the collection (Fig. 2a). While there are subtle variations in CP IDP morphology, they are distinct from CF IDPs.[8] The CP IDPs show a Gaussian size distribution with mean rms = 10 µm (Table 6) which includes 86% of all CP IDPs. About 75% of these IDPs are on the LACs. The CP IDP distribution shows a distinct enhancement on the collectors W7017, U2001, L2005 and L2006 (Fig. 3a).

Glazed CP IDPs

Glazed CP IDPs (Fig. 2b) show a bimodal size distribution of two populations with a Gaussian distribution at mean rms = 17.5 and 34.5 µm (Table 6). The first population includes 77% of these IDPs. All glazed CP IDPs on the U2001, U2011, W7029 and W7027 collectors belong to the first population while 83% of these IDPs on W7013, U2015, U2022 and U2034 belong to the first population. They are equally distributed among both populations on the LACs. The highest abundance of glazed CP IDPs in the lower stratosphere occurs on SACs W7013 and U2001 (Fig. 3b). The excessive abundance on SAC U2011 should be taken with caution due to the unusual character of this particular collector.

Table 6
The root-mean-square mean size (microns) and two-sigma standard deviation size range for two populations of non-spherical chondritic aggregate IDP subtypes

IDP Subtype	population 1		population 2	
	mean	range	mean	range
CP IDPs	10.0	3-17	--	--
glazed CP IDPs	17.5	7.5-27.5	34.5	24-45
CF IDPs	13.0	7-19	27.5	21-34
glazed CF IDPs	18.5	8.5-28.5	33.5	28-39
CS IDPs	19.0	10-18	30.5	20-41

CF IDPs

The CF IDPs (and glazed CF IDPs) are a catch-all subtype, *i.e.* non-spherical chondritic aggregate particles that neither have the unique CP morphology[8], nor the distinct CS or CR morphology (Fig. 2c). The CF IDPs include aggregates, smoke-like particles, and rare foliated low-porosity particles. In general, CF IDPs tend to be coarse-grained, coherent aggregates that seem to lack the matrix of granular units typical of CP IDPs.[9&10] They show two Gaussian size distributions at mean rms = 13 μm and a second, poorly-developed, distribution with mean rms = 27.5 μm (Table 6). The first population includes 83% of CP IDPs. But for a single IDP, this population comprises only particles from the collectors W7017, W7013, W7029, W7027, U2011, L2005, L2006 and L2011. About 50% of these IDPs are on the LACs (Fig. 3c).

266 NASA Johnson Space Center Cosmic Dust Collection

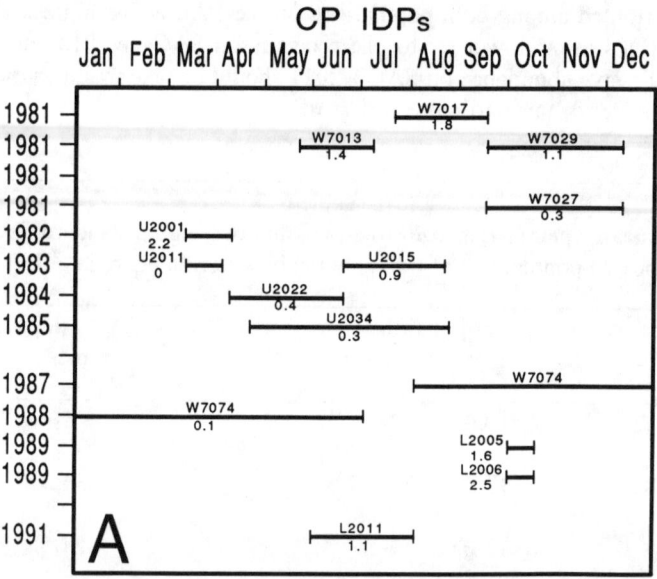

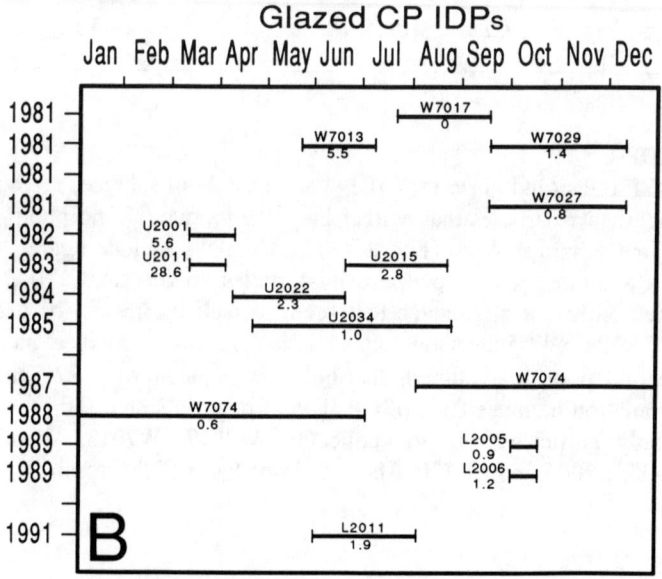

Figure 3: Normalized distributions shown below the bars for non-spherical chondritic interplanetary dust subtypes as a function of collection time in the lower stratosphere: (a) CP IDPs, (b) glazed CP IDPs, (c) CF IDPs, (d) glazed CF IDPs, (e) CS IDPs and (f) CR IDPs.

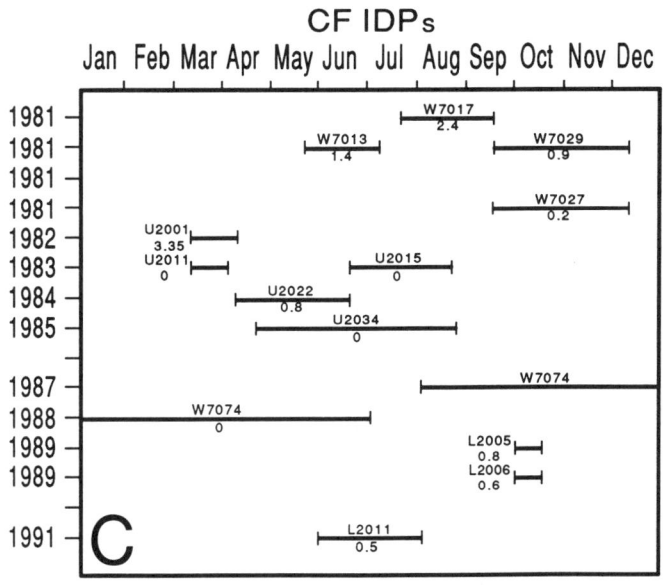

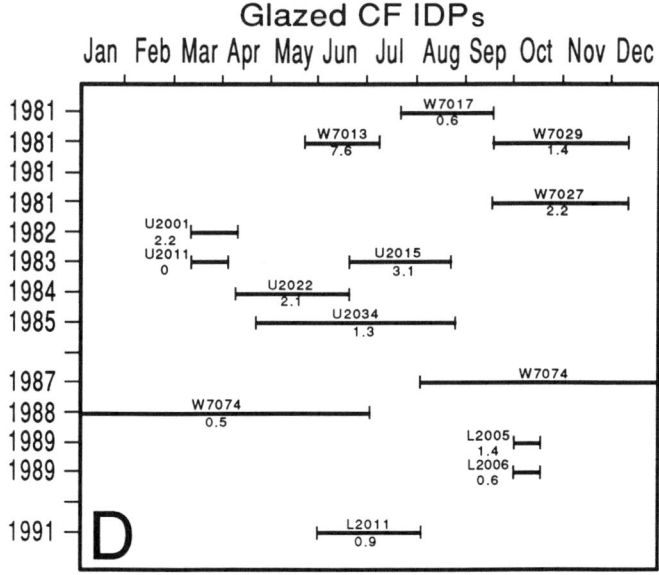

Figure 3 (cont.): Normalized distributions shown below the bars for non-spherical chondritic interplanetary dust subtypes as a function of collection time in the lower stratosphere: (a) CP IDPs, (b) glazed CP IDPs, (c) CF IDPs, (d) glazed CF IDPs, (e) CS IDPs and (f) CR IDPs.

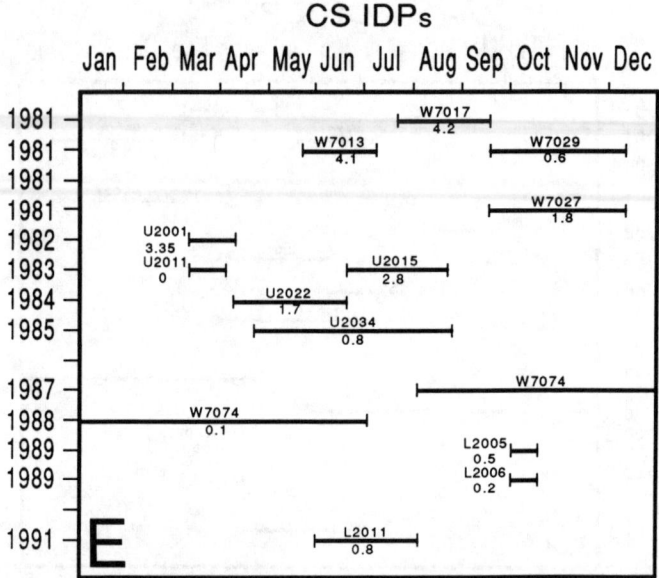
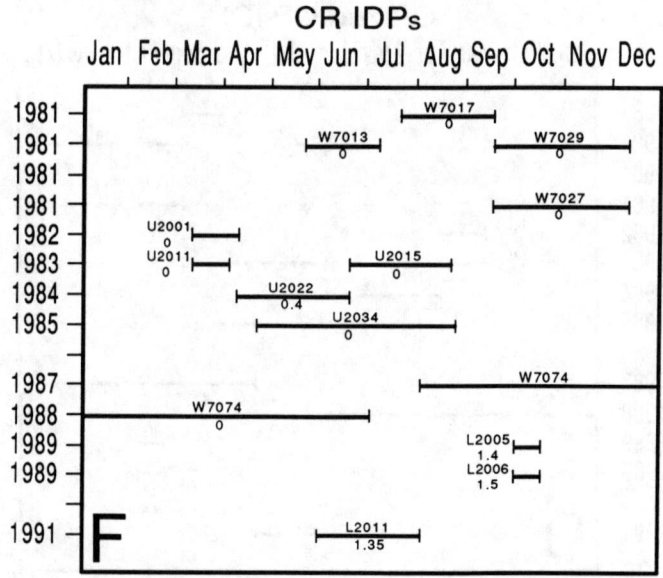

Figure 3 (cont.): Normalized distributions shown below the bars for non-spherical chondritic interplanetary dust subtypes as a function of collection time in the lower stratosphere: (a) CP IDPs, (b) glazed CP IDPs, (c) CF IDPs, (d) glazed CF IDPs, (e) CS IDPs and (f) CR IDPs.

Glazed CF IDPs

The glazed CF IDP (Fig. 2d) (cf. previous section) show two populations with Gaussian size distribution at mean rms = 18.5 and 33.5 µm (Table 6). There are two small glazed CF IDPs on W7027 and U2015, two large IDPs on L2005, and one on L2011. The first population includes 85% of these IDPs. The first population is defined by 90% of these particles on the LACs. Their distributions of as a function of collection time in the lower stratosphere is shown in Fig. 3d.

Non-Aggregate Particles
CS IDPs

These particles generally have a smooth, featureless surface (Fig. 2e). Their rms sizes show a Gaussian distribution with mean rms = 19 µm that comprises 61% of CS IDPs. A second, poorly-developed population has a Gaussian distribution with mean rms = 30.5 µm and incorporates ~30% of these particles (Table 6). The dataset is small but it appears that each collector contains particles from both populations. High amounts of CS IDPs are present on the SACs W7017, W7013, and U2001 (Fig. 3e).

CR IDPs

These particles are morphologically distinct from the aggregate particles and CS IDPs (Fig. 2f). The CR IDPs have a distinctive rough-textured surface. Their rms sizes range from 11.3 to 56.6 µm. Their sizes show a Gaussian distribution with mean rms = 28.7 µm for 95.5% of these IDPs (Table 6). Fifty-four CR IDPs are present on the LACs and two particles occur on SAC U2022 (Table 7).

Table 7
Number of CR IDPs (n) and total number of non-spherical chondritic IDPs (N) on four collectors and calculated abundances, $[c.100n/N]/a$.

Collector	n	(N)	$[C.100n/N]/A$
W2022	2	40	0.37
L2005	23	86	1.39
L2006	18	64	1.46
L2011	13	50	1.35

Sulfur-Free Spherical CF IDPs

While outside the scope of this paper (but for completeness sake), we draw attention to a distinct group of sulfur-free spherical CF IDPs in the NASA/JSC cosmic dust collection. They are only abundant on SAC U2015. There is one spherical CF IDP on SAC U2022 and two of them on LAC L2005. Their rms sizes are between 1.2 and 15.6 μm. They display a skewed, normal distribution with rms_{mean} = 6.2 μm and rms_{mode} = 3 μm (skewness = 0.78). Spherical CF IDPs are among the smallest IDPs on the collectors. They occur individually, in clusters[11&12], and in aggregates wherein they are associated with CP-like material. Their typically low sulfur and zinc abundances are suggestive of reheating of these particles after formation.[12] It was previously suggested that spherical CF IDPs were added to the stratosphere by a single, unusual event within the months preceding the collection on SAC L2015.[12] These unique IDPs reach the lower stratosphere between late-June and mid-August. The high 1983-abundance together with their presence on LAC L2005 supports a 5-6 year periodicity. Recent developments in IDP survival rates as a function of atmospheric entry velocity and particle size[13] suggest that the spherical CF IDPs might be from a high velocity (> 15 km sec^{-1}) meteor stream.

DISCUSSION

General

As indicated elsewhere in this paper there are many reasons why the our exercise might have been doomed from the start. However, despite uncertainties in collection and curation procedures, and the distribution of individual collectors within the first decade of lower stratospheric sampling, we maintain that it is still justified to make a preliminary attempt that explores temporal variations in chondritic IDP subtype distributions. Yet, the conclusions warrant some caution. The distributions of chondritic, mostly non-spherical, IDP subtypes in the lower stratosphere are clearly non random. They potentially contain information on both intra- and inter-annual variations in IDP subtype fluxes. This information linked with models of orbital evolution of dust from different sources within the solar system may contain important clues on the origin and evolution of protoplanets. The unique spherical CF IDPs are an excellent example to indicate where efforts of quantification of temporal variations in the IDP subtypes might be headed. Similarly, size differences among aggregate particles might constrain their different responses to atmospheric entry heating and to stratospheric settling rates. On the average, glazed aggregate IDP are larger than their CP and CF counterparts (Table 6). This difference could result in subtle differences in their mass, or size (cross-section) - density product.[14]

The distinction between CP and CF IDPs is partially artificial. Still, there are at least two types of aggregate particles in the collections.[3&8] The majority of non-spherical chondritic IDPs in the JSC CDC fit one of to six morphological subtypes, viz. CP and glazed CP IDPs, CF and glazed CF IDPs, CS IDPs, and CR IDPs. A few trends emerge from the subtype distributions as a function of collection time in the lower stratosphere. For example, CP IDPs are slightly more common among the aggregate particles. The average ratio of (CP + glazed CP IDPs) to (CF + glazed CF IDPs) is 1.3 (1 stand. dev. = 0.7).

IDP Subtype Distributions

The CR IDPs were uniquely present in the lower stratosphere from early April through mid-October (Fig. 3f) during 1984, 1989, and 1991. The actual character of their inter-annual distributions might have been constrained by stratospheric samplings during 1986 and 1990. These opportunities regrettably did not materialize. Thus, any periodicity in a unique CR IDP-producing event cannot be ascertained but the data seem to exclude their presence prior to 1984. Among the CR IDPs are at least two CM meteorite-like IDPs: L2005T12[15] and L2011O3 [FJMR, unpubl. data]. If indeed the CR IDPs include a sizable amount of type CM meteorite debris, the fraction of CM chondrite-like IDPs among phyllosilicate-rich chondritic IDPs could be much higher than the 10% abundance suggested[16] for this type of dust. Also, their absence prior to 1984 might be significant to constrain their origin. The distributions of the other non-spherical chondritic IDP subtypes is less evident.

The CP IDP distributions indicate three maxima for this subtype in the lower stratosphere during March (U2001), from mid-May to mid-September (W7013 & W7017), and during the first-half of October (L2005 & L2006) (Fig. 3a). One might be tempted to conclude that the maxima were limited to events during 1981/82 and 1989. Yet, the collectors W7029 (1981) and L2011 (1991) indicate that at least two maxima are intra-annual events during a ten-year period rather than represent isolated, interannual events. The effects of these two events in the lower stratosphere peak during late-July/August and first-half of October. The CF IDPs show similar distributions but a maximum in the first-half of October is much evident. A linear correlation (corr. coeff. = +0.62) between the CP and CF subtype abundances on the collectors (cf. Table 4) supports their co-occurrence in events that were sampled in the lower stratosphere.

Glazed CP IDPs show two distinct maxima, viz. (1) during March (U2001), and (2) from mid-May to early-July (W7013). In the absence of adequate lower stratospheric samplings between 1984 and 1991, the data might be interpreted to support abundant glazed CP IDPs from 1981 till 1984 and a recurrence during 1991 (L2011). Any importance of the absence of this subtype on SAC W7017 is unclear. There may be a third maximum from mid-September through October.

The glazed CF IDPs distributions are quite similar to those of glazed CP IDPs as indicated by a linear correlation (corr. coeff. = +0.92) between both subtype abundances on the collectors (cf. Table 4). On the other hand, linear correlations between CP-glazed CP IDPs and CF-glazed CF IDPs show negative correlations, viz. corr. coeff. of -0.11 and -0.52, respectively. Despite the very poor correlations, they support small intra-annual offsets in the maxima of CP and CF aggregate IDPs and their glazed equivalents. Finally, the linear correlation (corr. coeff. = +0.94) between the glazed CF IDP and CS IDP subtype abundances on the collectors (cf. Table 4) support close temporal relationships among the glazed aggregate and chondritic-smooth IDPs in the lower stratosphere.

IDP Stratospheric Residence Time

When all the dust has settled and experimental uncertainties have been addressed, we find that it should be possible to quantify intra- and inter-annual variations in chondritic IDP subtype abundances in the lower stratosphere. As discussed above there is a rich area of stratospheric and tropospheric dynamical processes that is still wide open to the IDP-community. The residence time of individual IDPs in the stratosphere is another important issue that this community will have to address. Both issues are a prerequisite for efforts that link lower-stratospheric subtype maxima to the events that deliver these subtypes to the terminal velocity altitude at ~80 km. From this altitude at ~80 km, IDPs will settle in the Earth's gravitational field. To constrain IDP stratospheric residence times, or settling rates, we rely on analyses of volcanic ash size distributions at ~35 km altitude. This analysis shows that particles with an effective diameter $\{D_{eff} = (a+b+c)/3\}$ smaller than the mean free path of the atmosphere have a negligible stratospheric residence time. They settle according to the Stokes-Cunningham law for spherical aerosol particles.[17] When D_{eff} equals the mean free path of the atmosphere, a particle slows down and settles according to the Wilson-Huang law for non-spherical particles[18] which is a subtle function of particle D_{eff} and density. The changeover altitude for these settling laws is derived from the data of *U.S Standard Atmosphere (1979)*. For IDPs with known size and density[5], the stratospheric residence times are calculated between ~13-70 days.[17] For long residence times, turbulent transport and convective overturn, among others, may seriously affect the stratospheric residency of IDPs. Particle contamination may also become a problem but, more seriously, the link between subtype distributions at ~80 km altitude and the collection altitude (17-19 km) may be lost.

The density of chondritic IDPs is the single most important property that needs to be determined. Measured densities of chondritic IDPs range from 0.4-2.0 g cm^{-3} [5&19], 0.7-2.2 g cm^{-3} [20], and 3.4 g cm^{-3}.[19] Although difficult to ascertain from the published record, these densities seem to cover all subtypes of non-spherical chondritic IDPs. The truly aggregate CP particles[8] have a fractal matrix. They show a simple particle size-density relationship[7] which yields an average CP IDP density of 0.28 g cm^{-3}. The calculated density is comparable to the density in meteor streams, 0.01-1.06 g cm^{-3}.[21] It is slightly lower than the measured densities but still consistent with the experimental data.[20] The CP IDP density of 0.28 g cm^{-3} in the Wilson-Huang equations[18] results in a mean CP IDP stratospheric residence time of 480 days which is significantly longer than those calculated for the non-CP IDPs, viz. ~13-70 days.[17]

Recommendations

In this paper we highlighted parameters that are unknown or ill-constrained in the current collection program. They need to be determined in order to obtain a reliable database on temporal variations of non-spherical chondritic IDP subtypes in the lower stratosphere and to correlate maxima in their abundances with events that deliver IDPs to the Earth's upper atmosphere. Instead of a detailed wish-list, we would like to make a few first-order recommendations:

(1) to fill-in the December-March gap in the collections despite logistical constraints such as aircraft maintenance and re-scheduling,

(2) to deploy LACs for short periods (< 1 month)[22],
(3) to initiate a systematic program to sample the lower stratosphere each month with a high coverage factor during at least two, not necessarily consecutive, years to explore temporal variations in non-spherical chondritic IDP subtype abundances,
(4) to obtain good determinations of particle dimensions, shape and density,
(5) to analyze archived collectors dating back to 1981 to augment the current database of subtype abundances,
(6) to continue chemical and petrologic analyses of non-spherical chondritic IDPs of all subtypes in order to fine-tune the current subtype distributions that are on the basis of morphology alone, and
(7) to obtain detailed subtype characterization for all particles on collectors for which complete particle abundances are available.[23&24]

CONCLUSIONS

Not every investigator will agree with our subtype groupings of non-spherical chondritic IDPs that are on the basis of morphology alone. Our main purpose was to demonstrate that the abundances of non-spherical chondritic IDP subtypes in the lower stratosphere do not display a random distribution. There are indications for systematic intra- and inter-annual maxima in the IDP distributions since May 1981. The maxima for CP and CF IDPs, as well as for glazed CP, glazed CF and CS IDPs, are correlated as a function of time. The best indications for temporal variations during the first decade of interplanetary dust collection are shown by the CR IDP and spherical CF IDP distributions. Although many questions remain, the answers should, and can be, obtained. Amongst these are a more complete database of IDP properties and a better understanding of atmospheric processes that determine the stratospheric residence time of IDPs. This information is needed to correlate the subtype distributions that are sampled in the lower stratosphere with events that deliver IDPs to the Earth's atmosphere. Ultimately, collection times will be combined with the chemical and petrologic properties of the non-spherical chondritic IDPs. With IDP collections proceeding at a more frequent and systematic rate, studies of IDPs collected in the lower stratosphere might substitute for *in situ* samplings of outer belt asteroids and comet nuclei. At this juncture the stratospheric collections and a Cosmic Dust Collection Facility in low-Earth orbit become complementary experiments.[22&25] We appreciate the speculative, possibly naive, nature of our paper. Yet, we hope that our pioneering effort will inspire NASA to support a long-term systematic program of lower stratospheric dust collection. In that manner, this type of collection will not only provide information on the physical, chemical and petrologic properties of the unique ultra-fine IDPs but also reveal the dynamics of solar system processes that produce and transport this dust into Earth-crossing orbits and their sources.

ACKNOWLEDGMENTS

FJMR is supported by a grant from the National Aeronautics and Space Administration NAGW-3626. We appreciate the reviews by Don Brownlee and George Flynn. George has held a long-standing interest in this topic and his critical review clearly improved the final version. Fu Guofei translated the raw data into computer graphics. Fleur Rietmeijer-Engelsman provided technical support at the Electron Microbeam Analysis Facility at UNM.

REFERENCES

1. R.A. Barrett, A.L. Dodson, K.L. Thomas, J.L. Warren, L.A. Watts and M.E. Zolensky, *Cosmic Dust Catalog 13, NASA JSC* **#25980**, 339p (1992).
2. G.J. Flynn, Changes to IDP composition and mineralogy by terrestrial encounters. *This volume* (1994).
3. I.D.R. Mackinnon and F.J.M. Rietmeijer, *Rev. Geophys.* **25**, 1527 (1987).
4. J.P. Bradley, S.A. Sandford and R.M. Walker, In *Meteorites and the Early Solar System* (J.F. Kerridge and M.S. Matthews, eds), pp. 861-898. (Univ. Arizona Press, Tucson, 1988).
5. G.J. Flynn and S.R. Sutton, *Proc. Lunar Planet. Sci.* **21**, 541 (1991).
6. D.E. Brownlee, *Ann. Rev. Earth Planet. Sci.* **13**, 147 (1985).
7. F.J.M. Rietmeijer, *Earth Planet. Sci. Lett.* **117**, 609 (1993).
8. F.J.M. Rietmeijer, On the possibility of petrological classification of carbonaceous chondritic micrometeorites. *This volume* (1994).
9. F.J.M. Rietmeijer, *Trends Mineral.* **1**, 23 (1992).
10. J.P. Bradley, *Geochim. Cosmochim. Acta* **52**, 889 (1988).
11. G.E. Blanford, K.L. Thomas and D.S. McKay, *Meteoritics* **23**, 113 (1988).
12. G.J. Flynn and S.R. Sutton, *Proc. Lunar Planet. Sci. Conf.* **22**, 171 (1992).
13. D.E. Brownlee, D.J. Joswiak, S.G. Love, A.O. Nier, D.J. Schlutter and J.P. Bradley, *Meteoritics* **28**, 332 (1993).
14. F.J.M. Rietmeijer, *J. Volc. Geotherm. Res.* **55**, 69 (1993).
15. F.J.M. Rietmeijer, *Lunar Planet. Sci.* **XXIII**, 1153 (1992).
16. J.P. Bradley and D.E. Brownlee, *Science* **251**, 549 (1991).
17. F.J.M. Rietmeijer, *J. Geophys. Res.-Planets* **98**, 7409 (1993).
18. L. Wilson and T.C. Huang, *Earth Planet. Sci. Lett.* **44**, 311 (1979).
19. M.E. Zolensky, D.J. Lindstrom, K.L. Thomas, R.M. Lindstrom and M.M. Lindstrom, *Lunar Planet. Sci.* **XX**, 1255 (1989).
20. P. Fraundorf, C. Hintz, O. Lowry, K.D. McKeegan and S.A. Sandford, *Lunar Planet. Sci.* **XIII**, 225 (1982).
21. D.W. Hughes, *Eur. Space Agency* **SP-249**, 173 (1986).
22. I.D.R. Mackinnon, In *Trajectory Determinations and Collection of Micrometeoroids on the Space Station.* (F. Horz, ed.), pp. 70-71. LPI Tech. Rpt. 86-05 (Lunar and Planetary Institute, Houston, 1986.
23. M.E. Zolensky, D.S. McKay and L.A. Kaczor, *J. Geophys. Res.* **94(D1)**, 1047 (1989).

24. M.E. Zolensky and I.D.R. Mackinnon, *J. Geophys. Res.* **90(D3)**, 5801 (1985).
25. I.D.R. Mackinnon, In *Trajectory Determinations and Collection of Micrometeoroids on the Space Station.* (F. Horz, ed.), pp. 68-69. LPI Tech. Rpt. 86-05 (Lunar and Planetary Institute, Houston, 1986).

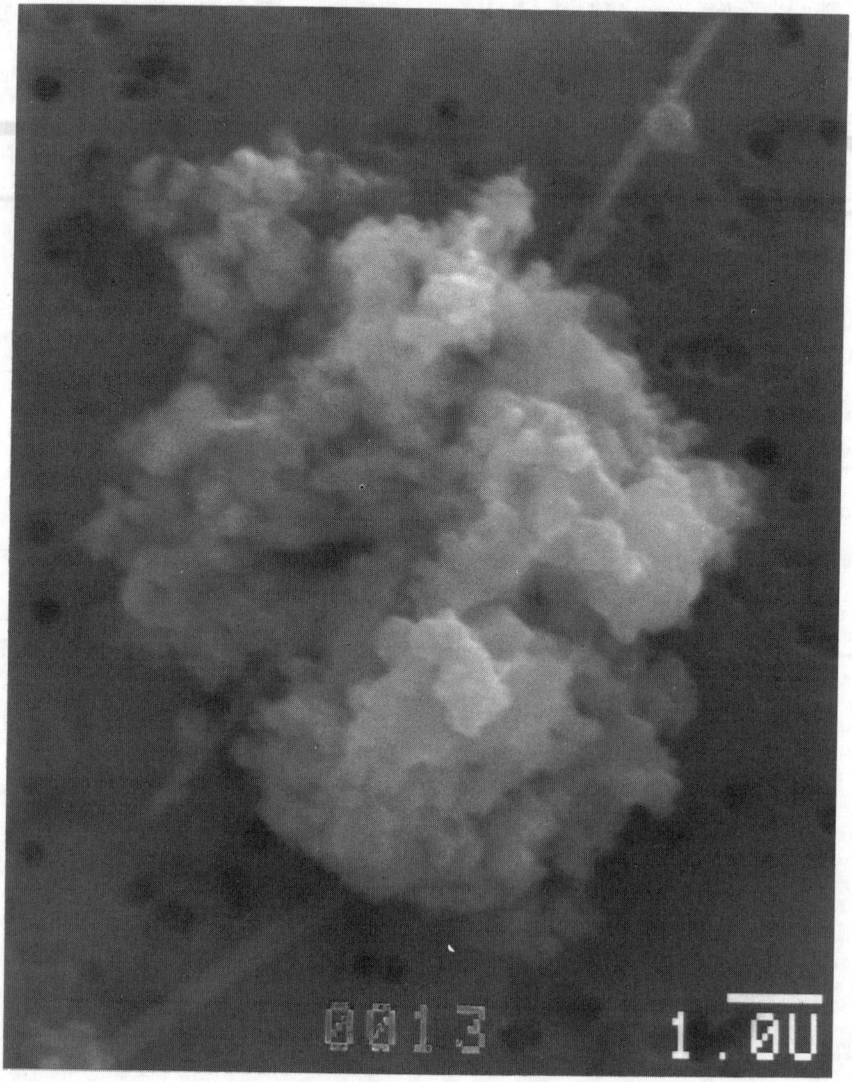

A secondary electron image of a chondritic interplanetary dust particle (#L2005 E4) collected in the stratosphere. This particle exhibits several enstatite crystals with an acicular or filiform morphology, which often results from rapid crystal nucleation and growth along screw dislocations (see the paper by John Bradley in this volume). The scale bar measures 1 μm. (NASA photo S90-38165)

COLLECTION AND CURATION OF IDPs FROM THE GREENLAND AND ANTARCTIC ICE SHEETS

M. Maurette and G. Immel
C.S.N.S.M., Batiment 108, 91405-Orsay, France

C. Hammer
Institute of Geophysics, University of Copenhagen, Haraldsgade 6
DK-2200, Copenhagen, Denmark

R. Harvey
Department of Geological Sciences, University of Tennessee,
Knoxville, TN 37996, USA

G. Kurat
Mineralogisch -Petrographische Abteilung, Naturhistorisches Museum
Postfach 417, A-1014 Wien, Austria

and
S. Taylor
CRREL, Hanover, NH 03755, USA

ABSTRACT

Upon melting or sublimating Greenland and Antarctica ice yields a sandy material which is very rich (up to ~10% by weight in the 50-100 µm size fraction) in unmelted and partially melted micrometeorites. These polar micrometeorites are remarkably unweathered despite having lost unknown amounts of soluble sulfates and carbonates and having "accreted" some trace elements during their settling time in the Earth's atmosphere. Although they mainly bear similarities with the rare class of CM and CR chondrites, some of the characteristics of their primary minerals strongly suggests that they are composed of a material not yet represented in meteorite collections. This paper describes improved methods for collecting micrometeorites in Greenland and Antarctica, and ways to study the past variations of the micrometeorite flux over a time scale > 200,000 years.

INTRODUCTION

In another paper in this volume, Warren and Zolensky[1] discuss interplanetary dust particles (IDPs) collected in the stratosphere. Here we describe the recovery of much larger unmelted to partially melted micrometeorites (MMs) from the Greenland and Antarctica ice sheet ("polar" MMs), and discuss problems arising in their collection and curation. The collection of MMs in other terrestrial environments such as deep sea sediments has been well covered by Brownlee[2], and will not be reviewed here.

Stratospheric IDPs and polar MMs are altered by "space weathering" (i.e. solar wind and UV irradiations, grains collisions, etc.) in the interplanetary medium, frictional heating upon atmospheric entry, atmospheric weathering (i.e. "scavenging" of various species of both the upper and lower atmosphere) as they settle in the atmosphere, terrestrial weathering in their host sediments ("hydrocryogenic" alteration in the ice). Collection procedures also alter the types of IDPs and MMs recovered. Comparison of particles collected from different environments and of a wide size range, is important for identifying those particles characteristics, that are independent of terrestrial processes.[3&4]

Polar micrometeorites, which are remarkably well transmitted throughout the atmosphere[3&5], are similar to the rare meteorite classes CM and CR. However, some features of MMs set them apart from both CM and CR chondrites.[5-9] The mineralogy of hydrous phases is not yet well characterized, but both serpentine and smectite seem to be present, as well as cronstedtite, ferro-brucite, and tochilinite. The low olivine/pyroxene ratio of MMs differs from that of CM chondrites but fits that of CR chondrites and of the unique chondritic breccia Kaidun, which contains CI, CM, CR, and enstatite chondrite lithologies.[10] The chemical compositions of olivines and pyroxenes in MMs are within the range of mineral compositions from CM and CR chondrites but the very low-Fe, refractory element-rich compositions are apparently missing. Also, the bulk chemical compositions of MMs differ from that of carbonaceous chondrites. These differences include: a depletion in Ca, Ni, Co and S, and sometimes also in Mg and Mn[5&11]; a higher content of carbonaceous material[12]; an enrichment in K, Br, Au, As and Fe.[13&14] The mineralogical and mineral chemical features of polar MMs are most probably of primary origin, and clearly set them apart from the CM chondrites. But some of their bulk chemical features appear to be likely due to a terrestrial origin, including the elemental depletions, which are probably related to the dissolution of their constituent Ca-Mg-Mn carbonates and Mg-Ni-Co sulfates[7,11,13], and some elemental enrichments, which probably reflect some form of weathering and/or condensation during settling in the atmosphere[15] (see also the paper by Flynn[16] in this volume for additional discussion of this controversial topic), and weathering in the ice. All these investigations strongly suggest that polar type MMs of all sizes are a common type of matter falling on the Earth today, and are not represented in meteorite collections.

Polar ices may be the most favorable collectors of large MMs. The ice is clean, and can be easily removed by melting. Furthermore, particles imbedded in ice are shielded from alteration processes operating on the ice surface, and their hydrocryogenic alteration is kept to a low value, as illustrated by the lack of detectable etch canals in chondritic barred spherules.[17] The major problem with ice is the low concentration of MMs. This concentration is computed assuming that the micrometeorite flux measured at ~1 AU in the interplanetary medium[18] is the same as that on the Earth's surface (100% transmission efficiency), and that ice flow models properly determine where the host ice formed, and thus, its accumulation rate, $V(cm/y)$. At the sites of our Greenland and Antarctic expeditions, $V \sim 50$ cm/y and

~20 cm/y, respectively, and would yield ~2 and ~5 micrometeorites (diameter > 100 µm) per ton of ice, respectively. At these low concentrations, a huge amount of ice has to be melted to get a reasonable number of MMs. One has thus to find methods to melt these amounts of ice in the field, or to discover zones where local processes would dramatically increase the concentration of MMs in the ice. Such "accumulation" processes fortunately operate on both the Greenland and Antarctica ice sheets.

COLLECTING MICROMETEORITES ON THE MELT ZONE OF THE GREENLAND ICE SHEET

In the melt zone of the west Greenland ice cap, small seasonal lakes are formed each year during the short Arctic summer. They are fed by the melting of a ~1m-thick layer of ice, over collection basins which have diameters of a few km. The amount of melt ice water flowing through a lake can reach ~10^8 tons. Aerial photographs showed dark patches of sediments (called "cryoconite") on the lake bottom and we hypothesized that as the melt water flowed through the lake, micrometeorites and terrestrial dust would settle to and be concentrated on the lake bottom.

If the lakes form at about the same locations each year (not an unreasonable assumption as the ice topography reflects the much steeper and invariant "hill/valley" structure of the basement rocks on which the ice flows), extraterrestrial particles may have been accumulating at the same site since the formation of the ice field. At the site of our first expedition (Blue lake 1), at ~20 km from the margin of the Sondrestromfjord ice field (latitude of 67°08' N), ice flow models indicated that the ice surface was about ~2000 years old. If our model is correct, in excess of 10^{10} tons of melt water have flowed through Blue Lake 1.

In July 1984, we used a small hand water pump to vacuum-up cryoconite first from a lake bottom (~10 kg), and then ~50 kg at other types of sites, including the bottom of the ~20 cm deep "cryoconite" holes that constitute most of the ice field surface. We returned to Greenland in July and August 1987 to sample cryoconite at 30 different locations between ~2 and 50 km from the margin of the same ice field, with the view of tracing back the past activity of the micrometeorite flux. In August 1984, 1985, 1987 and 1988, colleagues collected several ~1 kg sample of cryoconite at higher latitudes including: two sites at the Jakobshavn ice field (69°20' N; ~10 and ~25 km from margin), and; two sites near de Quervains Harbor (69°44' N; ~5 and ~9 km from margin). The expeditions yielded ~200 kg, of wet cryoconite from about 50 different samples.

The cryoconite at all locations is essentially composed of "cocoons" of filamentary siderobacteria[17] that tightly encapsulate a fine-grained mineral sand, predominantly terrestrial in nature but containing a minor component of melted (cosmic spherules) to unmelted micrometeorites. The >100 µm size fraction (and occasionally the 50-100 µm fraction) from ~100 g aliquots of the different samples was disaggregated using a stainless steel sieve and a hard nylon brush. Less than 10 g of mineral particles ("sand") per kg of wet cryoconite was recovered. The concentration of cosmic spherules in both this sand and its host cryoconite leads to the following results:

(a) About 800 cosmic spherules and ~200 unmelted micrometeorites >100 μm were found per kg of (wet) cryoconite. This figure is independent of the type of deposit, the latitude of the ice field, and its distance to the margin. As argued elsewhere[3], this "saturation" concentration corresponds to a model where cryoconite holes act as MMs collectors; they appear near the limit of the melt zone, at about 50 km from the margin, and move with the ice over a "lifetime" of ~250 y, until they are destroyed by events such as flooding and the opening of a crevasse.

(b) In contrast, the concentration of cosmic spherules, C_{cs}, measured *in the residual "sand" extracted from cryoconite*, is dependent on the location of the collection site. This trend reflects sharp variations in the abundance of terrestrial grains deposited by the winds. Thus, C_{cs} scales the "purity" of the sand in extraterrestrial grains. On a given ice field, C_{cs} decreases toward the margin, mainly because the amount of wind-borne dust increases. On the other hand C_{cs} dramatically increases if the ice field terminates at the ocean, because the winds have a smaller sediment load. Thus, the most favorable ice fields are those which terminate on the coast (Jakobshavn and de Quervain Harbor), and not 30 km inland (Sondrestromfjord).

(c) Although the absolute number of large spherules in cryoconite sharply decreases with increasing sizes (as expected from the mass distribution of the micrometeorite flux), their concentration in the "sand" markedly increases, and reaches a value >0.1 in the >300 μm fraction. This is the result of a much sharper drop in the number of large wind borne terrestrial particles. In contrast at all sites, with the exception of the de Quervains Harbor ice field, the <100 μm size fraction can hardly be used ($C_{cs} < 10^3$).

These trends in the "purity" of the cosmic dust component indicate that wind-borne dust from ice free lands is the dominating contaminant. This "size fraction" trend is just the opposite on the Antarctic ice sheet where coarse moraine debris is the major terrestrial contaminant, and the coarser size fraction is now heavily contaminated. *The Greenland and the Antarctica collections complement one another in this regard.*

Although it is easy to recover a large number of cosmic spherules and micrometeorites from cryoconite, two factors presently limit the representiveness and usefulness of the Greenland cryoconite collection for MMs <300 μm:

(a) The harsh mechanical disaggregation procedure used to break down cryoconite may destroy the most friable grains; these are the least thermally altered and possibly the most interesting MMs (the partially melted scoria type chondritic particles, that have also been extracted from deep sea sediments by magnetic raking, are much less friable). The present collection of Greenland cosmic dust

particles is thus biased toward cosmic spherules, scoria-type particles, and crystalline MMs;
(b) During their metabolism, the cryoconite siderobacteria release colloidal forms of poorly defined "hydrous" iron oxides, that infiltrate porous grains. Such colloids act very efficiently as "ion exchange resins" and by adsorbing trace elements from the melt water they contaminate the porous grains. Robin[19] showed that all porous Greenland MMs have a terrestrial REE pattern, while the crystalline MMs exhibit the typical chondritic pattern, suggesting that the least melted grains have been the most severely altered by these siderobacteria.

Despite these two problems related to cryoconite, there are still bright prospects in using the Greenland ice sheet to collect micrometeorites. First, if cryoconite-free ice fields are found in North Greenland (see the section below on "Future Prospects"), then they will probably be the best collector of "giant" MMs with size >400 µm. Moreover, cryoconite can also be used to probe the past activity of the MMs flux. Indeed the ice flow model of Neels Reeh suggests that a 10,000 year record of the types of extraterrestrial materials reaching the Earth is preserved in the Sondrestromfjord ice field. Possible changes in the composition of spherules, crystalline grains, and scoria type particles trapped in cryoconite could be detected by taking the ~30 cryoconite samples that we collected between ~2 and 50 km from the margin of the Sondrestromfjord ice field. Finally, a method, based on one already developed by geochemists to remove iron hydroxides from "rusty" clays, should be developed to remove such hydroxides from porous micrometeorites extracted from cryoconite.

COLLECTING MICROMETEORITES IN ANTARCTICA

The experience gained in the collection of Greenland micrometeorites, lead one of us (M.M.) to sample the cryoconite-free blue ice fields of Cap-Prudhomme, near the French Dumont d'Urville Station, Antarctica, with the logistical and financial support of the Institut Français pour la Recherche et la Technique Polaire (IFRTP).

A preliminary exploration took place from December 1987-January 1988 in collaboration with Michel Pourchet from Laboratoire de Glaciologie et de Géophysique de l'Environnement, Grenoble (LGGE). The idea was to melt ~100 tons of blue ice. Pockets of melt water would be formed by injecting a jet of 70°C water delivered by a steam generator into a ~2 m deep drill hole. Micrometeorites initially trapped in the ice would be released into the water which would be re-injected into the steam generator by an immersion pump placed at a shallow depth in the hole. The meltwater would be filtered using stainless steel sieves. With ~$10,000 we purchased two used steam generators, two used immersion pumps, one water pump, one used electrical generator, the necessary plastic tubing and stainless steel sieves (with openings of 50, 100, and 400 µm), and we left for Antarctica with this primitive ice melter.

After one week of unsuccessful attempts, we found an area where the ice was not too contaminated with moraine debris or full of open microcracks, through which water could be lost. In this ~2000 m² blue ice area we melted ~100 tons of ice. This was done by melting two, 2 to 3 m³ water pockets a day for 20 days.

The field observation of cosmic spherules in the glacial sand recovered in the sieves revealed a sharp decrease in terrestrial contamination with decreasing size fraction, leading to an amazingly high "purity" of the 50-100 µm size fraction (the opposite trend is observed in Greenland). Later work in the laboratory revealed that in this size fraction the concentrations of MMs was very high (~0.1), and the ratio of unmelted to melted MMs was unexpectedly large (>5). This last trend is contrary to what is predicted by atmospheric entry models, which estimate that >99% of MMs with sizes ~100 µm should be completely melted.[2] Although micrometeorites > 400 µm are still found, they occur in much smaller numbers. This reflects both an increase in contamination by moraine debris, a decreasing flux of larger MMs in the interplanetary medium, and a stronger frictional heating in the atmosphere. The total number of micrometeorites recovered from the 100 tons of ice exceeded 20,000.

In December 1990 and January 1991, we returned to the same area with Michel Pourchet, to collect MMs for EUROMET (the European Meteorite Consortium). About 260 tons of ice were melted over 25 days of field operation, and an additional sieve size was included to collect MMs in the 25-50 µm size fraction. We wanted to make a connection between the Antarctic and the stratospheric collections of interplanetary dust. The three years between the two expeditions allowed the ice sampled in 1987-1988, which has been punctured with ~40 holes and heavily polluted with our activities, to flow toward the sea and be replaced by fresh, ultraclean ice (the ice in this area moves at about 10 m a year and was formed in pre-industrial times ~50,000 years ago). We improved the EUROMET ice melter, by using 3 new steam generators, new pumps and fittings. We also hired an expert cartographer from LGGE (Christian Vincent), to both measure the ice flow at and around the collection site, and investigate aerial photographs. We hope to learn how to identify rare and favorable areas such as this one from these studies.

In 1987, Gunter Faure and Christian Koeberl examined samples of neogene-aged tills from the Walcott Névé area and found that they contained abundant cosmic spherules.[20] During the subsequent Antarctic Search for Meteorites (ANSMET) expedition, one of us (R.H.) collected several kg of similar samples (surficial aeolian and moraine debris) from various locations near local blue-ice meteorite stranding surfaces. High concentrations (>20 per g) of large (>250 µm) cosmic spherules were found in all samples. These deposits have probably formed when materials exposed by the ablation of the blue ice are moved by strong katabatic winds to local aeolian traps. Harvey and Maurette[21] searched these sediments for possible MMs, but had little success. In one favorable sample, more than a thousand >100 µm size spherules were recovered, while only half a dozen possible MMs were located.

This very small ratio of MMs to cosmic spherules (at least 100 times smaller than at Cap-Prudhomme), which was quite unexpected, makes the recovery of micrometeorites from these aeolian deposits very difficult and inefficient. This low ratio of unmelted to melted micrometeorites might be related to the abundance of coal in the samples, that might camouflage any dark irregular MMs. In addition, sufficient weathering may have occurred to disaggregate fragile MMs, moving their fragments to a smaller size fraction (<50 µm), which has not yet been carefully investigated.

The value of these sediments lies in their high concentration of >500 µm cosmic spherules and "mini-meteorites" (which can be directly obtained from R.H.), and in the relatively unweathered condition of the recovered specimens (as compared to those recovered from deep sea sediments). Similar samples are collected annually by ANSMET expeditions in the hopes of finding more unusual samples and perhaps higher concentrations of MMs.

To identify blue ice fields having the highest concentration of micrometeorites and to understand how these concentrations change as a function of depth within the ice (possibly modulated by past climatic conditions), we studied ~50 kg blocks of blue ice from Cap-Prudhomme and the Queen Alexandra range. These blocks were collected by D. Barnolla (LGGE), in January 1987, and by the ANSMET field party in 1990 and were cut with a wire saw. In 1987, the number of >50 µm size chondritic cosmic spherules in the Cap-Prudhomme block were counted at LGGE[22] to ensure that our December 1987 collection at this site would be fruitful. In 1992, with James Cragin from CRREL (Cold Regions Research and Engineering Laboratory), we measured the concentration of MMs and cosmic spherules with sizes >35 µm within the Queen Alexandra block[23], and we are presently determining the depth profiles of the concentration of various trace elements and/or micron-size aerosols in the same block.

So far, the striking result is that the concentration of >50 µm size chondritic cosmic spherules (~100 per ton) is very similar in these two very different ice fields, and about 20 times larger than the background level (~5 per ton). This background was determined by melting a ~2 m^3 pocket of melt water at a depth of ~5 meters at the same location where the ice block was taken at Cap-Prudhomme. Thus an efficient mechanism, still to be understood, markedly increases the concentration of melted and unmelted micrometeorites in the top layers of these two very different ice fields.

Some analyses of polar MMs point out limitations that we hope to overcome during our next expedition. These include: the lower S, Ca and Ni contents of MMs recovered from the ice, which are probably related to the dissolution of highly soluble carbonates and sulfates in melt water; the corrosion of the steam generator pipes (made of ordinary steel, as no stainless steel and/or copper pipe was available), that released a large amount of fine-grained rust, which heavily polluted the 25-50 µm fraction. Contaminant trace elements such as lead and uranium were also detected in some of the grains.[11&24]

CURATION OF POLAR MICROMETEORITES FOR EUROMET

The present curatorial facilities at C.S.N.S.M. have to be improved in order to store the EUROMET collection of polar micrometeorites properly, which includes:

(a) ~50 distinct samples of Greenland wet cryoconite, collected from July 1987 to August 1988, stored into polyethylene bags, and representing a total mass of ~200 kg.
(b) >40,000 micrometeorites from Cap-Prudhomme, Antarctica, distributed in 4 size fractions (25-50, 50-100, 100-400, and >400 µm), but mostly found in the 50-100 µm size fraction. This is by far the purest and the richest sample of MMs ever extracted from terrestrial sediments.

The ~60 g of Cap-Prudhomme sand extracted from 360 tons of ice, have been encapsulated in about 250 vials (made of either glass or polyethylene). There is a very sharp drop in the numbers of MMs with increasing size, as we only collected 10 partially melted MMs, and 93 cosmic spherules with size of >400 µm, from 260 tons of melt ice water. We also clearly showed[25] that it is feasible to collect a very pure 25-50 µm IDPs size fraction, at the condition of eliminating the rust grains.

Currently the "EUROMET" collection of vials and plastic bags is stored in a freezer. We need to repair and modernize a small dust free room, which will be used to store, handle, and preserve this collection. We have no SEM equipped with a fast EDX system (the basic instrument for curatorial work on micrometeorites). Fortunately, since 1989 we have been able to use the new SEM at the Naturhistorisches Museum in Vienna, which we utilize about 20 weeks a year. Very recently we received funding to improve the rapidity of the SEM analyses, by acquiring a faster automated analysis system.

Past problems have made us reluctant to distribute samples of bulk cryoconite, with just a note explaining how to extract MMs from these sediments. Instead, we prefer to distribute ~500 mg aliquots of the >100 µm size sand material extracted from ~100 g of wet cryoconite. These sand aliquots contain about 80 spheres and 20 micrometeorites of the scoria and crystalline types.

The major characteristic of the 1991 EUROMET collection of MMs, are outlined elsewhere.[26] To receive a few mg of the 50-100 mg sand, containing a few hundred MMs, or an aliquot of twenty MMs from the 100-400 µm size fraction, a one page proposal should be sent to Michel Maurette (fax: 0033 1 69 41 50 08).

Both the 25-50 µm fraction, and the few "giant" micrometeorites from the >400 µm size fraction are given only to groups with the best expertise in either the handling of stratospheric IDPs, or in the multidisciplinary microanalyses of >400 µm size grains. The 10 "giant" MMs recovered from 260 tons of melt ice water have not been allocated, but any short proposal will be kindly considered by EUROMET. During our forthcoming expedition at Cap-Prudhomme we hope to collect a very pure 25-50 µm size fraction. If so, we plan to lend about 1/3 of this material to the Curatorial Facility at NASA Johnson Space Center, which has developed considerable expertise in handling and curating micrometeorites in this size range.

FUTURE PROSPECTS : FROM THE HANS TAUSEN PROJECT IN GREENLAND TO MELTING ICE WATER AT THE SOUTH POLE AND DOME C

As part of the international Hans Tausen project in Greenland, Claus Hammer is organizing in 1995 a glaciological expedition to a high northern ice cap, which is considered to be extremely sensitive to climatic variations. At this northern latitude the Arctic summer is too brief to support the growth of siderobacteria, which make the bulk of cryoconite. Melt water, however, is still flowing right at the shallow margins of the ice cap. One of our major goal is to filter this water, and collect a cryoconite-free sand, in which we hope to find "mini-meteorites" >1 mm. Using ~10 stacks of sieves, about 100,000 tons of melt water could be filtered in one month. Simultaneously, we shall look for meteorites near the margin of the ice field.

In 1992, one of us (M.M.) received funding from IFRTP, to develop an all stainless steel ice melter. Besides some teflon pipes all components of this melter in contact with the ice or melt water (pipes of the steam generators, hot water pipes, immersion pumps, ice coring device, etc.) will be made of the same stainless steel used in the water pipes of French nuclear reactors. This steel is ranked as one of the best with regard to corrosion resistance in water. Thanks to the cooperation of IN2P3, one of us (G.I.) was authorized to work full time on this melter, which has been built and was transported to Dumont d'Urville in December 1992. It will be used during the next Antarctic summer, to obtain a new EUROMET collection of micrometeorites. These samples should be free of rust particles and of plasticisers released from the plastic tubing. They should yield a very pure 25-50 µm size fraction. We shall also attempt to collect a 10-25 µm size fraction.

Françoise Yiou, Grant Raisbeck and Célestine Jéhanno have shown that >50 µm size MMs and cosmic spherules (about 1 per ~10 kg of ice) can be extracted from Antarctica ice cores.[27] This difficult work has to be encouraged in the future, as it gives the exciting prospect of detecting possible changes in micrometeorite flux and compositions over a time period >200,000 years.

Moreover, we recently discovered that the external surface of cosmic spherules exposed for much shorter time to melt water show micron-size deposits (Figure 1), mainly composed of K and Ca-rich sulfates. These deposits might represent the "dry" residues of complex mixtures of stratospheric aerosols scavenged by the spherules during their settling time in the atmosphere, being thus related to the sulfate nanocrystals already observed by Rietmeijer[28] in a much smaller ~10 µm size IDP (W7029 E5). Such "sulfate-coated" cosmic spherules appropriately recovered from ice cores might also help in the investigation of the past "activity" of such aerosols on a similar time period.

We are planning to build for Ralph Harvey a "mini-melter" (a smaller version of the EUROMET melter, fixed on a Nansen sledge) for the collection of MMs from remote meteorite stranding surfaces during future ANSMET seasons. This will permit the investigation of the micrometeorite content of ice that is relatively free of terrestrial debris, as well as the collection of MMs from ice of different ages. In addition the meteorite and MMs content of a parcel of ice can be compared.

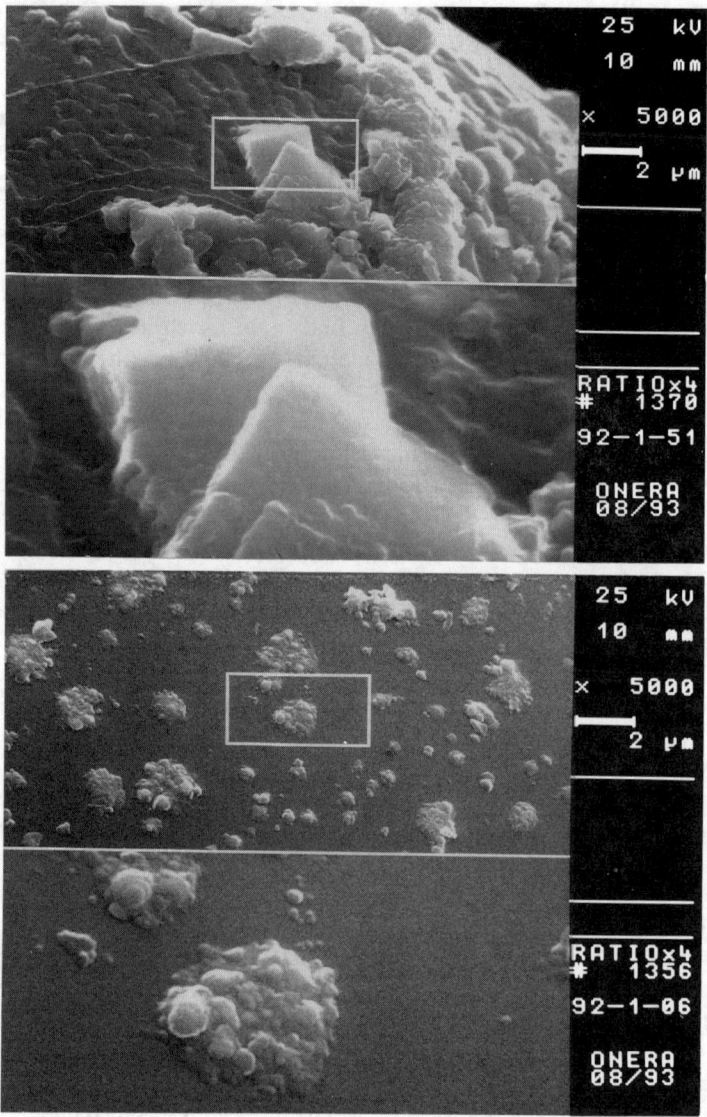

Figure 1. SEM micrographs showing the external surfaces of a chondritic "barred" spherule (top), and a chondritic vesicular glassy spherule (bottom). These cosmic spherules have been found in a ~50 kg block of blue ice recovered from the Queen Alexandra range (Antarctica). This block was melted in about 30 minutes with a microwave oven in order to reduce the exposure of the grains to melt ice water (still at 0°C). Both spherules are sprinkled with bright sulfate-rich deposits (Courtesy of Daniel Boivin, Division des Matériaux, ONERA).

Susan Taylor and Ralph Harvey have a plan to retrieve MMs from the bottom of the new water well at the United States South Pole Station. Over the lifetime of the well, 5 to 10 years, a cylinder 15 m in diameter and 100 m deep will be melted in the clean firn and ice and thousands of MMs, released from the ice, can be collected on the bottom. This is equivalent to sampling a core 15 m in diameter. As the age of the melted ice is well known, by sampling yearly it will be possible to detect changes in the type and flux of MMs as a function of time. We also hope to exploit the small surface snow melter which will be built at the future Italian-French station "Concordia", to be constructed at Dome C.

Finally Claus Hammer and Michel Maurette are considering drilling a ~100 m deep core in Greenland specifically dedicated to a search for both the "ashes" of the 1908 Tunguska event, and the "residues" (deposited on cosmic spherules) of stratospheric aerosols from the major volcanic eruptions observed since 1908.

ACKNOWLEDGMENTS

Our two first expeditions in Greenland were made possible by grants from INSU in France, and the Commission for Scientific Research in Greenland. The two expeditions at Cap-Prudhomme have been entirely funded and logistically supported by IFRTP (Institut Français pour la Recherche et la Technique Polaire). One of us (MM) also acknowledge both the EEC Program "SCIENCE (Twinning and operations)" (Contract SC1-CT91-0618 SSMA), the support of IN2P3, the invaluable technical help of Michel Pourchet, and the generous scientific cooperation of D.E. Brownlee, M. Christophe Michel-Levy, C.T. Pillinger and R.M. Walker. R. Harvey received support from NSF (grant DPP 8314496) for field work and analytical studies. The analytical work of G. Kurat is funded in Austria by contract P8125-GEO of FWF

REFERENCES

1. J. Warren and M.E. Zolensky, Collection and curation of interplanetary dust particles recovered from the stratosphere by NASA, this volume (1994).
2. D.E. Brownlee, Cosmic dust: Collection and Research, *Ann. Rev. Earth Planet. Sci.* **13**, 147 (1985).
3. M. Maurette, C. Hammer, M. Pourchet, Multidisciplinary investigations of new collections of Greenland and Antarctica micrometeorites, in *From Mantle to Meteorites*, eds K. Gopalan, V.K. Gaur, B.L. Somayajulu (Indian Academy of Sciences, Bangalore, 1989).
4. S. Taylor, D.E. Brownlee, Cosmic spherules in the geologic record, *Meteoritics* **26**, 203 (1991).
5. M. Maurette, C. Olinger, M. Christophe, G. Kurat, M. Pourchet, F. Brandstatter, M. Bourot-Denise, A collection of diverse micrometeorites recovered from 100 tonnes of Antarctic blue ice, *Nature* **351**, 44 (1991).
6. I.M. Steele, Olivine in Antarctic micrometeorites: comparison with other extraterrestrial olivine, *Geochim. Cosmochim. Acta* **56**, 2923 (1992).

7. G. Kurat, F. Brandstatter, T. Presper, C. Koerberl, M. Maurette, Micrometeorites, *Geol. Geofiz.*, in press (1993).
8. M. Christophe Michel-Levy and M. Bourot-Denise, Mineral compositions in Antarctic and Greenland micrometeorites, *Meteoritics* 27, 73 (1992).
9. T. Presper, G. Kurat, M. Maurette, Preliminary composition of anhydrous minerals phases in micrometeorites from Cap-Prudhomme, *Meteoritics* 27, 278 (1992).
10. A.V. Ivanov, The Kaidum meteorite: composition and history, *Geochem. Internat.* 26, 84 (1989).
11. G. Kurat, C. Koerberl, T. Presper, F. Brandstatter, M. Maurette, Bulk composition of Antarctic micrometeorites: Nebular and terrestrial signature, *Meteoritics* 27, 246 (1992).
12. M. Perreau, C. Engrand, M. Maurette, G. Kurat, Th. Presper, C/O atomic ratios in micrometer-size crushed grains from antarctic micrometeorites and two carbonaceous meteorites, *Lunar Planet. Sci.* XXIV, 1125 (1993).
13. T. Presper, G. Kurat, C. Koerberl, H. Palme, M. Maurette, Elemental depletion in Antarctic micrometeorites and Arctic cosmic spherules: Comparison and relationships, *Lunar Planet. Sci.* XXIV, 1177 (1993).
14. G.J. Flynn, S.R. Sutton, W. Klöck, Unmelted Polar Micrometeorites: Compositions, Mineralogies, and Similarities/Differences with IDPs and Meteorites, *Proc. NIPR Symp. Antarct. Meteor.* in press (1992).
15. E.K. Jessberger, J. Bohsung, S. Chakaveh, K. Traxel, The volatile element enrichment of chondritic interplanetary dust particles, *Earth Planet. Sci. Lett.* 112, 91 (1992).
16. G. Flynn, Changes to IDP composition and mineralogy by terrestrial encounters, this volume (1994).
17. G. Callot, M. Maurette, L. Pottier, A. Dubois, Biogenic etching of microfractures in amorphous and crystalline silicates, *Nature* 328, 147 (1987).
18. E. Grün, H.A. Zook, H. Fechtig, R.H. Gleen, Collisional Balance of the Meteroid Complex, *Icarus* 62, 244 (1985).
19. E. Robin, *Des poussières Cosmiques dans les Cryoconites du Groénland: Nature, Origine et Applications*. PhD Thesis, Université de Paris-Sud, Centre d'Orsay (1988).
20. C. Koerberl and E.H. Hagen, Extraterrestrial spherules in glacial sediments from the Transantarctic Mountains, Antarctica: structure, mineralogy and chemical composition, *Geochim. Cosmochim. Acta* 53, 937 (1989).
21. R.P. Harvey and M. Maurette, The origin and significance of cosmic dust from the Walcott Névé, Antarctica, *Proc. Lunar Planet. Sci. Conf.* 21, 569 (1991).
22. M. de Angelis, L. Ferenbach, C. Hammer, C. Jéhanno, M. Maurette, Search for extraterrestrial grains in polar ices, *Lunar Planet. Sci.* XV, 98 (1984).
23. M. Maurette, J. Cragin, S. Taylor, Cosmic dust in ~50kg blocks of blue ice from Cap-Prudhomme and Queen Alexandra Range, Antarctica, *Meteoritics* 27, 257 (1992).
24. D.J. Lindstrom, W. Klöck, Analyses of 24 unmelted Antarctic Micrometeorites by Instrumental Neutron Activation Analysis, *Meteoritics* 27, 250 (1992).

25. M. Maurette, D.E. Brownlee, S.R. Sutton, D. Joswick, Antarctic micrometeorites smaller than 50μm, *Lunar Planet. Sci.* XXIII, 857 (1992).
26. M. Maurette, M. Pourchet, M. Perreau, The 1991 EUROMET micrometeorite collection at Cap-Prudhomme, Antarctica, *Meteoritics* **27**, 473 (1992).
27. F. Yiou, G.M. Raisbeck, C. Jéhano, The micrometeorite flux to the Earth, during the last ~200,000 years as deduced from cosmic spherules concentration in Antarctic ice cores, *Meteoritics* **26**, 311 (1991).
28. F.J.M. Rietmeijer, The bromine content of micrometeorites- Arguments for stratospheric contamination, *J. Geophys. Res.* **98**, N E4, 7409 (1993).

Recovery of the Long Duration Exposure Facility (LDEF), by the Space Shuttle Columbia, following 69 months in orbit. This bus-sized satellite exposed thousands of materials and experiments to the harsh environment of space (meteoroids, spacecraft debris, temperature excesses, radiation, etc.). The planet Earth lies in the background. See papers in this volume by Berthoud and Mandeville, and Zolensky et al. (NASA photo S32-8539)

METEOROID INVESTIGATIONS USING THE LONG DURATION EXPOSURE FACILITY

Michael E. Zolensky, Friedrich Hörz
Solar System Exploration Division
NASA Johnson Space Center, Houston, TX 77058, USA

Thomas See, Ronald Bernhard, Claire Dardano, Ruth A. Barrett, Kimberly Mack, Jack Warren
Lockheed Engineering and Science Company
2400 NASA Rd 1, Houston, TX 77058, USA

and
William H. Kinard
LDEF Science Office
NASA Langley Research Center, Hampton, VA 23665, USA

ABSTRACT

The Long Duration Exposure Facility was recovered in January, 1990, following 5.7 years of continuous exposure in low-Earth orbit. The gravity-stabilized (non-spinning) nature of LDEF permits the spatial resolution of the flux and trajectories of impacting meteoroids and spacecraft debris particulates. We have completed the acquisition of high-resolution, stereoscopic video imaging of all large impact features on the entire LDEF, and present here the preliminary results of our efforts to analyze these digitized images, and extract critical data. We present results of detailed crater surveys of LDEF frame intercostal members, and find an unusual local variation in the impact frequency. In a discussion of impactor fluxes derived from LDEF results we explore apparent directionalities for impacting particulates which are not accounted for in current models. We also describe current efforts to characterize meteoroid residues recovered from the impact craters, and we have found that a low, but significant, fraction of these residues have survived in a largely unmelted state.

INTRODUCTION

Most of the papers in this book concern the study of Interplanetary Dust Particles (IDPs) collected in and below the atmosphere. However, there is considerable value in sampling these materials before atmospheric entry; before IDP velocity differences can lead to the selective destruction of cometary particles in the atmosphere and consequent sampling bias in the stratosphere. If we wish to learn the true flux of IDPs of all different sources then we must look above the stratosphere. Recognition of this fact motivated studies of returned portions of Earth-orbiting satellites like the Solar Maximum[1] and Palapa satellites. However, these spacecraft were not designed to serve this purpose, and the flux data they yielded did not really permit particle trajectories to be well-determined. However, these studies did enable the design, testing and development of new techniques for the characterization of IDP impact features and the particle residues they sometimes contained. Laboratories and workers were thus well-prepared for the return of the Long Duration Exposure Facility (LDEF).

The LDEF was recovered in January, 1990, following 5.7 years of exposure of 130 m^2 of surface area (about the size of a large city bus) in low-Earth orbit (LEO, 250-179 nautical miles altitude). The LDEF spacecraft was an open-grid, 12-sided, cylindrical structure on which a series of rectangular trays used for mounting experiment hardware were attached. These trays faced in 14 directions, 12 along the sides ("rows") and 2 ends (Earth- and space-facing). In addition, portions of the LDEF frame, tray attachment clamps and half of the tray lip (flanges) faced into directions between each of the 12 experiment rows. Since the LDEF was gravity stabilized, elements of the LDEF continually faced in 26 different directions that remained fixed relative to the spacecraft's velocity vector. In addition to the large area-time of space exposure (two orders of magnitude greater than previously returned spacecraft surfaces[1]), LDEF also affords the opportunity to obtain information about the directionality of the meteoroid and debris fluxes. This data can then be related to the sources of meteoroids and orbital debris. This information is needed to deduce the asteroidal versus cometary abundance of impacting meteoroids. Results from LDEF thus build on and complement previous studies of such materials returned from space as blankets and louvers from the Solar Maximum Satellite.[1]

The LDEF was host to many individual experiments designed to characterize aspects of the meteoroid and space-debris environment in LEO. It was realized from the beginning, however, that the best and most efficient way to accomplish this goal was to exploit the impact record of the entire LDEF spacecraft. Therefore, the Meteoroid and Debris Special Investigation Group (M&D SIG) was organized to achieve this goal. The basic role of the M&D SIG is to complement the specialized research being performed by the principle investigators (PIs) on their dedicated experiments. The membership of the M&D SIG has grown as other disciplines have recognized the potential value of its work.

One of the first M&D SIG activities was to assist in the initial documentation of the flown spacecraft in the Spacecraft Assembly and Encapsulation Facility clean room at the Kennedy Space Center (KSC). In this paper we describe the results of this survey effort, as well as post-KSC activities of the M&D SIG to further characterize the meteoroid and space debris environment through continued analyses of LDEF surfaces. In particular we present the results of detailed impact feature scanning of aluminum intercostal members from LDEF's frame, which has the attribute of being a single, homogeneous material continuously exposed in all 26 facing directions of the LDEF for its entire 5.7 year lifetime. We briefly summarize efforts to better understand the flux of particulates in LEO, based upon LDEF experiments, and efforts to obtain refined crater morphological data from stereo images. The LDEF Meteoroid and Debris Database is described briefly. Finally, we present the preliminary results of efforts to characterize the mineralogy and structural state of meteoroid residues recovered from LDEF impact craters.

DATA-ACQUISITION PROCEDURES

During a three month period from February through April, 1990, members of the M&D SIG monitored all LDEF deintegration activities, documenting the meteoroid and debris impact features present on its surface. At that time the M&D SIG photodocumented all impact features measuring >0.5 mm in diameter present on structural surfaces, and all penetrations >0.3 mm in diameter in thin materials such as thermal-control blankets. The dual size threshold was employed because the ratio of crater- to projectile diameter is generally larger than that of penetration hole size to projectile diameter. By employing two different threshold values, we assured complete overlap in mass frequency of project-iles responsible for penetrations. We made a visual survey, only, of all smaller but visible impact features. The photo-documentation was required due to the destructive nature of most of the analyses planned for LDEF surfaces. Subsequent analyses of LDEF surfaces by experiment PIs have indeed destroyed much of the impact record.

Our impact-feature documentation was performed with three stereomicroscope imaging systems, each built around a Wild Leitz M8 body. A beam splitter directed 50% of incoming light to camera systems (Nikon F3-HP 35-mm cameras or Sony XC-711 CCD video cameras). The microscope/camera system was attached to a fully articulated surgical floor-stand. This integrated system provided complete mobility of the microscope/camera system and permitted the microscope to be moved into virtually any position.

Output from the video cameras was carried to an NEC Portable Powermate 386 SX portable computer, to which was added a Data Translations frame grabber/digitizing board, a Data Translations encoder/multiplexer board, two Javelin video monitors, and two Storage Dimensions External Laser WORM disk drives. To permit the measurement of impact feature locations on experiment trays and LDEF frame, we utilized electronic Coordinate Registration Systems (CRS) fabricated by Prototype Machine Corporation. Each CRS was paired to an LDEF experiment-tray stand for use with the stereomicroscope imaging systems. In some cases a metric tape measure was used to determine the coordinates of impact features when the CRS could not be used.

M&D SIG scientists surveyed all LDEF components as they were removed from the satellite, harvesting a specific set of data for all large impacts, which included: (1) the size, type, location, and additional characteristics of all qualifying impact features (a total of approximately 4,500 features), (2) digitized, stereo, color imaging of these same impact features, and (3) an accounting of all impact features large enough to be observed visually (with the naked eye), but too small to warrant detailed documentation (approximately 30,000 impact features). In addition, we collected any other information on impact features which could be gathered visually. These KSC activities and the resulting data are well described by See et al.[2]

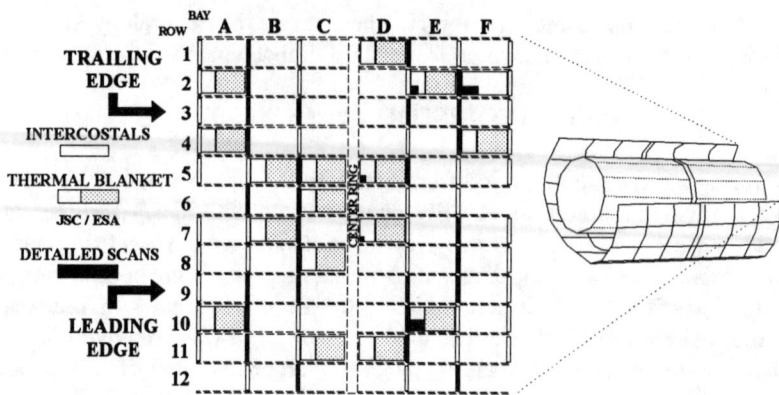

Figure 1. Illustration showing the multi-sided nature of the LDEF spacecraft and associated numbering scheme. The Earth- (G) and space-facing ends (H) would be to the left and right, respectively, in this illustration. Sixty eight, of a possible 72 peripheral intercostals reside at JSC and are being scanned in detail for craters <500 µm in diameter; the results of these detailed scans on 28 intercostals are included in this report. One third of all thermal blankets also reside at JSC

The M&D SIG survey of LDEF's frame was conducted following the removal of all of the experiment trays and thermal panels from the spacecraft. The purpose of this survey was to identify and photodocument features of interest residing on the longerons and intercostals which composed the skeletal framework of the LDEF spacecraft. This survey became necessary because portions of all LDEF frame members were exposed between tray flanges (*i.e.* the LDEF experiment trays did not completely cover the underlying frame). All LDEF frame members consisted of 6061-T6 chromic-anodized aluminum, and space-exposed surfaces were polished. Because of their polished surface, all intercostal frame members were given to the M&D SIG for more detailed scanning (see below).

All experiment trays were affixed to the LDEF frame by means of anodized aluminum clamps. These clamps also faced in all LDEF directions, although the total surface area of the clamps (3.5 m^2) was considerably less than that presented by the LDEF frame (15.4 m^2), and they are not polished to the same degree as the frame members. All clamps were surveyed for large impact features at KSC. Approximately one-half of these clamps have been retained by the M&D SIG for further analysis; the remaining clamps are in the possession of the LDEF Materials SIG.

Although all of the digitized images of LDEF impact features will be reduced to provide precise diameter and depth information, we optically measured impact feature diameters so that preliminary reports would be possible. These diameter measurements should be accurate to within 10%. All crater measurements presented here are center-of-rim to center-of-rim diameters. For more information regarding hypervelocity impacts and the morphologies presented by impact features consult Anderson[3] and Kinslow.[4] When highly asymmetric rim shapes were present, the minimum measurement is reported here.

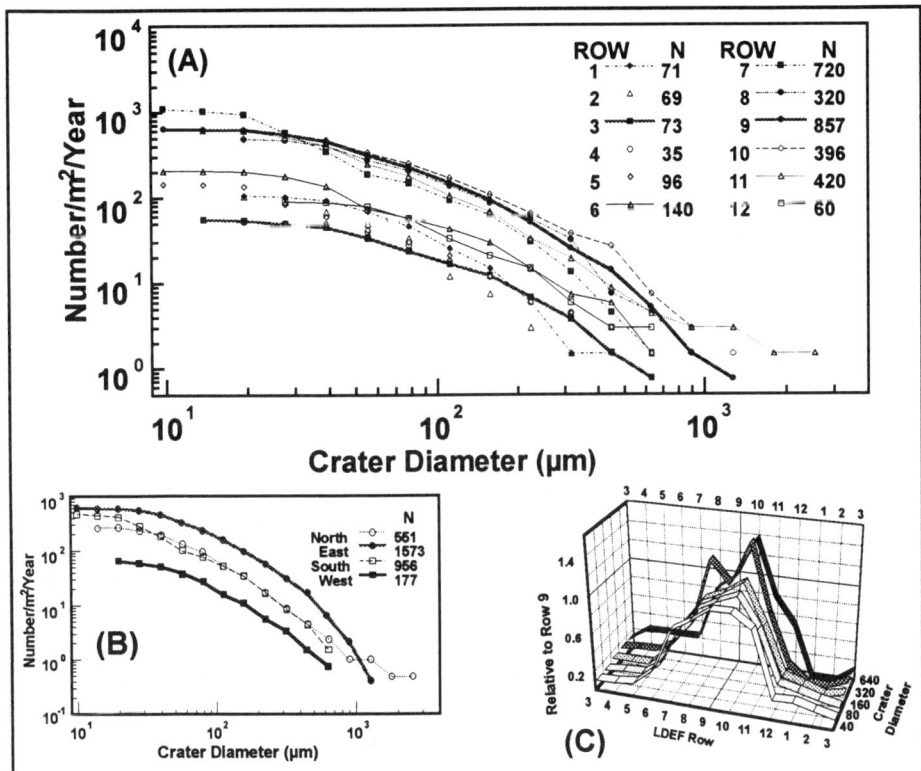

Figure 2. (A) Particle flux as a function of pointing direction for the LDEF intercostal frame members; Row 9 was the leading-edge or velocity-vector pointing direction. (B) Average particle flux for the four major pointing directions of LDEF (North equals the average flux of Row 11, 12, 1, etc.) (C) Relative impact frequency for selected crater diameters with respect to Row 9. N (40-640) is the crater diameter associated with each line.

To foster continued studies, we have carefully selected a large variety of materials from LDEF displaying impact features, and returned them to the Curatorial Facility at the Johnson Space Center (JSC). We are currently re-surveying curated LDEF surfaces in order to document smaller features than those documented at KSC during LDEF deintegration activities. All of the LDEF surfaces obtained by the M&D SIG are available for allocation to qualified investigators For information on the allocation of LDEF samples contact the Office of the Curator at JSC.

RESULTS AND DISCUSSION

LDEF Intercostal Survey

The total surface area of the exposed LDEF frame amounted to 15.4 m^2. The aerial density of impacts >0.5 mm on these surfaces varied from ~0 to 78.5 impacts / m^2. We have been re-surveying the intercostal frame members from LDEF because they were removable from the structure and are polished, permitting recognition and characterization of all impact craters down to diameters of 30 μm (10 µm in favorable conditions, but the coverage is *not* complete below 30 µm). All scanning has been performed at the JSC Facility for the Optical Inspection of Large Surfaces (FOILS Lab), on a very stable, translatable optical table.

The cumulative size frequency distributions and spatial densities of craters for 28 intercostal frame members are shown in Figure 2a. This figure includes data from at least two intercostals from each row. These data are in good agreement with our earlier results[5,6], with the highest cratering densities being present on rows 8-10. Similar size-frequency slopes for each individual intercostal suggest that the overall ratio of small-to-large particulates remains constant for all directions.

However, something different is observed for intercostal F07F02 (Row 7, Bay F, clamp #F02; see Figure 1 for location), which exhibits a high density of craters <30 µm in diameter (Figure 2a). In fact, intercostal F07F02 exhibits an order of magnitude more of these small impact craters than does the other intercostal from the same row which we examined. Humes[7] has found that this large density of small impact features extends to an adjacent experiment tray.

We have satisfied ourselves that the very localized, unusually high density of impact features found on one end of Row 7 of LDEF is real, and not due to scanning bias, or other measurement errors. The correct explanation might lie in an unexpectedly high variation in the spatial density of impacting particulates, such as might be expected from local contamination from the Space Shuttle which placed LDEF in orbit. During the documentation of intercostal F07F02 we noted that an unusually high number of impact craters contained impactor residues, suggesting relatively low velocities for these impacts. However, chemical characterization of these residues by scanning electron microscopy-energy dispersive X-ray analysis (SEM-EDX) techniques revealed no single source of impactors (instead we found 23% meteoroids, 46% debris (including paint and electrical materials), 5% ubiquitous LDEF Si-Ca contamination (from the outgassing of lubricants, adhesives, epoxies and other volatile materials), and 25% of inconclusive origin; see Figure 3).

There is a significant increase in the relative number of paint-type residues as compared to residues found on other LDEF surfaces, which could lead one to posit an occurrence such as an impact on the Space Shuttle's remote manipulator arm during LDEF deployment or recovery. However, aside from this disparity, the distribution of impactor types for intercostal F07F02 appears similar to the LDEF as a whole, lending no support to a local, *homogeneous* source for the impacting particulates. We have formed no conclusions regarding the explanation of this apparent paradox, and are investigating the situation further. This phenomenon is described in greater detail elsewhere.[6]

Figure 2b shows simplified trends for intercostal data, as we have averaged crater size-frequencies for the four cardinal LDEF side-facing directions (N, S, E and W). As expected, forward facing rows (8-10, *i.e.* E) show the highest crater frequencies (due to the windshield effect), while the rearward-facing rows (2-4, *i.e.* W) show the lowest. The northern-(1, 12 & 11) and southern-facing rows (5-7) have essentially identical crater frequencies, despite the well-documented 8° skew of the velocity vector of LDEF towards Row 10 from Row 9.[8]

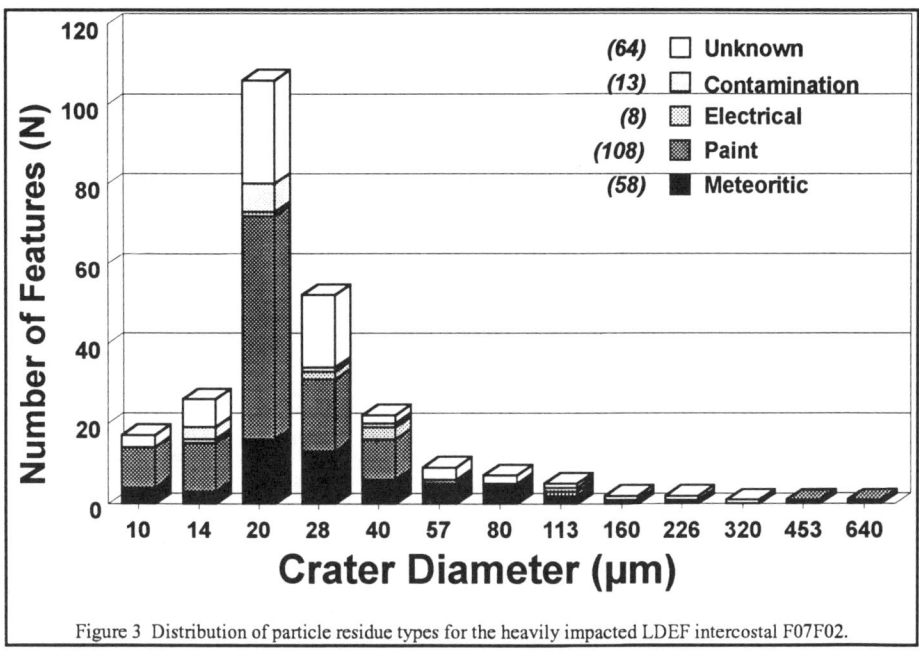

Figure 3 Distribution of particle residue types for the heavily impacted LDEF intercostal F07F02.

Figure 4 shows the impact frequency for all frame intercostals displayed in polar form. In our earlier work with only the largest craters (>0.5 mm in diameter), we found the highest impact frequency on Row 10.[9] However, with a data set extending to much smaller craters, Row 9 now has the higher impact density, as was expected. Note, however, that the influence of many small craters on intercostal F07F02 gives Row 7 a high apparent impact density.

Well before LDEF's recovery, Zook[10] theoretically deduced, under a "randomness" assumption, that from 6 to 9 times more meteoroids per unit area were expected to strike an LDEF leading-edge surface than would impact a trailing-edge surface; and, further, that this ratio depended on the velocity distribution with which meteoroids approached the Earth. These leading- to trailing-edge ratios of fluxes were due solely to LDEF orbital motion. When meteoroid impact velocities and cratering equations are also taken into account, relative spatial densities (leading- to trailing-edge) of meteoroid impact craters on LDEF can be calculated[11-12]; these ratios are found to range from 10 to 30, depending on the meteoroid velocity distribution and the meteoroid size distribution used. Kessler et al.[13] similarly deduced theoretical ratios to be expected for orbital debris.

A Gaussian curve which best fit the LDEF frame impact frequency data has a maximum at Row 9, the nominal leading-edge direction, and a leading to trailing impact frequency ratio of 10:1 is derived. A similar fit to the intercostal data solely for large impacts (≥0.5 mm) results in a ratio of only 4:1. However, it should be kept in mind that only 38 of the 2147 impacts (<2%) included in this study were ≥0.5 mm in diameter, resulting in a relatively large error for the 4:1 ratio. Nevertheless, it is interesting that the leading to trailing impact frequency increases as smaller impact features are considered. A similar relation has been found by other LDEF investigators examining impact penetrations through foils. McDonnell[14] found that the leading to trailing impact ratio climbs to approximately 50:1 for penetrations on the order of 5 μm in diameter. These measured ratios of leading-trailing impact frequencies for diminutive particulates are being used to augment current modeling of meteoroids or spacecraft debris, which have till now only dealt with larger particulates.[15-16]

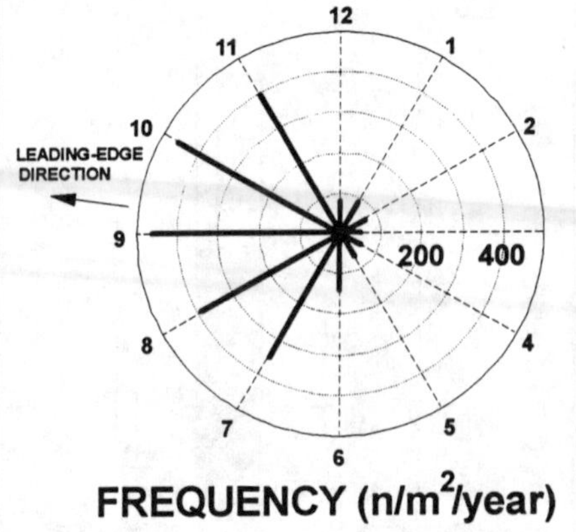

Figure 4 Polar coordinate diagram of the intercostal cratering frequencies for the 12 LDEF rows for craters of all sizes.

Particulate Flux Estimates

The largest impact crater on LDEF was 0.57 cm in diameter and was probably caused by an object about a millimeter in diameter. This is greatly helping to bridge the observational gap between the radar data (now estimated to reach down to about 1 cm diameter) obtained from ground stations and data returned from direct observations in space. With this new data shielding against meteoroids and debris needed to protect satellites from damage can now be better estimated; this is especially important for a Space Station where there is considerable concern over impact shielding.[17]

Analyses of residues in impact craters on gold surfaces that were facing the trailing direction of LDEF[5] has produced the very interesting result that approximately 15% of craters containing analyzable impactor residue were attributable to spacecraft debris. This result was surprising because before LDEF recovery it had been predicted by Kessler[13] that no debris would hit the backward-facing LDEF surfaces. The only way these surfaces can be struck is for particles to catch up to LDEF from behind, which implies that these particles must be in highly elliptical orbits. Modeling implicates debris sources in geosynchronous transfer orbits as the responsible agent.[16]

Although there are probably a number of debris impacts in the population of *large* (diameter 0.5 mm) impact craters, the ratio of leading-to-trailing crater spatial densities also appears consistent with meteoritic impacts alone.[11] On an aluminum surface facing about 50° off the leading edge, Bernhard and Hörz[5] found that orbital debris impacts start to become more numerous than meteoroid impacts for impact craters smaller than about 100 μm in diameter. Below 50 μm in diameter, orbital debris appears to dominate the crater populations on leading-edge LDEF surfaces. This result is also found by other LDEF investigators.

The time variation of the flux striking LDEF is also a strong indicator of the origin of the impacting particles. The only "active" meteoroid experiment on LDEF was the Interplanetary Dust Experiment (IDE, Singer et al.[18-19]) which recorded, for the first year, the time of impact triggered discharges of Metal-Oxide-Silicon (MOS) capacitors placed around LDEF. This experiment recorded over 15,000 impacts that penetrated either 0.4 μm- or 1.0 μm-thick dielectric layers of SiO_2. The IDE sometimes sensed multi-orbit "streams" of particles, where the impact rate would greatly increase for a few minutes on every orbit. The only reasonable interpretation of such multi-event sequences is that LDEF passed through the orbital plane of a debris cloud associated with still unidentified spacecraft. Also, the impact rate on IDE was elevated for the first few days of the mission, presumably caused by contaminant particles from the Shuttle that had deployed the LDEF.

In summary, analyses of impact craters and the time history of impacts on LDEF is giving us a much better picture of both the meteoroid and space debris populations in near-Earth orbit. We have become especially aware of new features of the orbital debris populations. Some debris clouds are concentrated into orbital planes and do not dissipate into the background as fast as expected from modelling. More debris is impacting trailing-edge surfaces than was expected, probably indicating that geosynchronous transfer orbits are well populated with debris (see Berthoud and Mandeville[20] in this volume for further support of this position). This phenomena has the unfortunate effect of peppering spacecraft with debris on all sides, including those for which it was once believed that only meteoroids would hit.

Mineralogy of Meteoroid Residues

LDEF impactor residues are being characterized to establish the nature and abundance of meteoritic and orbital debris materials in the LEO environment. We have developed simple techniques for the study of selected chondritic (containing Si, Mg, Fe, +/- Al, Ca, S, Mn, and Ni in appropriate amounts) impactor residues in shallow craters in gold plates, from the LDEF experiment A0178. A detailed structural and compositional analysis of several of these impactor residues was performed utilizing transmission electron microscopy, energy dispersive spectroscopy, and electron diffraction. The immediate goal of this continuing work has been to determine the shock effects exhibited by chondritic meteoroids (AKA Interplanetary Dust Particles or IDPs), and to compare the impactor residues to chondritic IDPs collected from the stratosphere.

Residues from the interior of several meteoroid impact craters were removed with a tungsten needle, mounted in EMBED-812 epoxy, and ultramicrotomed into 90 nm thick sections. Observation of the sections on carbon-coated copper grids was done by transmission electron microscopic techniques using JEOL 100CX and 2000FX analytical electron microscopes. Chemical analyses were performed with a PGT System 4, and an energy dispersive X-ray spectrometer and reduced with the PGT dedicated software. The structural state of all analyzed materials was assessed by electron diffraction, which proved to be a critical step, considering the non-crystalline nature of many materials observed.

We examined the mineralogy of residues from three impact features: nos. 102, 121, and 295. Impact residue 102 has abundant, very finely-divided, crystalline augite ($En_{55-59}Wo_{36-40}$) and orthopyroxene (En_{84-96}) showing abundant evidence of intense shock, these being planar deformation features, mosaicism, and, in some instances, evidence of recrystallization (120° grain intersections). The matrix

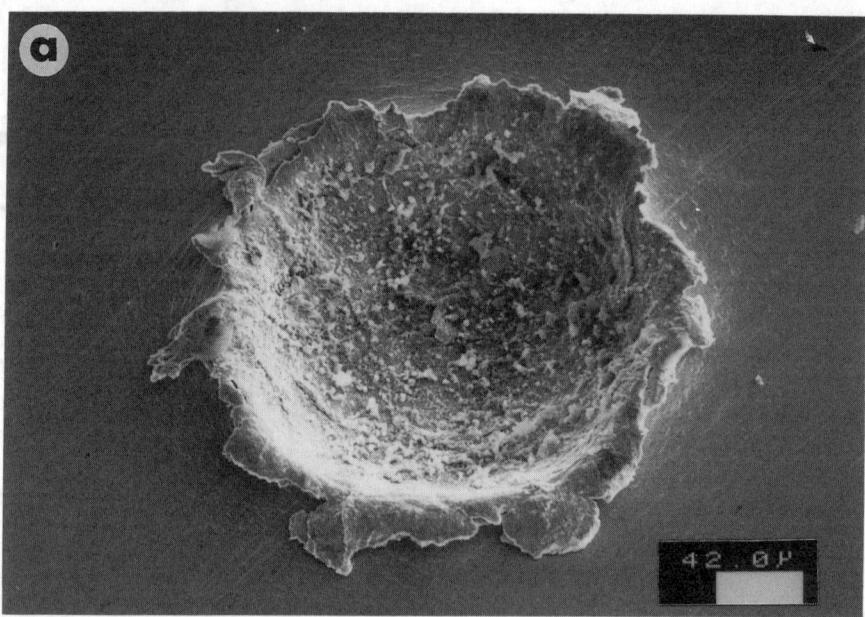

Figure 5 Impact crater 102, with IDP residue. (a) SEM image of the impact crater in gold; residue is visible inside. Scale bar measures 42 μm. (b) High-magnification SEM image of the impactor residue from the center of crater 102; mineral fragments and frothy glass are visible. Scale bar measures 5 μm.

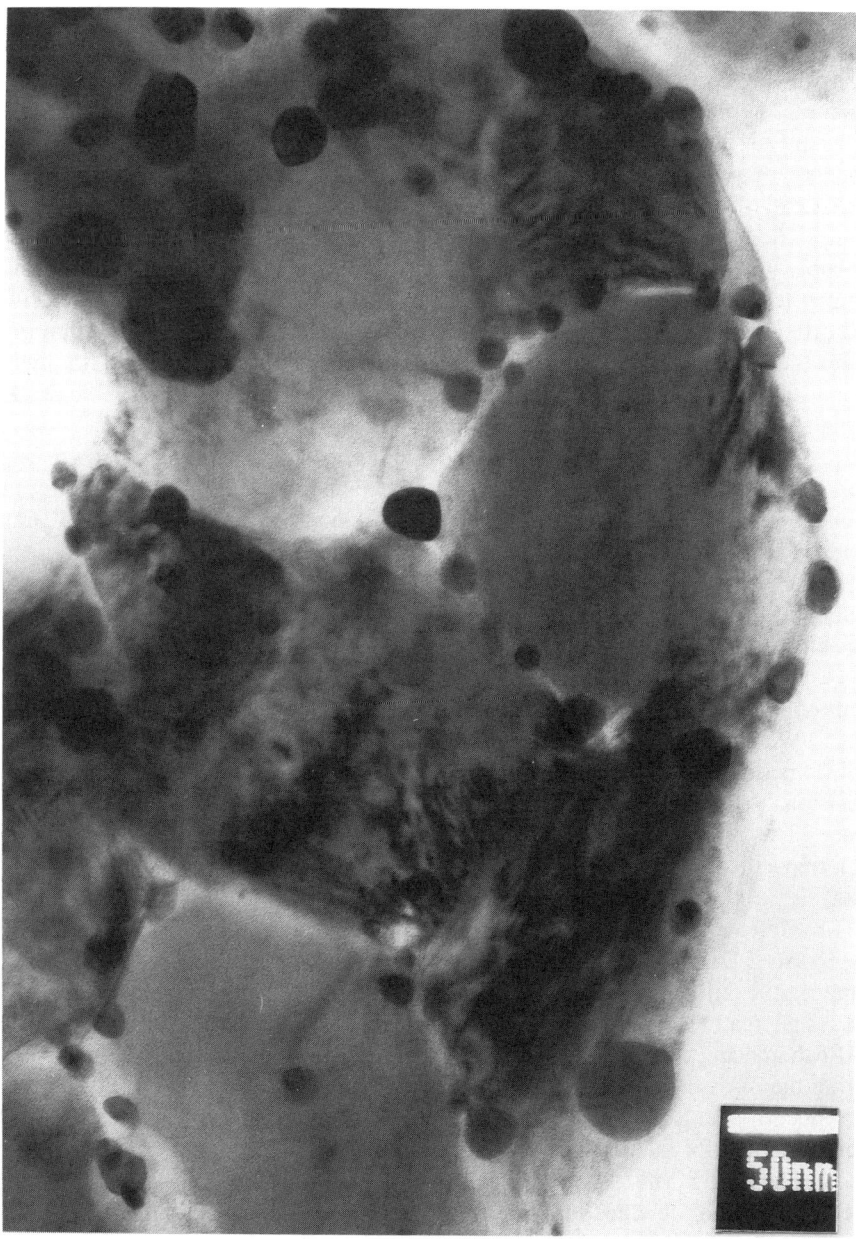

Figure 5 Impact crater 102, with IDP residue. (c) TEM image of a residue grain. Scale bar measures 50 nm. Abundant very finely-divided crystalline augite and orthopyroxene showing abundant evidence of recrystallization (120° grain intersections). The matrix consists of frothy ferromagnesian glass. Spherical bodies of Fe-Ni metal (some glasses) and pyrrhotite abound locally, particularly at grain boundaries.

consists of frothy ferromagnesian glass. Spherical bodies of Fe-Ni metal and pyrrhotite abound locally, particularly at grain boundaries (see Figure 5). Some Fe-Ni grains are metallic glasses (J. Bradley, personal communication, 1993). Impact residue 121 contains fragmental grains of olivine (Fo_{57-67}), orthopyroxene (En_{63-64}), Fe-Ni metal, and abundant glass. The olivine and pyroxene grains show abundant evidence of shock (see above for criteria). Impact residue 295 contains shocked, fragmental olivine (Fo_{56-71}) and orthopyroxene (En_{71}), pyrrhotite, and glass.

The pyroxenes in residue 102 have Fe-poor, restricted compositions, which probably identifies the impactor as a hydrous IDP (see Zolensky and Barrett[22] in this volume to make the comparison to IDPs). However, the compositions of olivines and orthopyroxenes in the other residues characterized in this study are equilibrated compared to anhydrous chondritic IDPs, and also Fe-rich compared to hydrous chondritic IDPs. They are also Fe-rich as compared to ferromagnesian silicates from partially melted chondritic IDPs[21], which are typically on the order of Fo_{90} and En_{90}. The presence of equilibrated and shocked ferromagnesian minerals, recrystallization textures, glass, and melted metal and sulfide bodies decorating grain boundaries, is indicative of varying degrees of shock metamorphism in all impact residues we have characterized. Our failure to locate any magnesian olivines or pyroxenes in these particular residues is illustrative of the pervasive shock metamorphism they experienced. We are continuing to characterize additional IDP impactor residues.

Reduction Of Digital Impact Crater Images

We collected digitized stereo images of all large impact features on LDEF at KSC to permit later analysis of the surface structure of each crater to a high degree of precision. As a minimum, we wish to determine the true depth and diameter of each crater. This information is critical to the extraction of particulate mass and size from each impact feature.

LDEF Meteoroid and Debris Database

An LDEF M&D database has been developed and is maintained at the Curatorial Facility at JSC. Five linked data tables contain information about individual features, digitized images of selected features, and inventory data for LDEF hardware curated at JSC. About 4500 impact features were characterized during the disassembly of the satellite at KSC, while an additional 4500 have been subsequently identified at JSC. The database also contains all information which has been submitted for inclusion by members of the LDEF PI community.

The LDEF M&D database may be accessed via Decnet, Internet, or modem. The capability for downloading results of searches to users' local computers via FTP, Kermit, or Mail is in place. Image files may be downloaded via FTP and, less efficiently, via Kermit. The image files do not stay on-line, but may be made accessible on request. For information regarding access to the LDEF M&D database contact the senior author of this report at the Office of the Curator. Stereo images of all LDEF impact features documented by the M&D SIG are available from the same source on compact disks.

SUMMARY

It is now clear that we have the capability to bring a large variety of analytical tools to bear on the characterization of impact features and the impactor residues they may contain. For example, we can usually distinguish between IDP and spacecraft debris residues. We can also, in favorable circumstances, distinguish the remains of hydrous from anhydrous IDPs, by examination of the compositional range of olivines and pyroxenes. LDEF experiments have already driven development of a new generation of IDP capture cells, which promise collection of less-altered particles (see Borg et al. in this volume.[23]).

The spatial density of impact craters and penetration holes on LDEF is only broadly consistent with traditional models of IDPs and man-made debris. As predicted, the forward-facing directions of non-spinning satellites do display much higher crater densities than trailing-edge surfaces, yet the ratio of trailing- to leading-edge crater densities varies between 4 and 50, depending on crater size; it is highest for the smallest craters (<20 μm), suggesting a preponderance of man-made debris. Real time impact rates measured by the IDE experiment reveal man-made debris particle clouds in distinct orbits. Also, chemical analyses of impactor residues suggest debris sources in highly elliptical orbits to be much more prolific than predicted.

It is clear that LDEF is living up to its promise of providing an unparalleled view of the meteoroid and orbital debris complexes in LEO. However, it is also clear that much additional work characterizing both particulate populations will be required to bring this view into sharp focus. It is also evident that LDEF has provided us with a detailed picture of the IDP environment in LEO, against which future changes in the environment will be compared.

ACKNOWLEDGMENTS

The members of the LDEF Meteoroid and Debris SIG are Martha Allbrooks, Dale Atkinson and J. D. Mulholland, (POD Assoc.), Don Brownlee (Univ. Washington), F. Buhler (Univ. Bern), Ted Bunch (NASA ARC), Vladimar Chobotov (Aerospace Corp.), Gunther Eichhorn (Space Telescope Science Institute), Miria Finckenor (NASA MSFC), Fred Hörz, Don Kessler, Michael Zolensky and Herb Zook (NASA JSC), Donald Humes and William Kinard (NASA LRC), Jean-Claude Mandeville (CERT-ONERA), J. A. M. McDonnell (Univ. Kent), Michael Mirtich (NASA LRC), Charles Simon and Jerry Weinberg (Inst. Space Sci. & Tech.), Thomas See (Lockheed), Robert Walker and Ernst Zinner (Washington Univ.)

REFERENCES

1. Warren, J.L., Zook, H.A., Allton, J.H., Clanton, U.S., Dardano, C.B., Holder, J.A., Marlow, R.R., Schultz, R.A., Watts, L.A. and Wentworth, S.J., In: *Proceedings of the 19th Lunar and Planetary Science Conference*, eds. G. Ryder and V. Sharpton, Lunar and Planetary Institute, Houston, pp. 641-657 (1989).

2. See, T., Allbrooks, M., Atkinson, D., Simon C. and Zolensky, M., *Meteoroid and Debris Impact Features Documented on the Long Duration Exposure Facility*, Planetary Science Branch Publication 84, NASA JSC, 565 p (1990).
3. Anderson, C.E., In: *Hypervelocity Impact, Proceedings of the 1986 Symposium*, Pergamon Press, Oxford (1987).
4. Kinslow, R., *High-Velocity Impact Phenomena*, Acad. Press, New York (1970).
5. Bernhard R. and Hörz, F., In: *Second LDEF Post-Retrieval Symposium Abstracts*, NASA Conf. Publ. 10097, p.46 (1992).
6. See, T.H., Mack, K.S., Warren, J.L., Zolensky M.E. and Zook, H.A., In: *Proceedings of the Second LDEF Post-Retrieval Symposium*, pp.313-324 (1992).
7. Humes, D., private communication (1992).
8. Peters, P.N. and Gregory, J.C., In: *Second LDEF Post-Retrieval Symposium Abstracts*, NASA Conf. Publ. 10097, p. 3 (1992).
9. Zolensky, M., Atkinson, D., See, T., Allbrooks, M., Simon, C. Finckenor, M. and Warren, J., *J. Spacecraft and Rockets* **28**, 204-209 (1991).
10. Zook, H.A., In: *Lunar and Planetary Science VIII*, pp. 1138-1139 (1987).
11. Zook, H.A., In: *LDEF-69 Months in Space*, NASA Conference Publication 3134, ed. A.S. Levine, pp. 569-579 (1991).
12. Humes D.H., In: *LDEF-69 Months in Space*, NASA Conference Publication 3134, ed. A.S. Levine, pp. 399-418 (1991).
13. Kessler, D.J., Reynolds, R.C. and Anz-Meador, P.D., *Orbital Debris Environment For Spacecraft Designed To Operate In Low Earth Orbit*, NASA TM 100 471 (1989).
14. McDonnell, J.A.M. and the Canterbury MAP Team, In: *Proceedings of the Lunar and Planetary Science Conference 22*, LPI, Houston, pp. 185-193 (1992).
15. Zook, H.A., In: *Second LDEF Post-Retrieval Symposium Abstracts*, NASA Conf. Publ. 10097, p. 52 (1992).
16. Kessler, D.J., In: *Second LDEF Post-Retrieval Symposium Abstracts*, NASA Conf. Publ. 10097, p. 53 (1992).
17. Zolensky, M.E., Zook, H., Atkinson, D., Coombs, C., Watts, A., Dardano, C., See, T., Simon, C. and Kinard, W., In: *Proceedings of the Second LDEF Post-Retrieval Symposium*, NASA Conf. Publ. 3194, pp. 277-302 (1993).
18. Singer, S.F., Stanley, J.E., Kassel, P.C., Kinard, W.H., Wortman, J.J., Weinburg, J.L., Cooke, W.J. and Montague, N.L., *Adv. Space Res.* **11**, 115-122 (1991).
19. Simon, C.G., Mulholland, J.D., Oliver, J.P., Cooke, W.J., and Kassel, P.C., In: *Proceedings of the Second LDEF Post-Retrieval Symposium*, NASA Conf. Publ. 3194, pp. 693-704 (1993).
20. Berthoud, L. and Mandeville, J.C., Analysis of remnants found in LDEF and Mir impact craters. This volume (1994).
21. Zolensky M.E. and Barrett R., *Microbeam Analysis* **2**, 191-197 (1993).
22. Zolensky, M.E. and Barrett, R.A. Olivine and pyroxene compositions of chondritic Interplanetary Dust Particles. This volume (1994).
23. Borg, J.,Bibring, J-P., Maag, C., Tanner, W. and Alexander, M.. Description of the COMRADE experiment. This volume (1994).

DESCRIPTION OF THE COMRADE EXPERIMENT

J. Borg and J-P. Bibring,
Institut d'Astrophysique Spatiale, Bat. 121, 91405 Orsay Cedex, France

C. Maag
Science Applications International Corporation; Glendora, Ca 91740, USA

and
W. Tanner and M. Alexander
Baylor University; Department of Physics; Waco, TX 76798, USA

ABSTRACT

The COMRADE experiment is designed to return minimally degraded particles to Earth along with complete in-situ information concerning mass, velocity and trajectory of encountered particles. The objectives of the program are very diverse. A set of flight-tested active detectors will be combined in an array to identify some of the physical properties of an incident grain, e.g., velocity vector, momentum, mass. The use of passive detectors gives access to the chemical and isotopic properties of the grains in the micron size range. We are also concerned simultaneously with a destructive capture, using metallic collectors, and a nondestructive capture, using a new low density target in which the impacting grains are captured, practically intact.

INTRODUCTION

The main scientific interest in the analysis of Interplanetary Dust Particles (IDPs), is due to the fact that most of these particles should be of cometary or asteroidal origin and contain information on the origin of the solar system. The smaller size fraction (<10 μm in diameter) is supposed to be enriched in grains of cometary origin.[1] In addition to IDPs, orbital debris (paint flakes, aluminum oxide spheres, etc.) is also encountered by a vehicle in low-Earth orbit (LEO), with the latter having much larger fluences than extraterrestrial grains at some size ranges.

The COMRADE experiment (Collection Of Micrometeorites, Residue And Debris Ejecta) has been proposed in order to gain information on all sizes of particles present in LEO, including submicron grains.[2] We are concerned simultaneously with a destructive capture of orbiting grains, using metallic collectors and a non-destructive capture, using new low-density targets in which the impacting grains can be captured in a relatively unmodified form permitting mineralogic, chemical and isotopic analyses to be performed. The advantage of the first type of capture is twofold: it allows us to gain information on the smallest size fraction, and to detect the presence of light elements such as carbon.[3] The interest in the second type of capture is to allow the extensive study of intact IDPs. Grains a few microns in size can be stopped in these low density materials and recovered for further studies.[4] Our COMRADE experiment will purposely couple the two techniques, so that the proposed investigation, which will collect micron/submicron particles with a minimum of particle degradation, will at the same time measure the dynamic particle parameters (determination of its mass, velocity, trajectory) with a high degree of confidence.

SCIENTIFIC OBJECTIVES

The primary objectives for this mission are: i) to identify the particle remnants of the micron sized grains having impacted on carefully designed metallic collectors, for complete and detailed chemical, isotopic and organic analysis thereby determining grain main composition as well as the existence of organic and inorganic molecules, to be related with the possible cometary origin of the grains showing an extraterrestrial signature; ii) to return captured particles to Earth for complete and detailed chemical, isotopic, spectral, mineralogical and organic analysis thereby determining grain composition as well as the existence of organic and inorganic molecules; iii) to capture micron/submicron dust grains in a manner that insures minimal particle degradation and guarantees confidence in state of the art measurements of the in-situ particle parameters including trajectory, velocity vector, mass and flux distributions.

CAPTURE OF REMNANT PARTICLES FOR CHEMICAL ANALYSIS

High purity metallic surfaces are used for the collection of all grains down to submicron sizes. During the impact of a high density impactor, a characteristic crater is formed, with rounded habits and a depth to diameter ratio characteristic of the encountered metal; the particle is destroyed and the remnants are mixed with the target material, concentrating in the bottom of the crater or on the surrounding rims. Gold and nickel are metals suitable for such collectors; the evaporation of 50 nm of gold on the exposed surface increases noticeably the identification of the impacting position, as the tear of the film creates a decoration of the impact position, allowing its easier discovery (Fig. 1).

The chemical and isotopic properties of the impactor can be identified, by analyzing the rim material. The major strength of the metallic collectors lies in the fact that many analytical techniques can be applied without modification to craters ranging from tens of nanometers up to millimeters in size, limited only by the thickness of the plate. Also, identification of carbon and organic material is made possible; this is essential for the study of extraterrestrial material.

Following exposure in space, the collectors are returned to the clean rooms of the lab in clean, sealed boxes. There, the impact positions of the grains are identified directly on the surface of the collectors, either by optical microscopy for the larger ones, or by using a scanning electron microscope for the micron-sized impacts, at a magnification that permits identification of crater features down to diameters of

1 mm. We can thus analyze the size distribution of the impact features, down to these sizes, permitting evaluation of the incident microparticle flux in the near Earth environment. In a second step, it is possible to determine, for each selected crater, the chemical composition, for elements as light as C, of the impacting particles. For grains identified on a chemical basis as being of extraterrestrial origin, this analytical step is to be followed by a high resolution analytical protocol including FESEM (Field Emission Scanning Electron Microscopy) imagery, molecular and isotopic characterization.

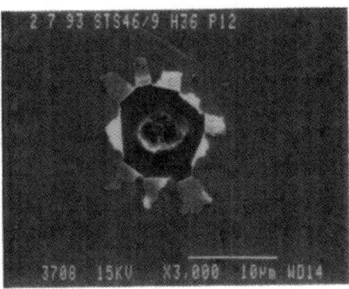 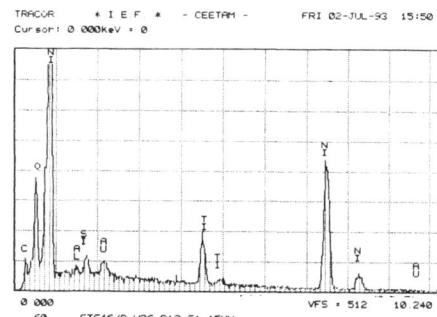

Figure 1. Typical crater observed in a Ni target, on which was evaporated 50 nm Au, torn through the impact. This collector was exposed to space during the STS 46 shuttle flight (July 1992). The impactor was an orbital debris grain, less than 5 mm in diameter and enriched in titanium oxide.

CAPTURE OF HYPERVELOCITY IMPACT PARTICLES

The return of extraterrestrial material to the laboratory is a primary goal of this investigation. The proposed investigation intends to retrieve relatively unshocked material by impacting three types of "underdense" capture devices. Shuttle experiments (STS 41-B, 41-D and 61-B) have shown that both organic foam and silica aerogel materials can be successfully used to capture, in some cases, intact particles.[4] These underdense capture devices will also be complemented by the well established technique of multilayer capture cells.

The impact of a hypervelocity projectile (>3 km s^{-1}) is a process which subjects both the impactor and the impacted material to a large transient pressure distribution. The resultant stresses cause a large degree of fragmentation, melting, vaporization and ionization (for normal densities). The pressure magnitude, however, is directly related to the density relationship between the projectile and target materials. As a consequence, a given impactor will experience the lowest level of damage on a low density target.

Historically, there have been three different approaches towards achieving the lowest possible target density. The first employs a projectile impinging on a foil or film of moderate density but whose thickness is much less than the particle diameter. This results in the particle experiencing a pressure transient with both a short duration and a greatly reduced destructive effect. A succession of these films, spaced to allow nondestructive energy dissipation between impacts, will reduce the impactor's kinetic energy without allowing its internal energy to rise to the point where compete destruction of the projectile mass will occur.

A second alternative uses an aerogel as the capture medium. This material, in its silicon-oxide form, is commonly produced for the nucleonics industry as a Cherenkov radiator, and has in fact been used in both space (HEOS, Ulysses, EURECA and STS-32) and balloon instrumentation. Other, analytically more desirable, compounds may also be processed to form this microporous structure, and these are currently being investigated. In comparison with polymer foams however, the extremely low densities (0.035 g cm^{-3}) cannot be achieved in any aerogel without producing great fragility. However, their transparency makes the recovery of the incident particles easier.

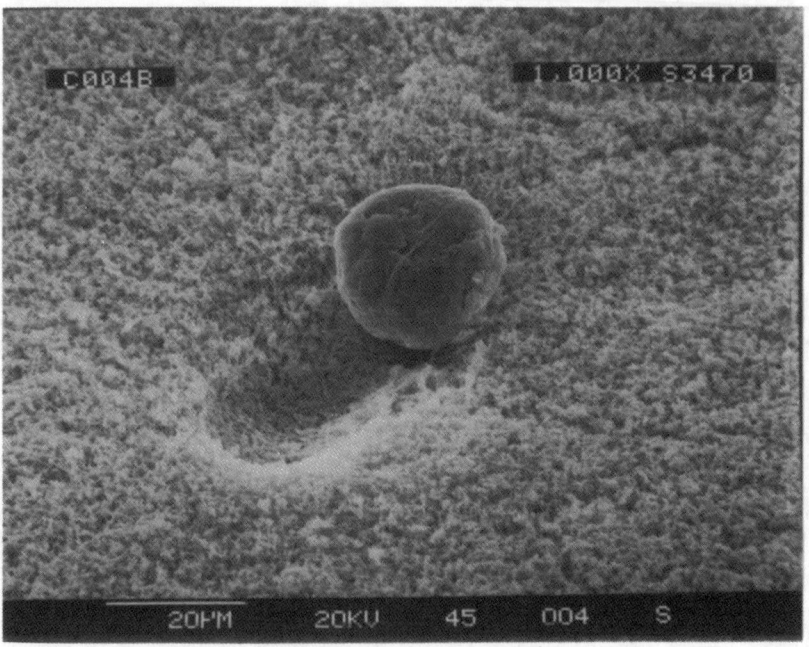

Figure 2. Alumina sphere of ~ 20 mm in diameter, recovered in polymer foam (flight STS41). It is encrusted with pyrolized foam.

Polymer foams have been employed as a third method of capturing particles with minimum degradation. The manufacture of extremely low bulk-density materials is usually achieved by the introduction of voids into the material base. It must be noted, however, that a foam structure only has a true bulk density of the mixture at sizes much larger than the cell size, since for impact processes this is of paramount importance. The scale at which the bulk density must still be close to that of the mixture is approximately equal to the impactor. When this density criterion is met, shock pressures during impact are minimized which in turn maximizes the probability of survival for the impacting particle. Polymer foam has been used currently as collecting material on various STS flights).[5] Intact high-density refractory particles, mainly of terrestrial origin, as small as 0.4 µm in size have been recovered. They are often recovered encrusted by pyrolized foam which is easily removed, leaving an undegraded specimen (Fig. 2).

MEASUREMENT OF FUNDAMENTAL PARTICLE PARAMETERS

The reliable determination of the trajectory of each individual dust particle is a high priority of the proposed investigation. Particle trajectories (as well as particle time of flight) can been determined using the thin film/plasma technique, which is based on the fact that a dust particle which impacts an extremely thin film will create a minute plasma cloud. The collection of this plasma cloud then allows for the analytic determination of dynamic particle parameters (Fig. 3). The use of multiple thin films thereby yields a method whereby particle trajectories and time of flight can be determined.

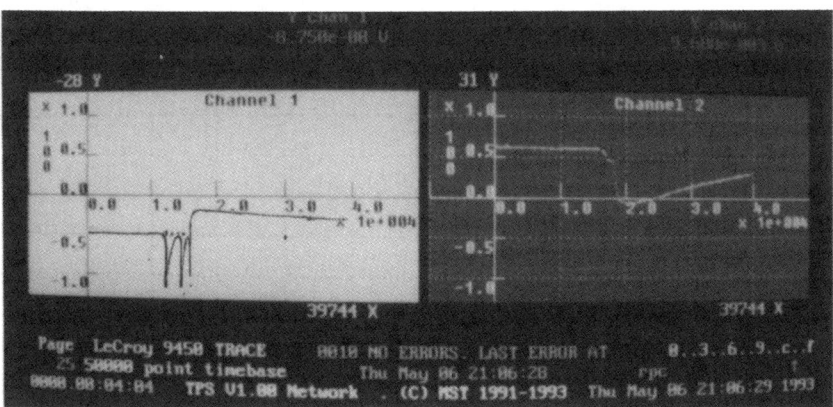

Figure 3. Indication of signals produced when a hypervelocity Fe sphere penetrates a thin film and is captured in an underdense foam. The signals on channel 1 and 2 correspond to the charge collected via an ultra low noise circuit connected to grids near the impact sites. In channel 1, they also indicate the charge induced on free standing grid wires. The time-of-flight of the dust grain can be ascertained by knowing the separation distance of the grids. The timebase of 20 ms/division implies that the dust grain was traveling at 4.5 km s^{-1}. (The simulation was performed at the plasma accelerator of the University of Kent, Canterbury U.K.)

In addition to the particle trajectory, it is vital that dynamic particle parameters also be measured with a high degree of reliability. The basic parameters which the proposed experiment will measure and/or determine are particle velocity and mass.

CONCLUSIONS

It is in the context of research of cometary particles down to submicron sizes that our proposal of exposing materials in LEO must be perceived. The detection of these particles requires the combination of active detectors, to determine the physical properties of the grains in LEO, with types of passive detectors (metallic collectors and low density material). Any collection facility designed for a long term exposure should contain some high purity metallic targets for chemical and isotopic identification of particles. The coupling of metallic collectors and low density material provides a powerful opportunity to obtain information on all sizes of grains from submicron sizes to a few microns.

All grains down to submicron sizes can be collected on our metallic targets. Because of their high relative velocity (>5 km s^{-1}), the impacting grains are physically destroyed, leaving a melted remnant that is mixed with the crater material. This process is more favorable for the smaller grains, in the micron size range; the larger grains may vaporize, leaving a remnant analyzable only by ion probe. Our previous results have shown that gold and nickel collectors are favorable for the collection and analysis of the small sized-grains orbiting around the Earth. For the less frequent larger grains, their collection is possible on large surfaces of low-density material.

The analysis of the grains, either remnants or complete, will be performed with the high resolution instruments to which we have access to (optical microscopy, Scanning electron microscopy - SEM, Energy Dispersive X-ray Spectrometry, Field Emission SEM, ion probe). By the time the collectors are returned from space, new techniques will have been developed and will be accessible for our analysis: for instance, IR spectroscopy of individual grains, or double laser probe, which are promising techniques for identifying organic molecular species present inside the grains.

The COMRADE proposal expands upon a program initiated a few years ago, with the COMET-1 experiment.[6] It provides for the collection of cometary dust and space debris as well as the characterization of their dynamic properties. It will involve exposing a variety of detectors and captors onboard spacecraft orbiting around the Earth. The opportunities in the future might include EURECA-2, LDEF-2 and possibly the Mir station to which an improved version of the COMET experiment could be attached. This latter flight opportunity would open the possibility to collect material at any given period and for any duration chosen.

REFERENCES

1. J.F. Bell, *Meteoritics* **26**, 4, 316 (1991).
2. J. Borg, C. Maag, J-P. Bibring, W. Tanner and M. Alexander, *Proceedings of First ESA Meeting on Orbital Debris* (ESA SD-01) 137-142 (1993).
3. J-P. Bibring, J. Borg, Y. Langevin, B. Rosenbaum, P. Salvetat and Y.A. Surkhov, *Lunar Planet. Sci.* XVI, 55-57 (1985).
4. C. Maag and W.K. Linder, *Hypervelocity Impacts in Space,* (McDonnell ed.), Canterbury, U.K., 186-190 (1993).
5. C.R. Maag, W.G. Tanner, T.J. Stevenson, J. Borg, J-P. Bibring, W.M. Alexander and A.J.Maag, *Proceedings of First ESA Meeting on Orbital dDbris* (ESA SD-01) 125-130 (1993).
6. J. Borg, J-P. Bibring, Y. Langevin, Ph. Salvetat and B. Vassent, *Meteoritics* **28**, (1993).

312 COMRADE Experiment

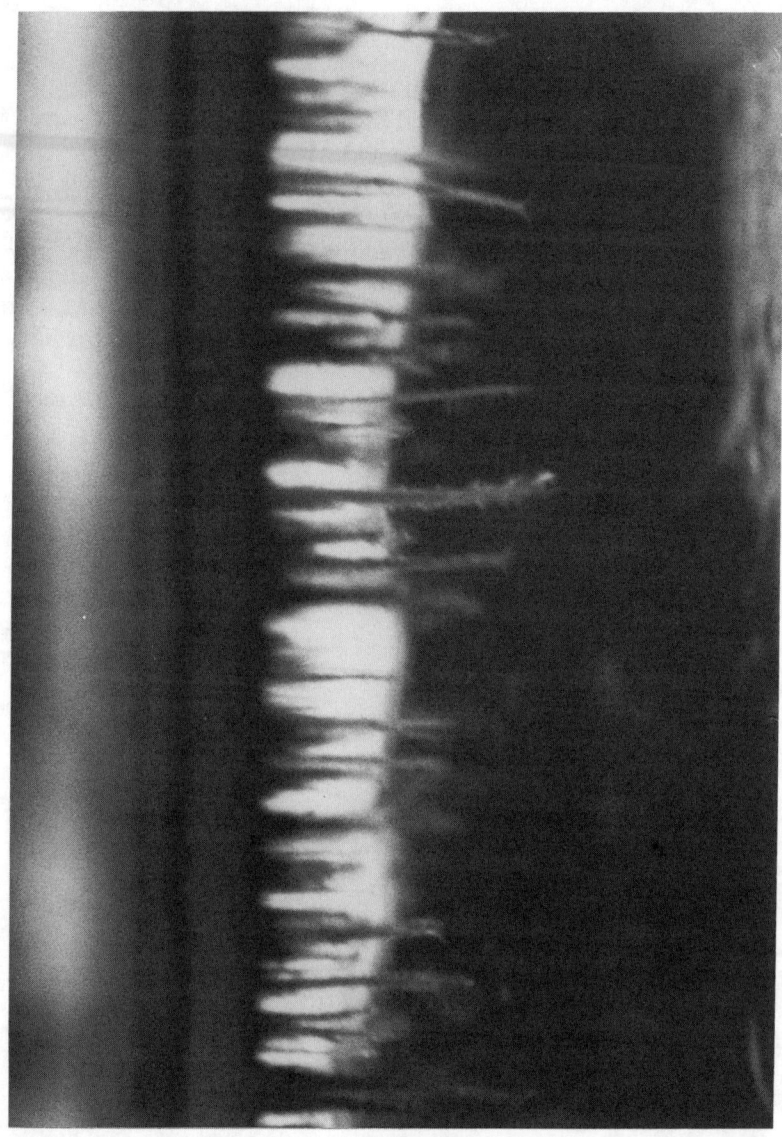

A transmitted light optical view of a block of silica aerogel which has served as a capture medium for mineral grains launched at high velocity. Many grains of olivine, pyrrhotite and pyroxene (individually measuring 100 μm in diameter) were fired into the low-density (0.04 g/cm^3) aerogel at a velocity of 7 km/sec; each captured grain bored a track to mark its passage (the particles entered the aerogel from the left). The captured particles (or their residues) lie at the base of each track as well as along its length. Workers have successfully extracted such captured grains from the enclosing aerogel, proving the utility of this technique for IDP capture on spacecraft. See the paper by Borg et al in this volume for discussion of these underdense media capture cell techniques. The longest tracks in this view measure 2 mm in length. (Photo by R.A. Barrett)

ANALYSIS OF REMNANTS FOUND IN LDEF AND MIR IMPACT CRATERS

L. Berthoud and J.C. Mandeville
CERT-ONERA/DERTS
2 Av. E. Belin, 31055 Toulouse cedex, France

ABSTRACT

This study is the further investigation of space-exposed samples recently recovered from the NASA satellite LDEF and the Franco-Russian 'Aragatz' mission on Mir. Some interesting impact features have been selected as examples to demonstrate various findings. Part of the objective of the experiments was to determine the nature and origin of particles in Low-Earth Orbit (LEO). Observations show that the 'Multiple Foil Detectors' appear to be an effective way of retaining impactor residues of larger particles. They provide a 'witness' foil which shows the shape and dimensions of the impacting particle. Several low velocity (< 4 km/s) oblique craters containing significant quantities of impactor residue have been identified. Energy Dispersive X-ray spectrometry (EDX) analyses of the residues show evidence of micrometeoroid and debris compositions. Similarities between the meteoroid signatures and those of chondritic interplanetary dust particles have been noted. Evidence of at least two different grain compositions in one impacting particle is shown. The discovery of debris impacts on the trailing edge of LDEF demonstrate that artificial debris may be found, not just in circular orbits, but also in elliptical orbits.

INTRODUCTION

The NASA Long Duration Exposure Facility (LDEF) was launched into a circular orbit with an initial altitude of 480 km and inclination of 28.5° in April 1984 and was retrieved from orbit after 5.7 years. The experiments were stored in trays fixed to the 5 m diameter, 10 m long frame[1]. These trays faced in 14 directions, 12 along the sides (called 'rows') and 2 on the ends. LDEF was 3-axis stabilized with the longitudinal axis continually pointed toward the center of the Earth, so that the rows remained at fixed angles relative to the velocity vector. The surfaces examined in this work include aluminum samples mounted on the French Cooperative Payload (FRECOPA) and aluminum clamps used to attach experiment trays.

The Mir space station has been in a circular orbit at an altitude of 350-450 km with an inclination of 51.6° since February 1986. The French experiment 'Echantillons' was deployed during the Franco-Russian mission 'Aragatz' on the 9th of December 1988 and was recovered on the 11th of January 1990, thirteen months later[2]. It was attached to the exterior of the core module of the Mir space station and consisted of a 1 x 1 x 0.2 m frame, upon which several experiments were mounted. Two of the experiments were designed to study the composition and distribution of dust particles in Low-Earth Orbit (LEO).

One of the objectives of retrievable experiments flown in LEO is the identification of the particles responsible for the formation of craters on surfaces exposed to the LEO particulate environment. In this work, particle identification was achieved by Energy Dispersive X-ray spectrometry (EDX) analyses of particle residues in and around the craters. Residues were classed as being of *natural* origin if they had a high Mg, Al, Si, Fe, Ca content. They were classed as being of *man-made* origin if they had high proportions of O (for spheres of Al_2O_3 fuel residues), Ti and Si (paint flakes), or Fe, Cr, Ni, Cu, Zn (stainless steel and alloys). Analysis efforts were handicapped by alloy inclusions, organic contamination containing Na, Cl, K, S (which may have occurred before, during or after space exposure) and the shadowing effect caused by crater geometry. For a summary of the observed meteoroid vs. debris proportions for several experimental surfaces and a discussion of the importance of particle impact parameters and experiment position (with respect to the leading edge of the spacecraft) see references 3-6.

IMPACT FEATURES

Multiple Foil Detectors

Multiple foil detectors mounted on the 'Echantillons' module and on LDEF FRECOPA acted as 'energy sorters' to collect the fragments of meteoroids of sizes between 1 and 20 µm (Figure 1). The principle of multiple foil detectors is to slow down particles without complete destruction so that some fragments remain on the thick target surface behind the foils (see case 2 in Figure 1). Examination of the space-exposed detectors from LDEF and Mir revealed several instances of fragmentation of the particle. These particles left perforations with a barely perceptible lip formation in the top foil. The perforations were the same shape and size as the particles, which avoids the necessity for conversion between crater size and particle diameter. An example is shown in Figure 2. The particle perforated the top foil (Figure 2a) and then fragmented, causing a star-shaped distribution of secondary craters on the target below (Figure 2b). EDX analysis of these secondary craters revealed the presence of Si, Na, Ca, Fe and Mg, indicating a particle of natural origin (Figure 2c).

LDEF Clamps

The clamps consist of plates of 6061-T6 aluminum (which have been treated with chromic acid) measuring 4.8 x 12.7 x 0.45 cm. The clamps examined here held trays onto Bay A, rows 1, 4, 5, 6, 7, 8 and 9 around the girth of the LDEF satellite (rows 3 and 9 were the satellite trailing and leading edges, respectively).

Figure 3 shows an oblique impact (depth to diameter (average) ratio P/D=0.41) on clamp A04 C05. The characteristic 'shelf' formation visible in this crater has been observed to occur at impact angles > 60° (to normal) in simulation experiments by the authors[7]. The jet of projectile ejecta produced by hypervelocity impact frequently leaves projectile fragments on these shelves. EDX analysis of the melt fragments revealed the presence of Al, Mg, Si and Fe - a classic signature for a particle of natural origin (see particle W7027F11 in reference 8). Figure 4 shows an impact on clamp A06 C06. Unmelted fragments of the impacting particle are clearly visible inside the crater. EDX spectra of different areas of the residue showed a wide range of proportions of the elements Al, Mg, Si, Ca, Fe and Cr (Cr probably comes from the chromic acid treatment). We deduce that the impactor was a 'fluffy' meteoroid consisting of an aggregate of at least two different minerals of different compositions.

LDEF FRECOPA (Trailing Edge)

FRECOPA was positioned on the trailing edge of LDEF (West side) on row 3. It consisted of 11 experiments mounted in 6 boxes. Two passive experiments were dedicated to the study of dust particles: A0138-1 and A0138-2.

Figure 5 shows a crater (P/D=0.3) found on the trailing edge of LDEF. EDX analysis of the melt residues shows a micrometeoroid signature - Al, Mg, Si, Fe, O, S, Ca and Cr.

Figure 6 shows an impact found on the trailing edge of LDEF. The impactor appears intact although fractured. EDX analysis revealed the presence of Al, Fe, Cr and V, indicating an origin as steel. This demonstrates that man-made debris exist in elliptical orbits around the Earth. While this circumstance was previously suspected, proof of the existence and magnitude of debris in non-circular orbit has come only from LDEF investigations by Fred Horz and Don Kessler[9], and the current study.

CONCLUSIONS

Chemical identification of residues is easier for low velocity impacts. This means that oblique impacts and multiple foil detectors, which both slow down the impacting particle, are more likely to retain impactor residues than normal impacts onto thick targets. Indeed, meteoroids travel at higher velocities than debris and are thus more likely to vaporize and leave no residue. These methods therefore offer a rare opportunity to observe and analyze micrometeoroids that have caused craters in experimental surfaces. They provide ideas for the design of future detectors.

The meteoroid impacts are all relatively shallow (P/D varying from 0.3 to 0.52), which is commensurate with the high lateral velocity component for oblique impacts and with the low density of the impactors (assumed to be around 2-3 g/cm^3 from the compositions). The examination of unmelted meteoroid fragments reveals their agglomerate nature and similarity to certain chondritic interplanetary dust particles.

ACKNOWLEDGMENTS

The authors would like to thank the following for their help: Christian Durin at CNES, Françoise Pichoir at ONERA-Chatillon, the Laboratoire de Metallurgie at ENSAE, Ronald Bernhard and Jack Warren from Lockheed ESC.

REFERENCES

(1) Clark, L.G. et al. (eds.) *NASA SP-473*.
(2) Mandeville, J. C. (1990) *Adv. Space Res. Vol. 10*, No. 3-4, pp. 3397-3401.
(3) Hörz, F., Bernhard, R.P. (1992) *NASA TM-104750*.
(4) Berthoud, L., Mandeville, J. C. (1994) *Proc. LDEF-3rd Post-Retrieval Symp.* (in press).
(5) Hörz, F., Cintala, M., Bernhard, R.P. and See, T.H. (1994) Penetration Experiments in Aluminum and Teflon Targets of Widely Variable Thickness. This volume.
(6) Zolensky, M.E., Hörz, F., See, T., Bernhard, R.P., Barrett, R.A., Mack, K., Warren, J. and Kinard, W.H. (1994) Meteoroid Investigations Using the Long Duration Exposure Facility. This volume.
(7) Berthoud, L. (1993) PhD thesis, ENSAE, Toulouse, France.
(8) Zolensky, M.E., et al. (1983) *Cosmic Dust Catalog 4. Planetary Materials Branch Pub. #65* (NASA Johnson Space Center, Houston) 170p.
(9) Zolensky, M.E., et al. (1992) *Part 2 LDEF-Second Post-Retrieval Symp.*, NASA CP-3194. pp 277-302.

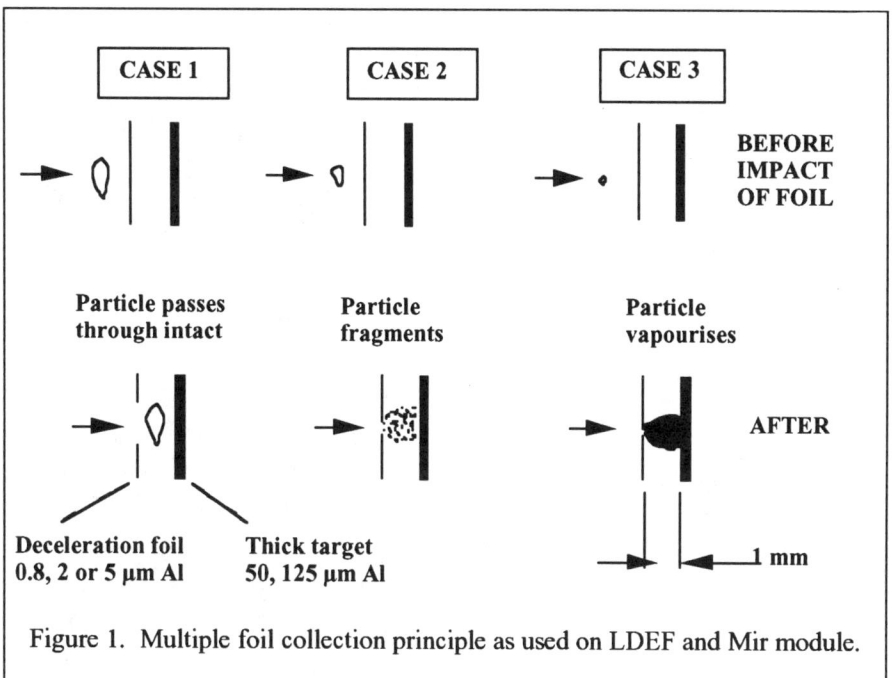

Figure 1. Multiple foil collection principle as used on LDEF and Mir module.

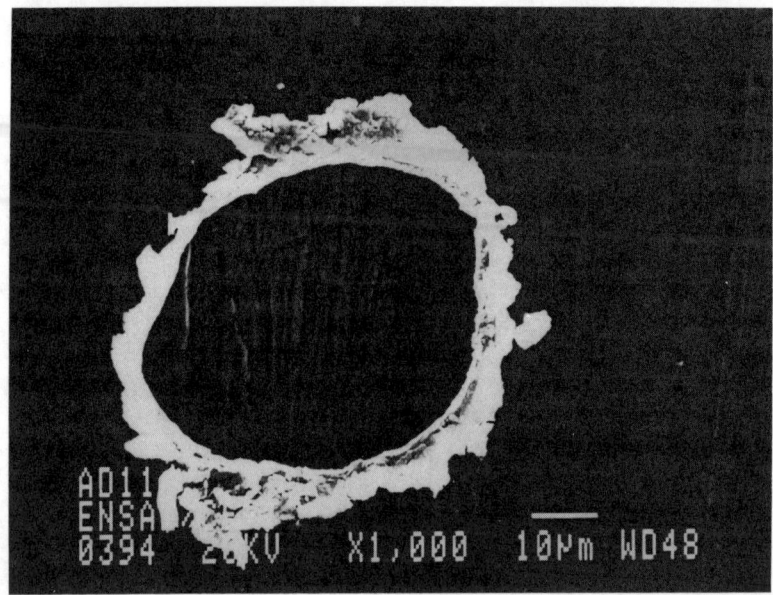

2a. Perforation (55 x 40 μm) of 5 μm thick Al top foil. Scale measures 10 μm.

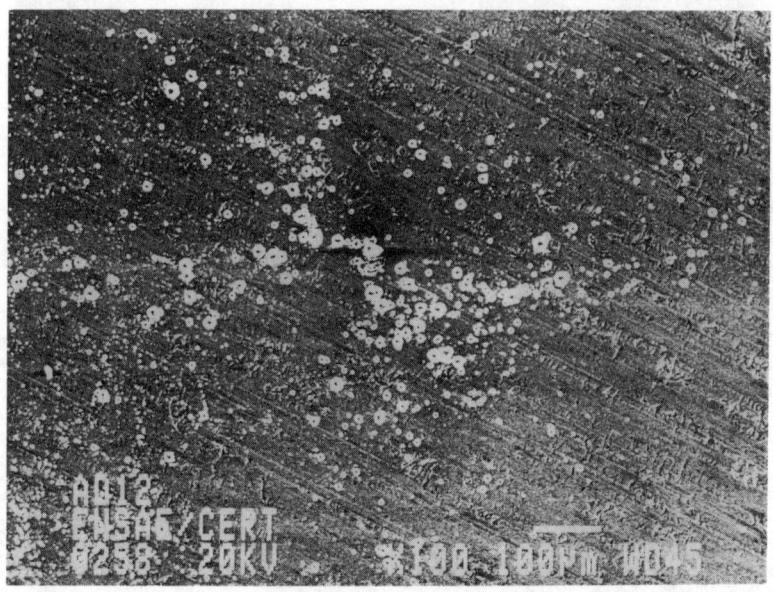

2b. Star-shaped crater distribution around crystalline fragment on 125 μm Al bottom foil. Scale measures 100 μm.

Figure 2. LDEF D11/2 Multiple Foil Detector

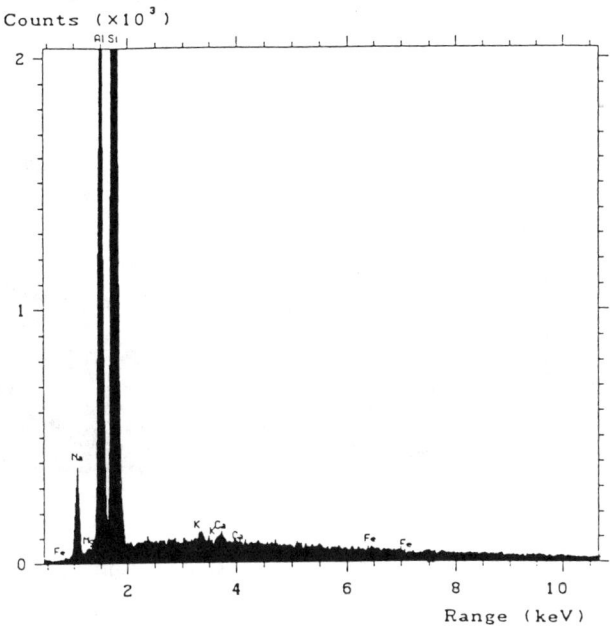

2c. EDX spectrum of crystalline fragment (20 keV).

Figure 2. LDEF D11/2 Multiple Foil Detector (cont.)

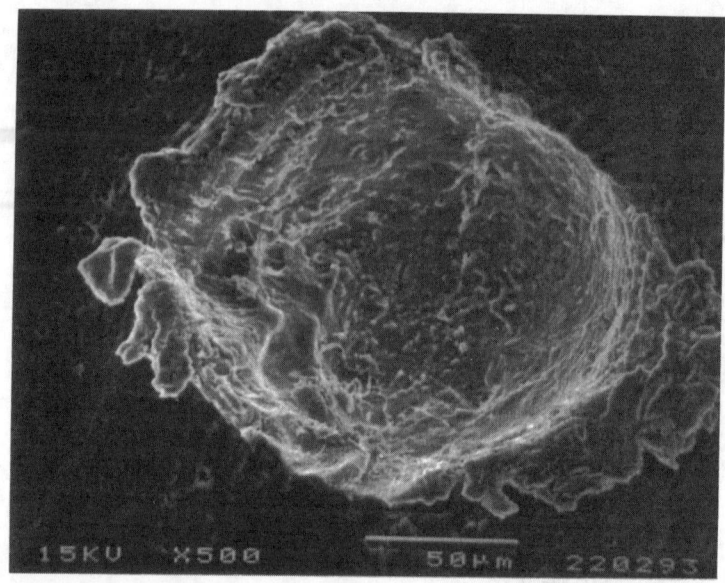

3a. Oblique impact (135 x 125 µm). Scale measures 50 µm.

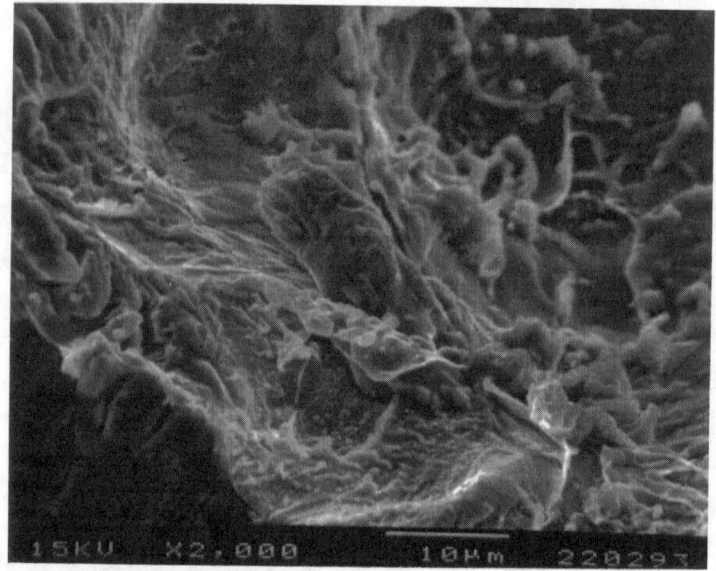

3b. Detail of molten residues on shelf. Scale measures 10 µm.

Figure 3. Meteoroid impact on LDEF clamp A04 C05.

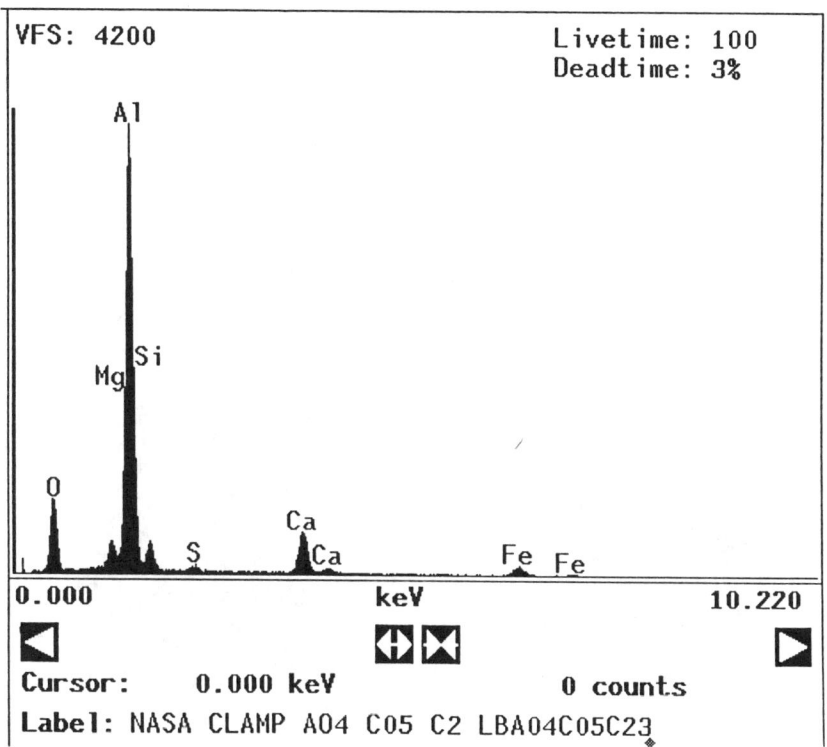

3c. EDX spectrum of residues (15 keV)

Figure 3. Meteoroid impact on LDEF clamp A04 C05 (cont.).

322 Analysis of Remnants Found in LDEF

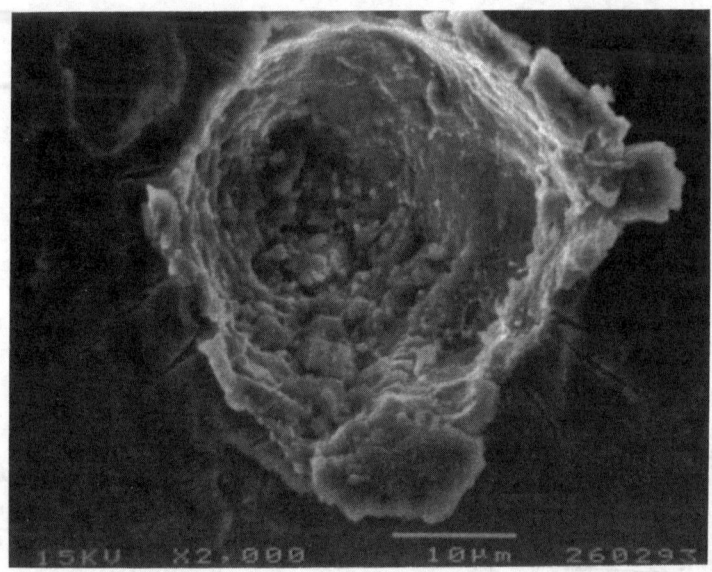

4a. Low velocity impact (33 µm) in LDEF clamp A06C06. Scale measures 10µm.

Figure 4. Low velocity impact (33 µm) in LDEF clamp A06C06 with spectra from different areas of the residues to show different grain compositions.

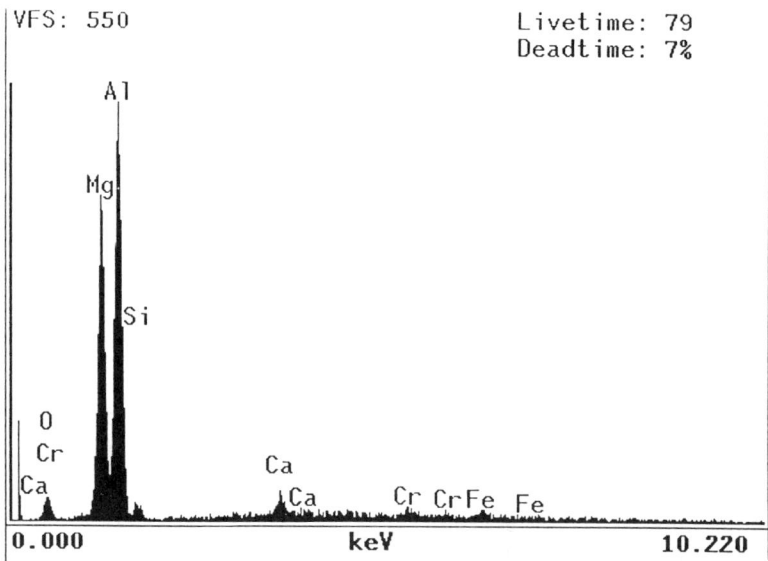

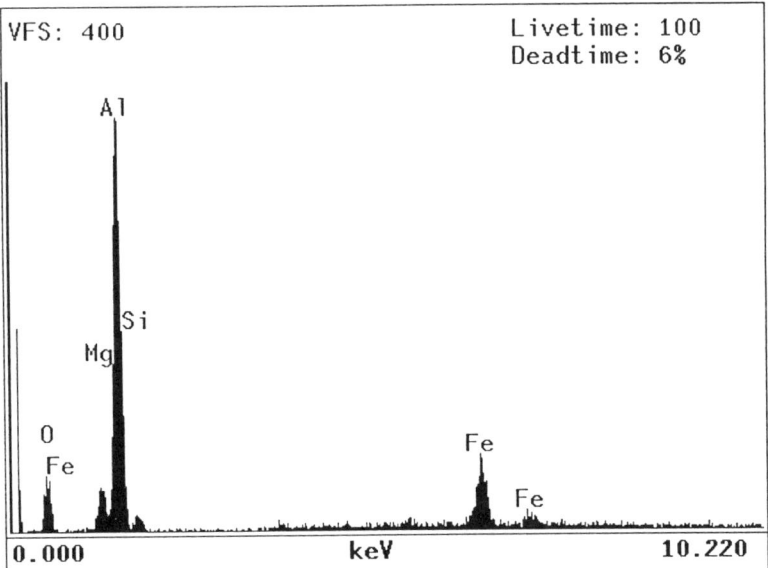

4b. Spectra from different areas of the residues to show different grain compositions.

Figure 4. Low velocity impact (33 μm) in LDEF clamp A06C06 with spectra from different areas of the residues to show different grain compositions (cont.).

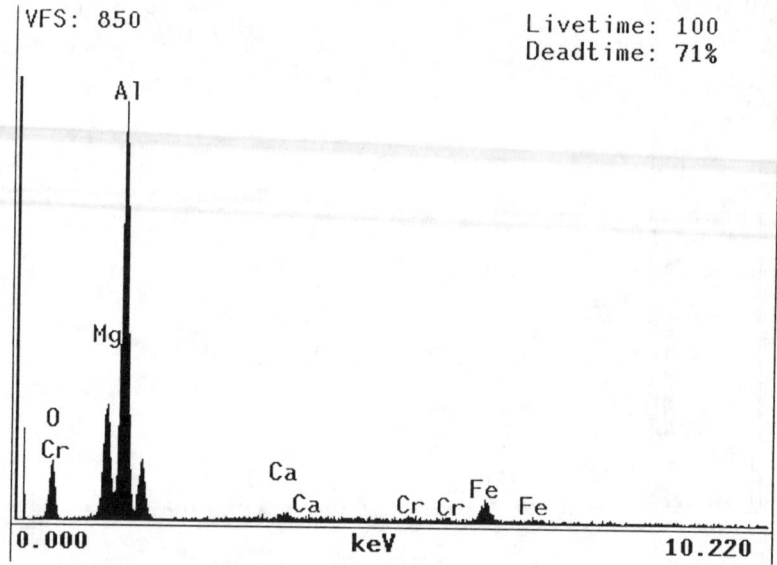
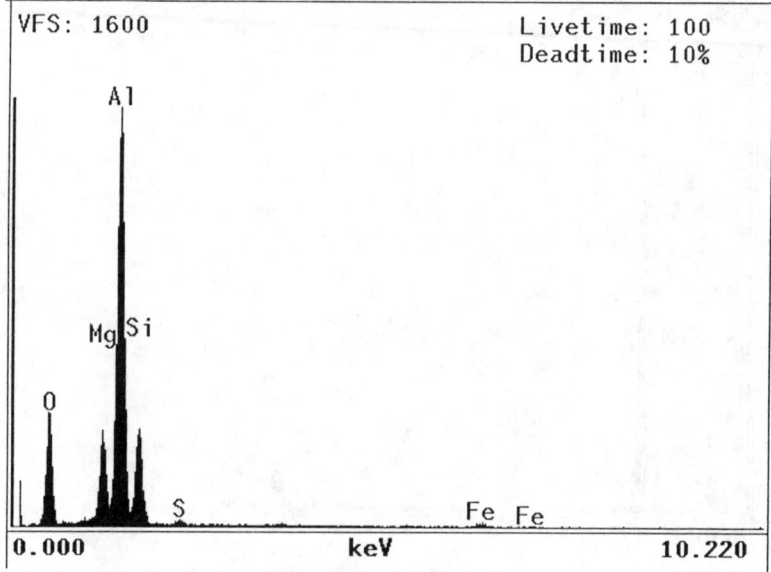

4b. Spectra from different areas of the residues to show different grain compositions (cont.).

Figure 4. Low velocity impact (33 μm) in LDEF clamp A06C06 with spectra from different areas of the residues to show different grain compositions (cont.).

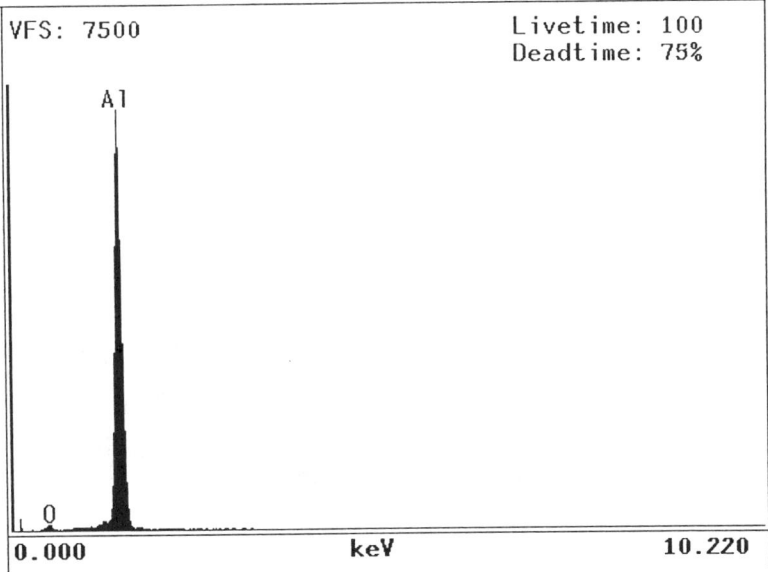

4b. Spectra from one area of the residue to show a final grain composition (cont.).

Figure 4. Low velocity impact (33 μm) in LDEF clamp A06C06 with spectra from different areas of the residues to show different grain compositions (cont.).

326 Analysis of Remnants Found in LDEF

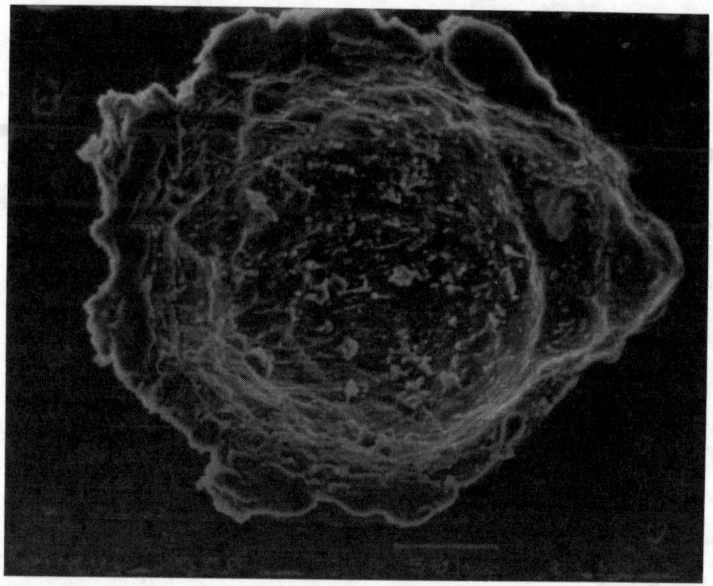

5a. Oblique impact (235 x 176 μm) on Al alloy experiment support surface. Scale measures 50 μm.

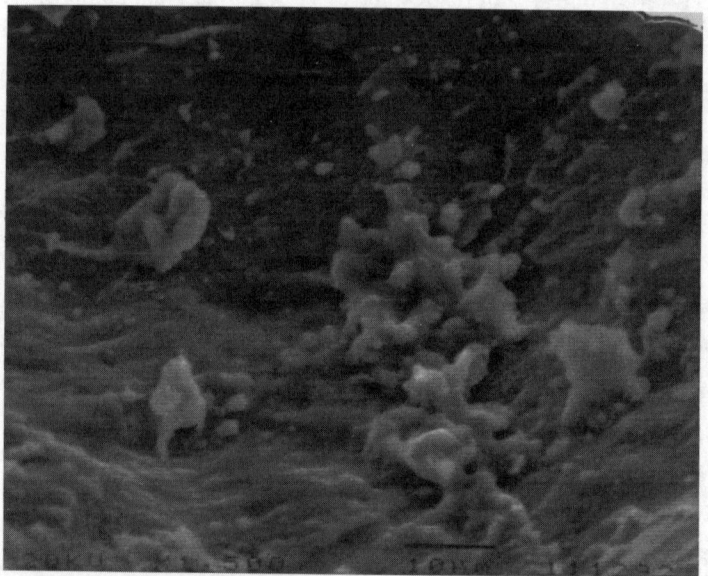

Figure 5. Meteoroid impact on LDEF FRECOPA (mounted on trailing edge) 'Ecran Rigide'.

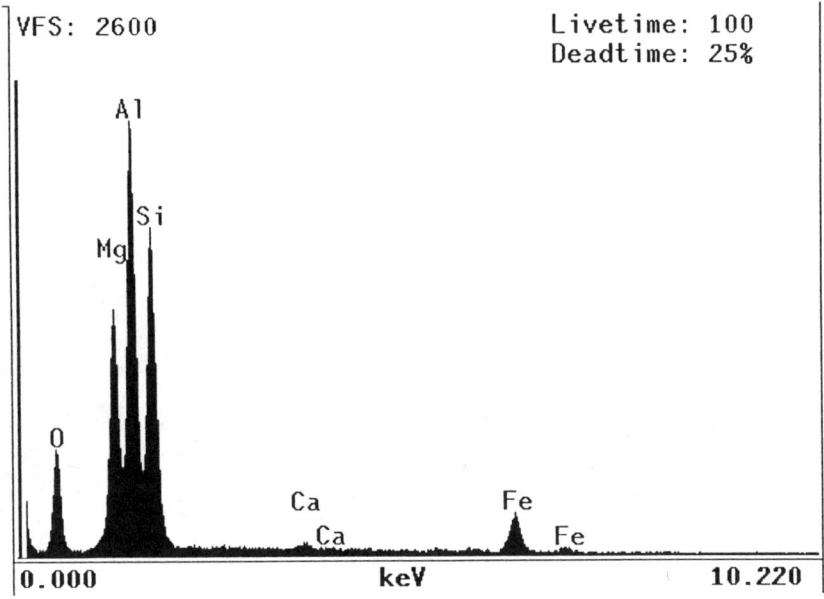

Figure 5c. EDX spectrum of residues (15 keV).

Figure 5. Meteoroid impact on LDEF FRECOPA (mounted on trailing edge) 'Ecran Rigide' (cont.).

328 Analysis of Remnants Found in LDEF

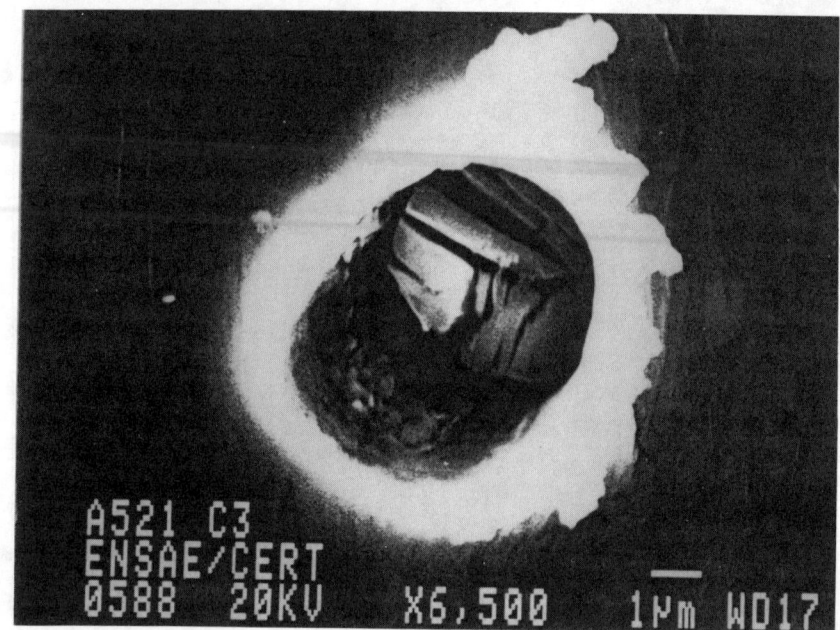

6a. Oblique impact (6.6 x 5.4 µm) on 99% pure Al experiment surface. Scale measures 1 µm.

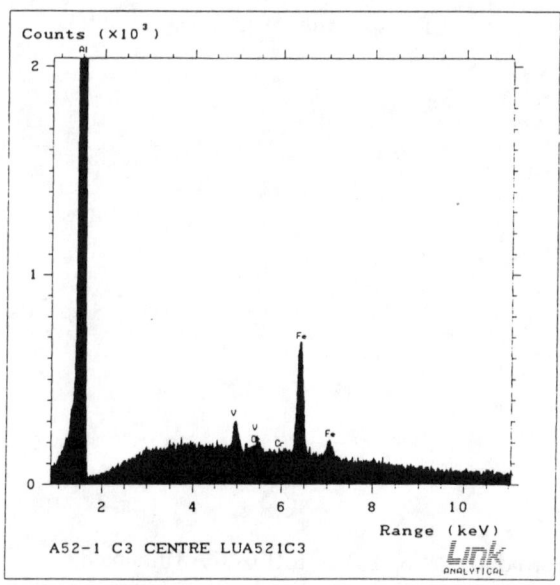

6b. EDX spectrum of residues (obtained at 20 keV).

Figure 6. Debris impact on LDEF FRECOPA.

PENETRATION EXPERIMENTS IN ALUMINUM AND TEFLON TARGETS OF WIDELY VARIABLE THICKNESS

Friedrich Hörz
Mark Cintala
Solar System Exploration Division
SN4, NASA Johnson Space Center
Houston, TX 77058, USA

Ronald P. Bernhard
and
Thomas H. See
Lockheed Engineering & Sciences Company
C23, 2400 NASA Road 1
Houston, TX 77058, USA

ABSTRACT

A 5 mm light-gas gun was used to fire spherical soda-lime glass projectiles from 50 to 3175 µm in diameter (D_p), at a nominal 6 km/s, into aluminum (1100 series; annealed) and Teflon (Teflon$^{TFE®}$). Targets ranged in thickness (T) from infinite halfspace targets (T ≅ cm) to ultra-thin foils (T ≅ µm), yielding up to four orders of magnitude variation in absolute and relative (D_p / T) target thickness. This experimental matrix simulates the wide range in D_p / T experienced by a space exposed membrane of constant T that is being impacted by projectiles of widely varying sizes.

Penetration hole size (D_h) decreases systematically with decreasing target thickness. Relative hole size (D_h / T) may be used to extract projectile diameter D_p from *individual* penetration holes in space-exposed surfaces, provided one assumes an impact velocity. The condition of $D_h = D_p$ mandates $D_p / T > 50$ in both targets. The ballistic-limit thickness (T_{BL}), at 6 km/s, occurs at $D_p / T = 0.29$ for our aluminum, and at $D_p / T = 0.16$ for the Teflon. While these thicknesses define the onset of physical perforation, they are not synonymous with the transition from cratering to penetration processes; this transition is gradual and occurs over a wide range of T. Consideration of the shock-pulse duration (t) in both the projectile (t_p) and target (t_t) identifies the condition of $t_p / t_t = 1$ as the real transition between cratering and penetration processes. This transition, at 6 km/s, takes place in our aluminum and Teflon targets at a $D_p / T = 0.83$ and 0.62, respectively (*i.e.,* at target thicknesses some factor of 2-3 thinner than the ballistic limit). Consideration of pulse duration is readily extended to impact velocities beyond those simulated in the laboratory; it may assist in understanding the velocity scaling of penetrative impact events.

© 1994 American Institute of Physics

INTRODUCTION

The morphologies and detailed dimensions of hypervelocity craters and penetration holes on space-exposed surfaces faithfully reflect the initial impact conditions. However, current understanding of this post-mortem evidence and its relation to such first-order parameters as impact velocity or projectile size and mass is incomplete. While considerable progress is being made in the numerical simulation of impact events, continued impact simulations in the laboratory are needed to obtain empirical constraints and insights.[1-3] This contribution summarizes such experiments with aluminum[4] and Teflon[5] targets, which were chosen in order to provide a better understanding of the crater and penetration holes reported from the Solar Maximum Mission (SMM[6]) and the Long Duration Exposure Facility (LDEF).[7-9] The ultimate objective of both the impact experiments and the dimensional analysis of the space-produced impact features is to obtain reliable mass-frequencies, fluxes and other dynamic properties of natural and man-made particles in low-Earth orbit. These parameters are of significance in their own right, yet they can also relate diagnostically to properties and processes of their parent objects, such as comets or asteroids.

Figure 1 illustrates the size distribution of penetration holes in the double-walled thermal louvers of the (A) SMM and (B) LDEF thermal-control blankets. The SMM louvers consisted of 125 μm thick sheets of aluminum (>99% Al, while the LDEF blankets consist largely of a 125 μm thick outer layer of Teflon, a mid-layer of vapor-deposited silver inconel, and a back 50-60 μm thick layer of organic binder and thermal protective paint.[7] These aluminum and Teflon foils rank among the most significant opportunities for investigations of particle populations in low-Earth orbit (LEO), with the largest holes approaching millimeter dimensions.

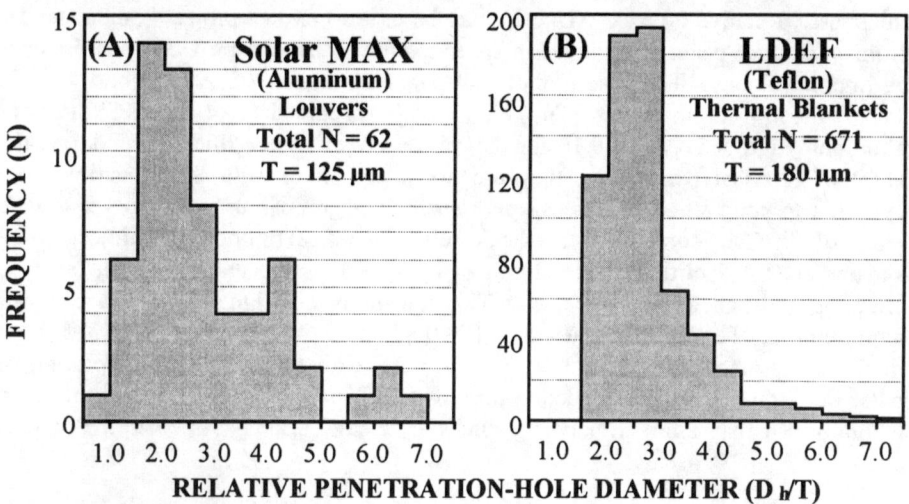

Figure 1. The frequency distribution of penetration holes normalized to target thickness (D_h / T) in the (A) Solar Maximum louvers[6] and (B) all penetrations holes >300 μm in diameter that were documented in the thermal-control blankets of LDEF.[7]

a

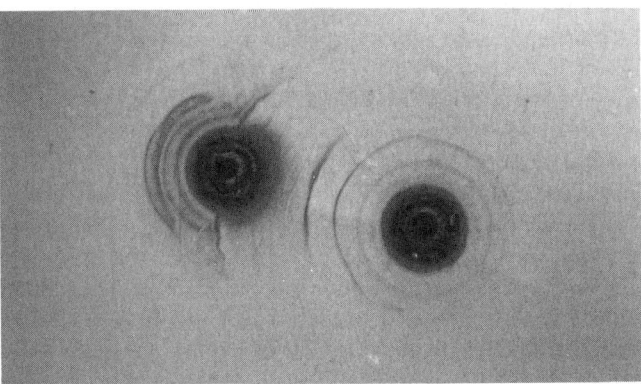

b

Figure 2. Photographs showing impacts into an LDEF thermal-control blanket. (A) Overall view illustrated by an on-orbit photo of Bay E10 showing the large number of penetrations (black dots) on a forward-facing blanket (NASA Photo # S32-78083). Vertical dimension of experiment tray equals ~1 m. (B) Close-up view of two penetrations in the thermal blanket occupying Bay D11 (Features 63 & 64; NASA Photo # S90-44968). The nature of the black dots is visible in this photo which reveals that these dark circular structures were, in reality, composed of numerous rings surrounding the penetration hole. The majority of penetrations through these blankets had such rings associated with them, as well as larger delamination zone (barely visible in photo). Field of view is ~ 8 cm across.

It should also be obvious that Figure 1 refers only to the most energetic impacts on these surfaces (*i.e.,* those capable of perforating a target of thickness T to form a penetration hole of diameter D_h). Both surfaces sustained a large number of non-penetrative, small impacts that resulted in hypervelocity craters, the majority with diameters of $D_c \ll T$.[6&10] As a consequence, interpretation of the complete bombardment record on most space-exposed surfaces requires a detailed understanding of both their cratering and penetration behavior (see reference 11). Therefore, we devised an experimental matrix that systematically varied target thickness from targets thick enough to sustain full fledged cratering events, to foils so thin that the size of the penetration hole approached typical projectile dimensions.

The above transition from cratering to penetration is traditionally characterized as the "ballistic limit" (T_{BL}) of the target. Given specific initial impact conditions, the latter specifies a minimum particle mass that is capable of physically perforating the target. By definition, targets of thickness T_{BL} will simultaneously sustain a full fledged cratering event of some characteristic, maximum depth P (P < T)[11-14], as well as incipient penetration holes of dimensions $D_h = 0$.[4] Applicable cratering and penetration formulas must yield identical particle mass for such threshold craters and penetration holes, yet analysis of the Solar Max[6] and LDEF impact features in aluminum targets[14&15] revealed that current cratering[16] and penetration formulas (*e.g.,* reference 17 and others) do not yield internally self-consistent results for the ballistic-limit case.[11&13] Typically, the penetration formulas yield smaller particle masses than the cratering formulas for penetration holes close to the ballistic limit. Therefore, special emphasis was placed on the transition from cratering to penetration processes in this work.

EXPERIMENTAL RESULTS

Procedures and Definitions

Detailed procedures of the impact experiments employing a 5 mm light-gas gun are given by Hörz et al.[4] All experiments were performed with spherical soda-lime-glass projectiles at normal incidence. Particles >0.5 mm were launched individually in serrated sabots; smaller projectiles were shotgunned. Velocities were measured by arrays of photo-diodes that either sense the occultation of an IR-LED laser beam in the free-flight chamber (four stations), or of an impact flash at the sabot separator and the target. All experiments were equipped with massive witness plates of aluminum (1100 series, 3.2 or 6.4 mm thick, and ~25 cm square) that were located ~10 cm to the rear of the penetrated target. We measured, using optical and Scanning Electron Microscope methods, the diameters of the crater (D_c; intercept of cavity walls with initial target surface); of the rim (D_r; rim crest to rim crest), the crater lip (D_l; average diameter of lip-periphery), the spall-zone (D_s; average extent of the lobate area of displaced mass), hole diameters (D_h; minimum diameter of physical opening) and spall diameter of the back surfaces (D_b).[4]

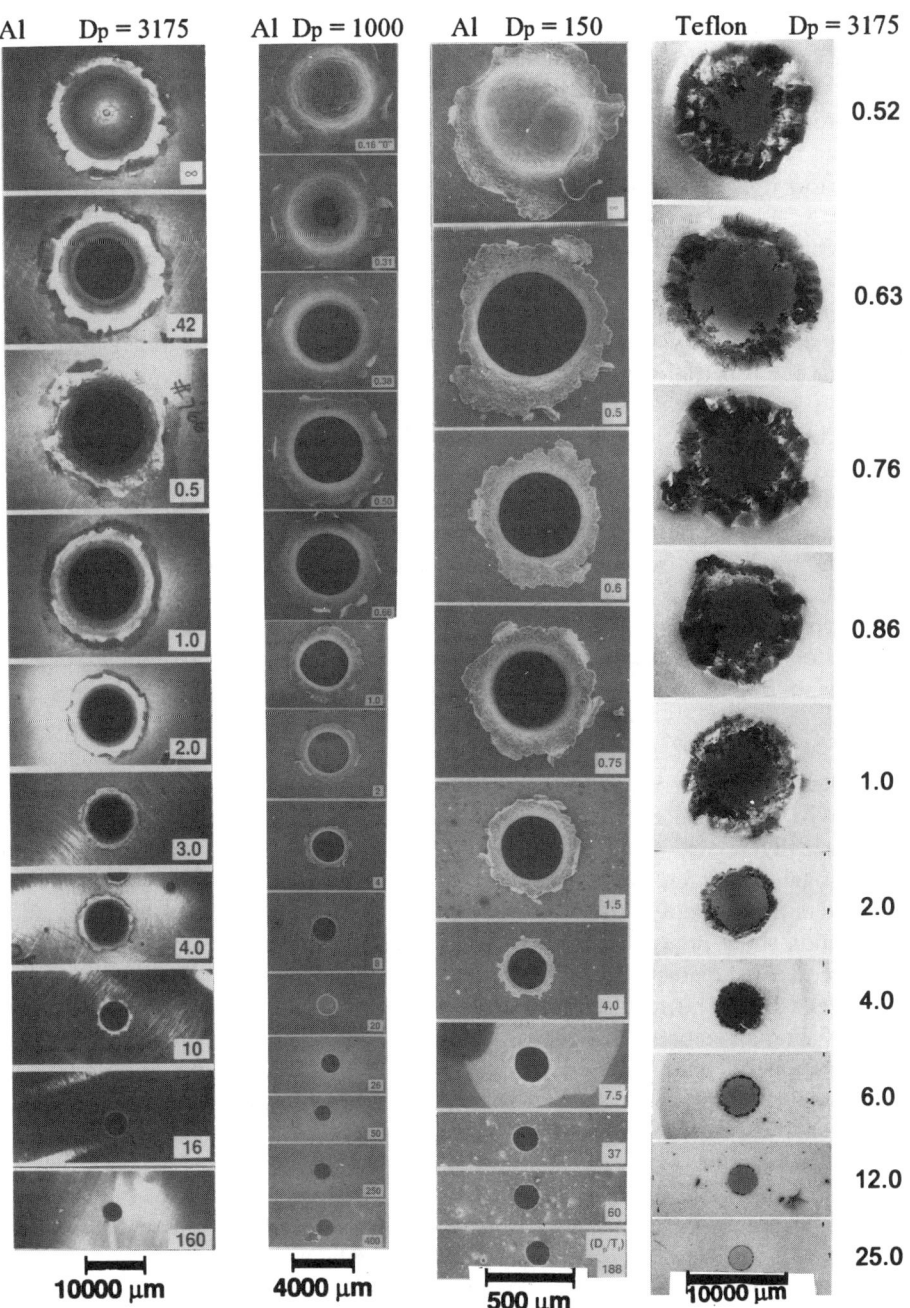

Figure 3. Craters and penetration holes resulting from soda-lime glass projectiles in aluminum and Teflon[TFE]® targets of highly variable thickness, identified by specific D_p / T. Each column illustrates a specific series of experiments at constant projectile diameter (D_p). Note that most holes in the thinnest foils approach the condition of $D_h = D_p$.

Crater Morphology

Typical craters and penetration holes produced in aluminum (1100 series) and Teflon^{TFE} targets are illustrated in Figure 3. Target thickness varies by essentially four orders of magnitude, from very massive specimens of essentially infinite thickness relative to the crater depth (*i.e.*, infinite halfspace-type target at top of each column) to foils as thin as 0.8 μm thick. Target thickness is given in normalized form (D_p / T) to facilitate convenient comparison with craters and penetration holes at different dimensional scales. The aluminum experiments employing projectiles of various sizes demonstrate that grossly similar morphologies are obtained at equivalent experimental conditions, no matter what absolute scale. All experiments show a strong and systematic relationship between hole size and target thickness. Note that the penetration holes in massive targets have dimensions typical for actual cratering events, yet hole size diminishes with decreasing foil thickness and approaches projectile dimensions only at very thin foils, typically of $D_p / T > 50$.

Figure 4 illustrates representative cross sections through relatively massive aluminum and Teflon targets, using the 3.2 mm projectiles, to specifically document the transition from pure craters to penetrative events. Although already visible in Figure 3, these cross sections best illuminate the substantially ductile deformation behavior of aluminum as opposed to the brittle response of Teflon. Nevertheless, many morphologic features, not just hole size, are part of a continuum that strongly depends on target thickness. For example, note the substantial, yet gradual increase in the depth of aluminum craters as the ballistic limit (at $D_p / T \cong 0.3$) is approached (Figure 4a), or the development of connective fractures leading to spallation at the rear surface of Teflon targets (Figure 4b), or the relative width of the crater-lip or of the spall-zones. (Figures 2 and 3).

The dependence of most morphologic features in Figure 3 and 4 is so systematically related to target thickness, that their detailed characterization and dimensional measurement should lead to diagnostic relationships between the parameters D_p, D_h, and T. As T is known for any space-retrieved surface, substantial information about projectile size may be extracted from the characterization of such quantities as hole diameter, relative rim-width, distribution of fractures, spall zones at the target's front and rear surfaces, etc. The current observations apply to some limited laboratory conditions, most notably a constant impact velocity of 6 km/s. Interpretation of space-exposed surfaces requires, as a minimum, the additional understanding of the velocity dependence of these relationships, and possibly other factors, such as projectile density and/or shape.

Returning to Figure 4, the physical perforation of both target materials is preceded by massive internal deformation and fractures close to the target's rear where reflection of the initial shock front results in a rarefaction wave and associated tensile forces.[18] This deformation proceeds to a stage where physical removal and spallation of substantial mass takes place at the target's rear, yet still without resulting in complete perforation (*e.g.*, at $D_p / T = 0.29$ in aluminum or at 0.149 in Teflon).

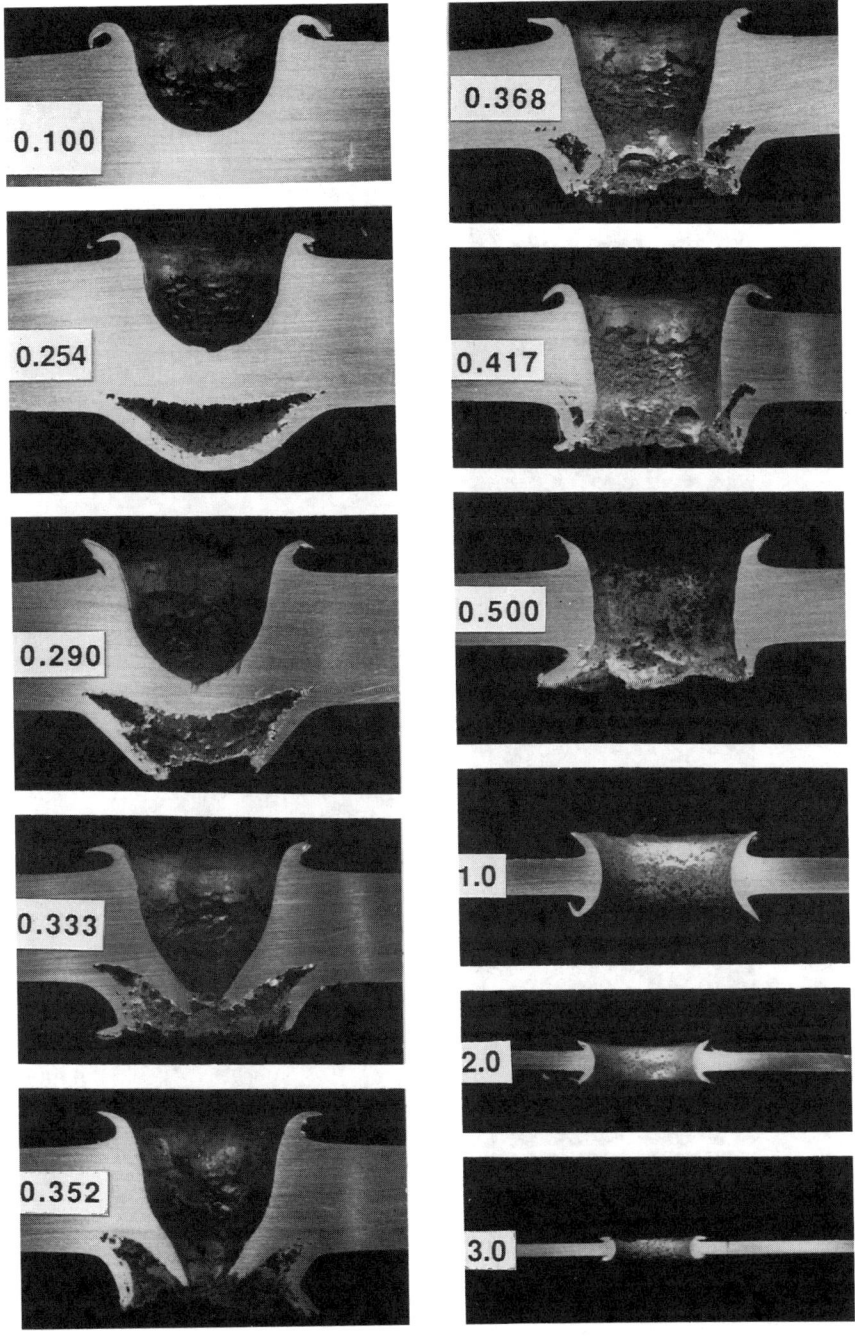

Figure 4a. Cross sections of a "standard" crater in an infinite-halfspace aluminum target (upper left) and the transition to genuine penetrations at $D_p / T > 2$ (V = 6 km/s; D_p = 3175 μm). For scale note that T = 3.2 mm at D_p / T = 1.0 (see text for discussion).

336 Penetration Experiments in Aluminum and Teflon

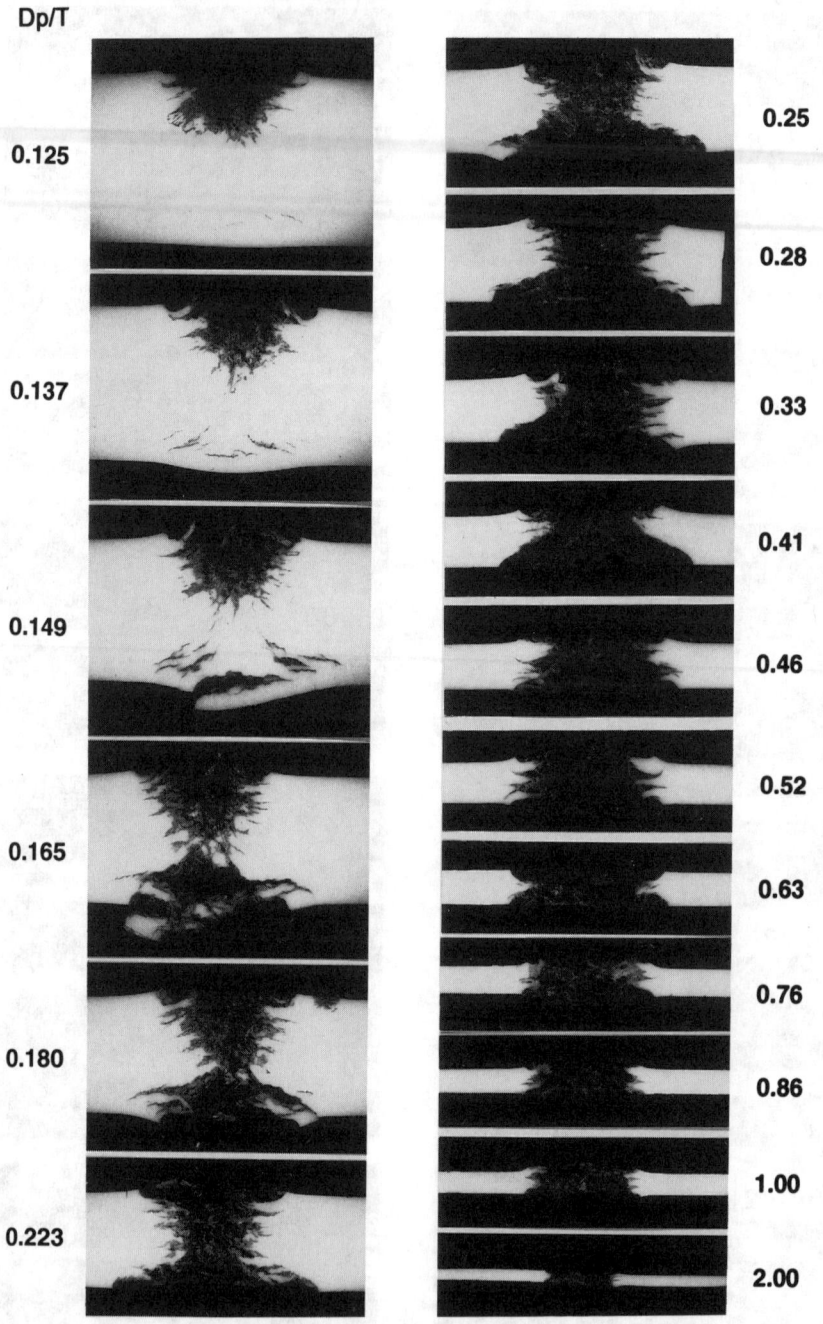

Figure 4b. Cross sections of Teflon targets (V = 6.3 km/s; D_p = 3175 μm).

Physical perforation is accomplished in aluminum (1100 series) at $D_p / T \cong 0.3$, whereas equivalent Teflon targets may be almost twice as thick ($D_p / T \cong 0.16$). Throughout these developments, the diameter of the crater and its rim or the extent of spall zones at the target's front side remain essentially that of the standard crater into infinite halfspace targets.

All experiments employed witness plates that monitored the evolving debris cloud towards the target's rear.[4] From the cross sections it seems self-evident that most debris in massive targets is dislodged from the target itself, and that the mass ratio of target and projectile residue must vary considerably as a function of T.[19&20] Detailed witness-plate analysis (Scanning Electron Microscopy and Energy-Dispersive X-ray) of aluminum experiments[4] reveals the first traces of projectile material at $D_p / T \cong 0.6$ (at 6 km/s). This seems surprising in view of the gaping holes produced at all conditions of $D_p / T < 0.6$, including the most substantial penetrations of the entire aluminum series, none of which revealed projectile traces on their witness plates. Not coincidentally, our observations reveal that projectile residue may exit a penetrated target only if the target thickness is smaller than the depth (P) of the associated standard crater (*i.e.*, if the condition of T < P applies; see Figure 4a). These observations imply that at all conditions of T > P, the projectile is totally entrained in cratering related material flows to eventually emanate from the evolving crater cavity as uprange ejecta. It is only at conditions of T < P that the cratering related flows[21] are sufficiently disturbed to have parts of the crater bottoms and projectile melts exit through the target's rear. This suggests that bona fide rim or spall phenomena at the target's front side should still resemble that of standard craters at these conditions, possibly at targets still thinner than P. Indeed, one may observe in Figures 2 and 3 that the cavity diameter at the initial target surface (D_c) is approximately that of the crater until $D_p / T \cong 1$ is reached. Rim details and spallation phenomena remain grossly similar to standard craters as well. Diameters D_c and D_h start to decrease significantly only in targets thinner than $T \cong D_p$.

Dimensional Measurements

Figure 4 illustrates measurements of D_c and D_h, which are the primary diameters and features to address projectile size. Figure 4 is purposely arranged such that it may function as a "calibration" curve to solve for the unknown projectile diameter D_p on space-exposed surfaces from the readily measured parameters D_h and T. Note, however, that these curves only apply to ~6 km/s cases, and that corresponding penetration experiments at different velocities are needed to understand the effects of velocity.

Returning to Figure 1 we see that most penetrations documented on the Solar Max and LDEF surfaces occurred at $D_h / T < 5$. This clustering around relatively low D_h / T values will be typical for all penetrated membranes retrieved from space, including those that are substantially thinner or thicker than a few hundred microns. The mass frequency of natural and man-made impactors is simply so dominated by "small" particles that all penetrated membranes will have a large population of genuine hypervelocity craters; penetrations of "massive" targets barely above the ballistic limit are the next most populous group of impact features, and the relative frequency of still larger events will rapidly decline with increasing D_h/T. The majority of penetration events will have dimensions that are truly intermediate between the genuine crater ($D_c = nD_p$) and the case of $D_h = D_p$ in the ultra-thin foil. This underscores, that the transition from cratering to penetration regimes must be well understood to interpret space-exposed surfaces correctly. If 6 km/s were indeed an applicable velocity for the data illustrated in Figure 1, the largest penetration holes would be due to impactors of ~500 μm in diameter for Solar Max and ~600 μm in diameter for LDEF surfaces.

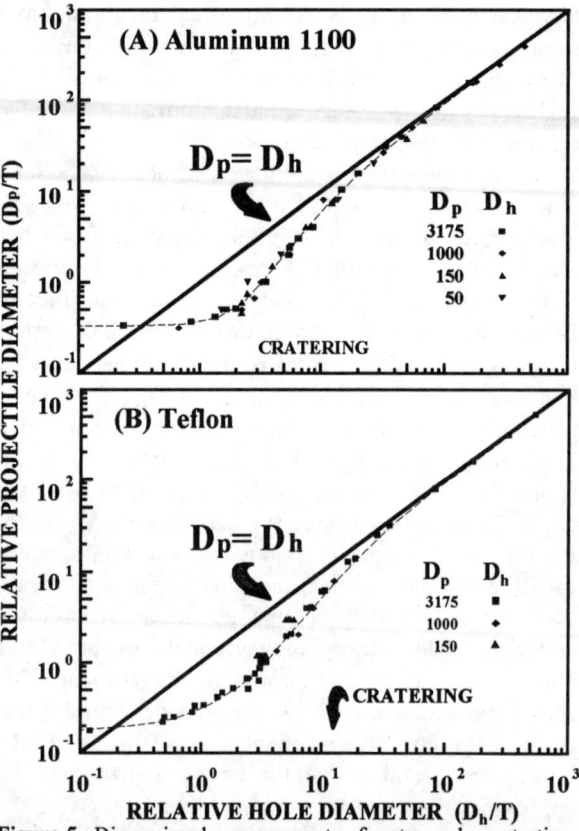

Figure 5. Dimensional measurements of crater and penetration-hole diameters for impact experiments into (A) aluminum and (B) Teflon targets at 6.0 and 6.3 km/s, respectively. Note that substantially more experiments were conducted and analyzed than could be illustrated in this short report, and that experiments of substantially different scales conform to a single curve. Unique solutions -- at some constant impact velocity -- for projectile diameter seem possible for any individual penetration hole that is characterized by the measurements of D_h and T.

Returning to Figure 4 we also see that experiments conducted with a wide range of projectile sizes combine to define a single curve that has rather similar characteristics for both target materials. This validates that dimensional scaling is a suitable approach to solve for first-order projectile dimensions from the measurement of D_h and T.[22] Hörz et al.[4] corroborate, however, that some scale-dependent behavior exists at very small projectile dimensions (<100 μm) as first described by Cour-Palais.[16]

We have performed various curve-fitting procedures to the data illustrated in Figure 4 to generalize the relationships of T, D_h and D_p. We found that a polynomial fit of the form

$$\log_{10} y = a_0 + a_1(\log_{10} x) + a_2(\log_{10} x)^2 + a_3(\log_{10} x)^3 + a_n(\log_{10} x)^n \qquad \text{Equation 1.}$$

best describes the experimental results with $y = D_p/T$ and $x = D_h/T$. Table 1 provides the associated coefficients for both aluminum and Teflon.

Table 1.
Polynomial fit of the experimental results illustrated in Figure 5.

	a_0	a_1	a_2	a_3	a_4	a_5	a_6	a_7	a_8
Aluminum (1100)	-0.458	0.175	1.008	1.199	-1.131	-0.800	1.152	-0.434	0.546
Teflon (FTE)	-0.485	0.667	0.562	-0.230	0.518	0.021	-0.661	0.415	-0.075

DISCUSSION

Aluminum alloys are among the best investigated materials in terms of cratering and impact behavior, due to their wide use as spacecraft materials as summarized by Carey et al.[17], Cour-Palais[16], or Hermann and Wilbeck[13], Christiansen[23], and Watts et al.[11], yet we are not aware of corresponding experiments and data for Teflon targets. Consequently, our present comparison with other investigations is limited to aluminum. The salient results are shown in Figure 5, where we compare our data with the penetration equations of other investigators, the latter suitably adjusted to our experimental conditions (velocity, densities, compressive strengths, etc.). We also extrapolated the equations to targets thicker than suggested and intended by the original workers, for illustrative purposes; many previous experiments were not necessarily meant to be extrapolated to very thin targets of $D_p/T > 10$.

The agreement among the various approaches is very good in the range of D_p/T of 0.5 to 10, in which most of the original data were gathered and for which most of the equations and applications were intended. Note, however, that we differ distinctly from all other workers at, or close to the ballistic limit. We not only observe experimentally, that hole size may be very small close to T_{BL}, but we also postulated that the ballistic limit satisfy the condition of $D_h = 0$. This causes a precipitous drop, over a very narrow range in D_p/T, of our experimental curve in Figure 6. While Carey et al.[17] do account for decreasing hole sizes in massive targets, they did not consider $D_h = 0$; yet this condition totally controls the curve shape over a very small interval in target thickness. Extrapolation of existing penetration formulas to the infinite halfspace case, as purposefully illustrated in Figure 5 and as performed by Warren et al.[6] or Humes et al.[15] and others, will always lead to projectile diameters smaller than those derived from cratering formulas. The ballistic limit obviously does not reflect the transition from cratering to penetration in the context of projectile size characterization based on the measurement of penetration-hole diameters.

We have emphasized throughout this report that the target's front-side morphology is either identical to, or strongly resembles that of the standard crater for targets as thin as $D_p/T = 1$ in aluminum or 0.6 in Teflon (see Figure 4). The cratering related material flow at or close to the target surface, as manifested by rim developments or spall zones and ultimately by a seemingly constant crater diameter D_c, does not seem to be significantly disturbed in targets thicker than these threshold thicknesses. As a consequence, penetrations of such targets must be thought of as craters when viewed (and measured) from the front side. The measurement of a hole diameter at some interior target location is incorrect and non-specific for the derivation of the projectile diameter in such massive targets.

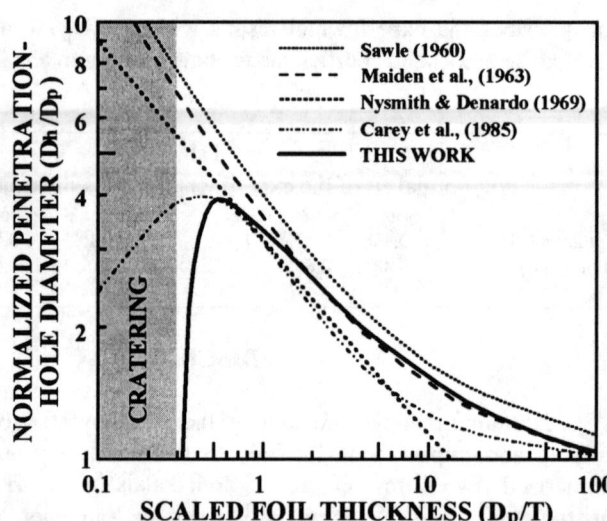

Figure 6. Comparison of our experimental aluminum impacts with diverse penetration formulas suggested by previous workers for metal targets if not specifically for aluminum. Note the fundamental difference between this work and all previous generalizations in the vicinity of the ballistic limit.

For the purposes of projectile-size characterizations, the transition from cratering to penetration occurs in targets much thinner than those which characterize the ballistic limit. This transition occurs when the feature diameter D'_c becomes measurably and distinctly smaller than the standard crater diameter D_c.

As detailed by Hörz et al.[4], each and every one of the scaled experiments reported here is uniquely characterized in terms of absolute and relative shock-pulse duration.[18] Pulse duration (t) is defined as the transit time for a shock front of speed U to reach the target's rear and to return as a rarefaction wave to the projectile/target interface where it will initiate decompression of the target. Pulse duration defines how long the compressive phase of a collisional event will last. It seems obvious that the dimensionally smaller member of an impacting pair is apt to control the pulse duration, in most cases. Cratering in infinite halfspace targets will be totally controlled by the pulse duration of the projectile (i.e., $t_p \ll t_t$). As long as $t_p < t_t$, the total duration of the event's compressive phase is precisely that of a standard crater, and corresponding material flows will be initiated (and largely sustained) by the target. However, as target thickness decreases, the condition of $t_p = t_t$ will be reached at some specific D_p / T; targets thinner than this threshold value imply that $t_p > t_t$ (i.e., that target pulse duration has become dominant). At all conditions where $t_p > t_t$, the material flow in the target will not be that normally associated with a standard crater.

We suggest that $t_t = t_p$ marks the transition from genuine cratering to penetration processes when extracting projectile sizes from space-exposed surfaces. All penetrations of $t_t < t_p$, when inspected in plan view, must be interpreted and *measured* as craters (i.e., a D_c reading must be taken rather than a hole diameter D_h).

Table 2.
Shock-stress and shock-wave velocities applicable to our experiments.

	Peak Stress (GPa)	Shock Velocity (km/s)	Velocity Ratio
Glass / Aluminum Impacts (6 km/s)			0.83
In Glass Projectiles	56	6.97	
In Aluminum Targets	56	8.43	
Glass / Teflon Impacts (6.3 km/s)			0.89
In Glass Projectiles	46	6.53	
In Aluminum Targets	46	7.35	

The condition of $t_t = t_p$ (and other pulse durations) are readily calculated if the equations-of-state (EOS) of both the projectile and target are known. Using the EOS for SiO_2, Al 2024, and Teflon^{TFE} [24] and the numerical method of Cintala[25], we calculated the shock-wave velocities (and peak pressures) applicable to our experiments (see Table 2). The ratio of these velocities corresponds directly to that D_p / T at which $t_p = t_t$ (*i.e.*, $D_p / T_{aluminum} = 0.83$ and $D_p / T_{Teflon} = 0.89$). These relative thicknesses are in excellent agreement with the experimental evidence illustrated in Figures 4a and 3b, as well as the data plotted in Figures 4 and 5.

In principle, shock velocities U_p and U_t may be calculated for any impact velocity, provided the EOS is known for both the target and projectile materials. This lead Hörz et al.[4] to calculate $t_t / t_p = 1$ for impact velocities beyond those simulated, and to determine the associated D_p / T values which mark the transition from cratering to penetration for any given model velocity. Example calculations are illustrated in Figure 6 for the glass/ aluminum pair. All impact features to the left of the $t_p = t_t$ line must be interpreted as craters. One may use applicable cratering equations and those for the ballistic limit in aluminum[16&17] to calculate any standard crater diameter (and associated D_c / D_p) and a ballistic limit thickness (expressed as D_p / T) as a function of velocity. All features to the left of the

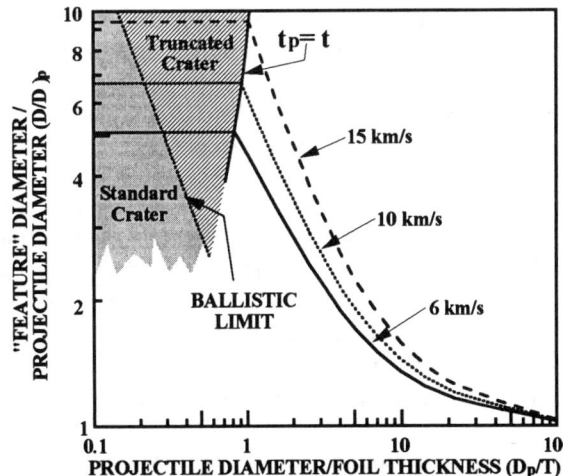

Figure 7. Possible extrapolation of experimental penetration results in aluminum targets to prevalent velocities in near-Earth orbit. The condition $t_t = t_p$ represents the demarcation between cratering and penetration regimes as calculated from equations-of-state data and associated shock velocities. Relative crater diameters (D/D_p) and ballistic limits were calculated from Cour-Palais[16] and Christiansen.[23]

ballistic-limit line are genuine craters, while the features between the ballistic-limit line and the pulse-duration line are actual penetrations, in part, substantial ones that must be interpreted as craters; all features of $t_t > t_p$ must be considered genuine penetrations. The detailed shape of all penetration curves at elevated velocities is schematic, as only the 6 km/s case is known. Nevertheless, all holes must decrease from crater diameter D_c[16&23] to the condition of $D_h = D_p$ at very thin foils. We assume that the latter is the case at $D_p / T = 100$ (Figure 6), and that this condition is essentially independent of velocity. Both assumptions are critical and the subject of ongoing laboratory investigations.

Figure 6 suggests that consideration of the shock-pulse duration may lead to an improved understanding of the velocity dependent aspects of hypervelocity penetrations in targets of arbitrary thickness. We consider the agreement between the experimental evidence and the pulse-duration arguments to imply that many penetrations should be viewed and measured as craters instead of penetrations. Furthermore, when interpreting space-produced penetrations, the transition from cratering to penetration formalisms may not occur at the ballistic limit, but at substantially thinner targets (*i.e.*, where $t_p / t_t = 1$).

SUMMARY

Employing constant impact velocities (*i.e.*, 6 km/s), our experiments encompassed a much wider range of target thicknesses than have previous studies.[11&13] This enabled us to demonstrate that the projectile size and target thickness totally control the dimensions of penetration holes in targets of arbitrary thickness. Indeed, craters and penetrations form a morphologic continuum that solely depends on the relative dimensions of the projectile and target, at otherwise identical impact conditions. Existing cratering formalisms and the conceptual approach illustrated in Figure 4 make it possible to extract projectile properties from every single impact in space-exposed surfaces, regardless of whether the events are craters or penetration holes. This new approach replaces the more traditional ballistic-limit method which only specify a single projectile, that of minimum size or mass, responsible for an entire population of penetrations on a given space-retrieved surface. Furthermore, we suggest that the discrepancies previously noted between cratering and penetration equations during analysis of Solar Max[6] and LDEF surfaces[15] are caused by extrapolating penetration formalisms to thick targets. Penetrations of massive targets at, or close to the ballistic limit must be thought of as truncated craters. The relative shock-pulse duration seems to be a suitable criterion to delineate the transition between cratering and penetration equations when extracting projectile sizes from space-exposed surfaces.

Similar experimental series need to be conducted employing higher and lower projectile velocities in order to understand the velocity-dependent relationships of projectile size, target thickness and penetration-hole diameter at LEO encounter velocities. Only then will it be possible to characterize the size-frequency distribution and fluxes of natural and man-made hypervelocity impactors with confidence.

ACKNOWLEDGMENT

We wish to acknowledge the contributions and skills of F. Cardenas, W. Davidson, G. Haynes and J. Winkler, of Lockheed Engineering & Sciences Company, in conducting the impact experiments. We also appreciate the many useful discussions and exchanges with H. Zook, A. Watts, D. Atkinson, J. Warren and M. Zolensky.

REFERENCES

1. Anderson, C.E., ed. (1987) Hypervelocity Impact, *Proceed. of the 1986 Symposium*, San Antonio, Texas, *Int. J. Impact Eng.* 5, 760 p.
2. Anderson, C.E., ed. (1990) Hypervelocity Impact, *Proceed. of the 1989 Symposium*, San Antonio, Texas, *J. Int. Impact. Eng.* 10, 635 p.
3. Anderson, C.E., ed. (1993) Hypervelocity Impact., *Proceed. of the 1992 Symposium*, Austin, Texas, *J. Int. Impact Eng.* 14, 891p.
4. Hörz, F., Cintala, M., Bernhard, R.P. and See, T.H. (1994) Dimensionally Scaled Penetration Experiments: Aluminum Targets and Glass Projectiles 50 µm to 3.2 mm in Diameter, *Int. J. Impact Eng.* 15, in press.
5. Hörz, F., Cintala, M., Bernhard, R.P. and See. T.H. (1993) Dimensionally Scaled Penetration Experiments to Extract Projectiles sizes from Space Exposed Surfaces, *Proceed. HVIS*, Austin, Texas, *Int. J. Impact Eng.* 14, 347-358.
6. Warren, J.L., Zook, H.A., Allton, J.H., Clanton, U.S, Dardano, C.B., Holder, J.A., Marlow, R.R., Schultz, R.A., Watts, L.A., and Wentworth, S.J. (1989) The Detection and Observation of Meteoroid and Space Debris Impact Features on the Solar Maximum Satellite, *Proc. 19th Lunar Planet. Sci. Conf.*, p. 641-657.
7. See, T.H., Allbrooks, M.K., Atkinson, D.R., Simon, C.G., and Zolensky, M.E. (1990) Meteoroid and Debris Impact Features Documented on the Long Duration Exposure Facility, *NASA JSC Report # 24608*, 561 p.
8. Levine, A.E., ed. (1992) *LDEF-69 Months in Space. First Post Retrieval Symposium, NASA Conference Publication, CP-3134*.
9. Levine, A.E., ed. (1993) *LDEF-69 Months in Space. Second LDEF Post Retrieval Symposium, NASA Conference Publication, CP-3194*.
10. McDonnell, J.A.M. (1991) Impact Cratering from LDEF's 5.75 Year Exposure: Decoding of the Interplanetary and Earth-Orbital Populations, *Proceed. 22nd Lunar Planet. Sci. Conf.*, p. 185-193.
11. Watts, A., Atkinson, D. and Rieco, S. (1993) *Dimensional Scaling for Impact Cratering and Perforation*, Contract Report by POD Associates, Albuquerque, New Mexico to Lockheed Engineering & Scineces Company, Houston, Texas, 84 pp.
12. Pailer, N. and Grun, E. (1980) The Penetration Limit of Thin Films, *Planet. Space Sci.*, 28, p. 321-331.
13. Herrmann, W. and Wilbeck, J. (1986) Review of Hypervelocity Penetration Theories, *Sandia National Laboratories Report, SAND-86-1884C*, 29 p.

14. McDonnell, J.A.M and Sullivan, K. (1992) Hypervelocity Impacts on Space detectors: Decoding the Projectile Parameters, in *Hypervelocity Impacts in Space*, J.A.M. Mc Donnell ed., University of Kent at Canterbury, p. 39-48.
15. Humes, D.H. (1992) Large Craters on the Meteoroid and Debris Impact Experiment, in *LDEF-69 Months in Space. First Post Retrieval Symposium*, A.E. Levine, ed., *NASA-CP-3134*, p. 399-419.
16. Cour-Palais, B.G. (1987) Hypervelocity Impact in Metals, Glass, and Composites, *Int. J. Impact Eng.* **5**, p. 681-692.
17. Carey, W.C., J.A.M. McDonnell, and D.G. Dixon (1985) An Empirical Penetration Equation for Thin Metallic Films Used in Capture Cell Techniques, in *Properties and Interactions of Interplanetary Dust*, R.H. Giese and P. Lamy, eds., Reidel Publ. Co., p. 131-136.
18. Gehring, J.W. (1970) Theory of Impact on Thin Targets and Shields and Correlation with Experiment, in *High-Velocity Impact Phenomena*, Kinslow, R. ed., Academic Press, p. 105-156.
19. Pietkutowsky, A.J. (1990) A Simple Dynamic Model for the Formation of Debris Clouds, *Int. J. Impact. Eng.* **10**, p. 453-472.
2). Stilp, A.J., Hohler, V. Schneider, E. and Weber, K. (1990) Debris Cloud Expansion Studies, *Int. J. Impact Eng.* **10**, p. 543-554.
21. Maxwell, D.E., (1977) Simple Z model of cratering, ejection and the overturned flap, in *Impact and Explosion Cratering*, Roddy., D.J., Pepin, R.O., and Merrill, R.B. eds., Pergamon Press, p.1003-1008.
22. Holsapple, K.A. and Schmidt, R.M. (1990) On the Scaling of Crater Dimensions, *J. Geophys. Res.* **87**, p. 1849-1870.
23. Christiansen, E.L. (1992) Performance Equations for Advanced Orbital Debris Shields, *AIAA Space Programs and Technologies Conference*, Huntsville, Alabama, *AIAA 92-1462*, 8 p.
24. Marsh, S.P., ed. (1980) LASL Shock Hugoniot Data, *University of California Press*, 658 pp.
25. Cintala, M.J. (1992) Impact-induced Thermal Effects in the Lunar and Mercurian Regoliths, *J. Geophys. Res.* **97**, p. 947-973.
26. Sawle, D.R. (1960) Hypervelocity Impacts in Thin Sheets and Semi-Infinite Targets at 15 km/s, *AIAA Hypervelocity Conf.*, Cincinnati.
27. Maiden, C.J., Gehring, J.W. and McMillan, A.R. (1963) Investigation of Fundamental Mechanism of Damage to Thin Targets by Hypervelocity Projectiles, *NASA TR-63-225*.
28. Nysmith, C.R. and Denardo, B.P. (1969) Experimental Investigation of the Momentum Transfer Associated with Impact into Thin Aluminum Targets, *NASA TN D 5492*.

Participants
NASA/LPI Workshop on Cosmic and Interplanetary Dust
May 15–17, 1993

Ruth Barrett
Lockheed Engineering &
  Sciences Company
2400 NASA Road One
Houston, TX 77058

Abhijit Basu
Lunar and Planetary Institute
3600 Bay Area Blvd.
Houston, TX 77058

Willet I. Beavers
Massachusetts Institute of Tech.
Lincoln Laboratory, Group 91
244 Wood Street

Lucinda Berthoud
Cert-Onera, France
2, Avenue Edouard Berlin
31055 Toulouse, France

George Blanford
University of Houston-Clear Lake
2700 Bay Area Blvd.
Houston, TX 77058

Janet Borg
I.A.S.
Batiment 121
91405 Orsay Cedex, France

John Bradley
MVA, Inc.
5500 Oakbrook Parkway #200
Norcross, GA 30093

Don Brownlee
University of Washington
Department of Astronomy
Seattle, WA 98195

Patti Jo Burkett
NASA, Johnson Space Center
Mail Code SN2
Houston, TX 77058

Bill Cooke
ISST
1810 NW 6th St.
Gainesville, FL 32609

Cassandra Coombs
POD Associates
2309 Renard Place SE
Suite 201
Albuquerque, NM 87106

Stanley F. Dermott
University of Florida
Department of Astronomy
211 SSRB
Gainesville, FL 32611

Tammy Dickinson
NASA Headquarters
Code SLC
Washington, DC 20546-0001

Joel Edelman
LDEF Newsletter
14636 Silverstone Dr.
Silver Spring, MD 20905

C. Y. Fan
University of Arizona
Department of Physics
Tucson, AZ 85721

Participants

Michel Faurette
C.S.N.S.M.
Batiments 108
91405 Campus Orsay, France

George J. Flynn
SUNY at Plattsburgh
Department of Physics
Hudson Hall-223
Plattsburgh, NY 12901

M. N. Fomenkova
University of California, San Diego
California Space Institute
La Jolla, CA 92093-0216

Evertt K. Gibson
NASA, Johnson Space Center
Code SN2
Houston, TX 77058

James Gooding
NASA, Johnson Space Center
Code SN2
Houston, TX 77058

M. S. Hanner
Jet Propulsion Laboratory
MS 183-601
4800 Oak Grove Drive
Pasadena, CA 91109

Friedrich Hörz
NASA, Johnson Space Center
Code SN4
Houston, TX 77058

Donald Humes
NASA, Langley
Mail Code 49B
Hampton, VA 23665-5225

A. A. Jackson
Lockheed Eng. & Sciences Co.
2400 NASA Road One, C23
Houston, TX 77258

James E. Keith
NASA, Johnson Space Center
Code SN3
Houston, TX 77058

Lindsay P. Keller
Lockheed Eng. & Sciences Co.
2400 NASA Road One
Houston, TX 77285

William Kinard
NASA, Langley
Mail Code 356
Hampton, VA 23665-5225

Wolfgang Klöck
University of Münster
Institut fur Planetologie
Wilhelm-Klemm Strasse 10
D-4400 Münster, Germany

Kimberly Leago
Houston, TX 77058

Wei Lee
University of Alabama, Birmingham
Physics Dept., 1300 University Blvd.
Birmingham, AL 35294-1170

David J. Lindstrom
NASA, Johnson Space Center
Code SN2
Houston, TX 77058

Participants

J. C. Liou
Department of Astronomy
University of Florida
221, SSRB
Gainesville, FL 32611

Carl R. Maag
SAIC
2605 E. Foothill Blvd.
Glendora, CA 91740

David S. McKay
NASA, Johnson Space Center
Code SN14
Houston, TX 77058

David Mendez
741 Flannery Street
Santa Clara, CA 95051

Scott Messenger
Washington University
McDonnel Center for Space
  Science, Box 1105
One Brookings Drive
St. Louis, MO 63130

Douglas W. Ming
NASA, Johnson Space Center
Code SN4
Houston, TX 77058

Alfred O. Nier
University of Minnesota
School of Physics and
  Astronomy
116 Church Street SE
Minneapolis, MN 55455

Kenji Nishioka
NASA, Ames Research Center
SETI Institute, M/S 239-12
Moffett Field, CA 94035-01000

Laurence E. Nyquist
NASA, Johnson Space Center
Code SN4
Houston, TX 77058

J. Oliver
ISST
1810 NW 6th St.
Gainesville, FL 32609

Doug Phinney
Lawrence Livermore, L-232
P.O. Box 808
Livermore, CA 94551

Frans Rietmeijer
University of New Mexico
Department of Geology
Albuquerque, NM 87131

Thomas H. See
Lockheed Eng. & Sciences
2400 NASA Road One, C23
Houston, TX 77058

Charles Simon
ISST
1810 NW 6th Street
Gainesville, FL 32609-3530

Frank Stadermann
University of Darmstadt
Fb 21-Material Sciences
Petersenstrasse 20
W-6100 Darmstadt, Germany

Stephen Sutton
Brookhaven National Laboratory
Building 815
Upton, NY 11973

William G. Tanner
Baylor University,
　Space Science Laboratory
Department of Physics
P.O. Box 97303
Waco, TX 76798-7303

Kathie Thomas
Lockheed Eng. & Science
NASA, Johnson Space Center-C23
Houston, TX 77058

Gary Toller
General Sciences Corp.
9364 Dewlit Way
Columbia, MD 21045

Anthony J. Tuzzolino
University of Chicago
Lab. Astrophysics & Space Res.
933 East 56th Street
Chicago, IL 60637

Robert M. Walker
Washington University
Dept. Physics, Box 1105
One Brookings Drive
St. Louis, MO 63130

Jack Warren
Lockheed Eng. & Science
2400 NASA Road One
Houston, TX 77058

Thomas J. Wdowiak
University of Alabama,
　Birmingham
Physics Department
Birmingham, AL 35294

Jerry Weinberg
ISST Space Astronomy Lab.
1810 NW 6th Street
Gainesville, FL 32609

Thomas L. Wilson
NASA, Johnson Space Center
Code SN3
Houston, TX 77058

Michael E. Zolensky
NASA, Johnson Space Center
Code SN2
Houston, TX 77058

Herbert A. Zook
NASA, Johnson Space Center
Code SN3
Houston, TX 77058

Author Index

A

Alexander, M., 305

B

Barrett, R. A., 105, 291
Bernhard, R. P., 291, 329
Berthoud, L., 313
Bibring, J-P., 305
Blanford, G. E., 165
Borg, J., 305
Bradley, J., 89
Brownlee, D., 5
Bustin, R., 173

C

Chang, S., 193
Cintala, M., 329

D

Dardano, C., 291
Dermott, S. F., 13

F

Fan, C. Y., 211
Flynn, G. J., 127, 223
Fomenkova, M., 193

G

Gibson, E. K., Jr., 173

H

Hammer, C., 277

Hanner, M. S., 23
Harvey, R., 277
Hörz, F., 291, 329

I

Immel, G., 277

K

Keller, L. P., 159, 165
Kinard, W. H., 291
Klöck, W., 51
Kurat, G., 277

L

Lee, W., 185
Liou, J. C., 13

M

Mack, K., 291
Maag, C., 305
Maurette, M., 277
Mandeville, J. C., 313
McKay, D. S., 159, 165

N

Nier, A. O., 115

R

Rietmeijer, F. J. M., 231, 255

S

See, T. H., 291, 329
Song, L.-G., 211
Stadermann, F. J., 51
Sutton, S. R., 145

T

Tanner, W., 305
Taylor, S., 277
Thomas, K. L., 159, 165

W

Warren, J. L., 245, 255, 291
Walker, R. M., 203
Wdowiak, T. J., 185
Wilson, T. L., xi, xiii, 33

X

Xu, Y.-L., 211

Z

Zhang, Y.-X., 211
Zolensky, M. E., xi, xiii, 105, 245, 291

SUBJECT INDEX

A

Achondritic Meteorites, 51, 66, 71, 74, 146, 155
Accretion, 5–6, 47, 89, 97, 100, 106
Aerogel, 307–308, 312
Aerosols, 154
Age determination, 214
Aggregates (see also Granular Units)
  Equilibrated, 47, 100
  Reduced, 47, 100
  Unequilibrated, 47, 100
Alumina, 137, 140, 154, 206, 305, 308, 314
Aluminum, 7, 10, 67, 71, 147, 149–151, 174, 198, 203, 254, 256, 294, 299, 314–315, 329–342
Amorphous Materials (including Glasses), xiv, 11, 29, 46–47, 52, 54, 60, 67, 72–74, 81–82, 93, 95, 112, 161, 170–171, 194, 196, 231–237, 300–302
Analytical Electron Microscopy, 52–84, 100, 138, 234, 238
Annealing, 112, 235–236
Antarctic Micrometeorites, 51–84, 205–206, 243, 277–287
Aqueous Alteration, 5, 7, 47, 51, 72, 74, 89, 101, 111, 161–162, 235–236
Argon, 133, 149
Arsenic, 149, 278
Asteroids
  General, 4–7, 11, 13–20, 33, 43, 47, 51, 53, 54, 61, 66, 72, 74, 84, 89, 92, 94, 96, 105, 111, 115, 118–119, 124, 133, 154, 162, 173, 198, 305–306, 330
  Hydrous, 100, 112–113
  Inner Belt, 7, 94
  Main Belt, 4–7, 11, 13–20, 94, 97, 100, 118, 128–130, 223–229
  Near-Earth, 11
  Outer Belt, 3, 92, 94, 236, 255
Atmospheric Entry Heating, 6, 11, 47–48, 53–54, 56, 59, 67, 69, 74, 89–90, 100, 118–119, 123–125, 130–133, 145, 154, 162–163, 170–171, 174, 186, 222–229, 238, 243, 245, 261, 270, 278, 291
Atmospheric Residence Time, 48, 127, 132, 134–137, 272–273, 277–278

B

Backscattered Electron Imaging, xiv, 75, 106–107
Barium, 57
Basalts, 231
Boron, 149, 211–221
Bromine, 48, 133, 137, 139, 147, 149, 152, 278
Brucite, 278

C

Cadmium, 54, 57, 141, 149
Calcium, 10, 57, 71, 111–112, 147, 149–151, 174, 203, 254, 256, 278, 283, 299, 314–315
Calcium Aluminum Rich Inclusions (CAIs), 51, 71–73
Carbides, 97, 162, 198, 203, 205–206
Carbon, 23, 49–50, 52, 54–55, 64, 111, 132–133, 149–150, 159–163, 165–171, 173–184, 185–191, 193–200, 203–204, 206, 211–213, 231–238
Carbonates, 53, 66, 132–133, 163, 170–171, 174, 176–178, 196–198, 277–278, 283
CAS Spheres, 10
Cerium, 57, 71
Chlorine, 137, 149, 198, 314
CHON Particles, 23, 196–197
Chondritic Meteorites
  Carbonaceous, 7, 23, 29, 48–49, 51–84, 89, 92, 97, 108, 115, 118, 132, 135, 145, 150, 153–155, 159, 160–161, 166, 173–184, 185–191, 194, 197–198, 203–208, 232, 271, 277–278
  Ordinary, 3, 51–84, 97
Chondrules, 5–6, 72

351

Chromite, 53
Chromium, 57, 71, 147, 149–152, 203, 314–315
Classification, 3, 47–50, 52, 90, 92, 94, 231–238
Cobalt, 57, 71, 149, 153, 155, 278
COBE Satellite, xi, 11
Collection of IDPs and Micrometeorites
  Balloon, 308
  Conventional Area Collectors, 245–251, 255–273
  Large Area Collectors, 141, 175, 245–251, 255–273
  Polar Ices, 48, 51–84, 130, 145, 206, 242–243, 245, 277–287
  Sea Floor, 48, 52–53, 130, 134, 145, 278, 280
  Spacecraft, 3, 11–12, 130, 155, 243, 245, 305–310, 312, 329–342
  Stratosphere, 3, 47–48, 52, 64, 72, 89, 92, 105, 122, 124, 130, 144–145, 150, 154, 165–170, 194, 223, 242–251, 255–273, 276–278, 291
Comets, 3, 6–7, 11, 22, 23–30, 33, 42, 47, 53–54, 61, 65–66, 89, 92–95, 97, 100, 105, 115, 118, 124, 128–129, 133–134, 154, 159, 173, 193–200, 203, 223–229, 236, 255, 291, 305–306, 330
COMET Experiment, 310
Composition (see also individual elements)
  Bulk, 3, 48, 71, 89–90, 105, 115, 119, 139, 145–155, 159–163, 166, 193–200, 229, 238, 305–306, 310
  Isotopes, 3, 7, 49, 51, 54, 72, 74, 90, 115, 148, 163, 197, 203–208, 211–221, 305–307, 310
  Noble Gases, 3, 47, 90, 116, 133, 142, 149, 205, 227
  Rare Earth Elements (REE), 149, 153, 149
  Refractory Elements, 6, 57, 72, 149, 153
  Trace Elements, 3, 54, 57, 71, 72, 145, 149–150, 152–153, 229, 277, 281, 183
  Volatiles, 6, 23, 47, 54, 57, 132–142, 145, 148–150, 165–171, 173–184, 227, 306
COMRADE Experiment, 305–310
Condensation, 5, 97, 99–100, 106
Contamination, 48, 134–142, 155, 243
Cooling Rate, 112
Copper, 54, 57, 139, 149, 314
Cosmic Rays, 23, 116, 211, 214
Cosmic Ray Tracks, see Solar Flare Tracks
Cratering Formulas, 329–342
Cronstedtite, 278
Cryoconite, 277–287
Curation, 119, 138–142, 145, 147, 154, 244–251, 277–287, 295, 302

D

Database, Meteoroid and Debris, 292, 302
Deep Sea Spheres, 48, 115, 134
Density, IDPs, 92, 118, 124, 130, 227, 229, 245
Deuterium, 49, 162, 185, 191, 196–197, 203–204, 208
Diagenesis, 236
Diamonds, 162, 194, 205–206
Diffusion, 66, 70
Dust, see IDPs
Dusty Plasma, 33–43
Dynamical Calculations, 11, 13–20, 33–43, 118, 270

E

Earth (as IDP collector), 11, 127–142, 242, 290
E-Layer, 48, 134–135
Electric Field, 33–43
Electron Diffraction, 61, 299
Electron Microprobe, 147
Emissivity, 130, 227, 229
Equilibrium, 66, 112, 235–236
EURECA, 243, 308, 310
Europium, 72, 149, 153

F

Feldspars, 235–237
Field Emission Scanning Electron Microscopy, 307, 310
Fischer-Tropsch Process, 97, 185
Flux
    General, 52, 255–273, 277–278, 287, 291–303, 306–307, 310
    Temporal Variations, 20, 255–273, 277–278, 287
Foams, 308–309
FOILS Laboratory, 12
FRECOPA Experiment, 313–327

G

Galileo Spacecraft, 4
Gallium, 139, 149, 152
Gamma Ray Observatory (GRO), xi
Garnet, 132
Gas-Phase Reactions, 89, 97, 99
Gehlenite (see also Melilite), 46
Germanium, 133, 139, 149, 152
Giotto Spacecraft, xi, 22, 61, 66, 94
Glasses, see Amorphous Materials
Gold, 149, 278
Granular Units (GU), 3, 47, 49, 52–65, 106, 108, 198, 231–237, 265
Graphite, 194, 203, 205–206
Graphitic Carbon, 133, 161, 163, 170–171, 176, 194, 197, 233, 237
Gravitational Field, 39, 42
Gravitational Segregation, 6, 118, 127–130, 223
Greenland Micrometeorites, 51–84, 206, 277–282, 284, 287

H

Halley (comet), 22–30, 193–200
Helium, 54, 115–125, 133–134, 149, 174, 211, 261
HEOS Spacecraft, 308
Hibonite, 46, 232
Hubble Space Telescope (HST), xi

Hydration, 11, 47, 113
Hydrocarbons, 133, 170–171, 174–184, 185–191, 196, 233, 237
Hydrocryogenic Alteration, 29, 198, 235–237, 278
Hydrogen, 7, 23, 49, 185–191, 194, 196–198, 200, 203–204, 207–208, 211–214
Hydroxide, 176–178, 281
Hypervelocity Impact, 12, 193, 200, 291–303, 305–310, 312–313, 329–342

I

Ida, 4
IDPs, Types
    Anhydrous, xiv, 3, 7, 23, 49, 51–84, 89–101, 105–112, 131–132, 139, 160–163, 165–166, 170–171, 176, 198, 204, 231, 233, 236, 302
    Basaltic, 3
    Carbonaceous, 231–238
    Chondritic, 3, 7, 11, 22–23, 30, 46–84, 105, 107, 112, 115–125, 130–132, 137, 140, 145–155, 159–163, 185–191, 192, 203–208, 210, 231, 255–273, 313, 315
    Hydrous, 3, 7, 29–30, 47–49, 51–84, 89–101, 105–113, 131, 159–163, 165, 170–171, 176, 198, 204, 210, 231–236, 271, 299–302
    Olivine, 24, 49, 51–53, 54, 57–58, 89–91, 100, 106, 159, 170–171, 204, 231–232
    Pyroxene, 29–30, 47, 49, 51–84, 89–91, 97, 106, 159, 170–171, 204, 231–232
    Refractory, 3, 46–47, 72, 232
    Saponite, 49, 51–84, 107, 231–232
    Serpentine, 49, 52, 61, 72, 108, 231–232
Impact Ejecta, 33
Impactor Residues, 291–303, 305–310, 313–327
Instrumental Neutron Activation

Analysis (INAA), 73, 105–106, 145, 148–150
Interstellar Particles, 5–6, 11, 23, 29, 49, 193, 196, 198, 203–206
Interstellar Medium, 185–191
Io, 34–36, 40, 42
Ion Imaging, 207
Ion Implantation, 136
Ion Microprobe, 57, 71, 203–208
IRAS Satellite, 11, 13–14, 223–224
Iron 7, 23–24, 29, 48, 53–83, 93–94, 133–135, 147, 149–153, 155, 174, 198, 203, 222, 254, 256, 278, 299, 301, 314–315

J

Jupiter, 5–6, 11, 14, 34, 36, 40, 42

K

Kaolinite, 235–236
Kerogen, 197
Kinetics, 236
Kuiper Belt, 3, 5–6

L

Laihunite, 132
Laser Microprobe, 148, 173–184
LDEF, 158, 243, 290–303, 310, 313–327, 330, 332, 338, 342
Levitation, Electrostatic, 33–43
Lexan, 246–249
Lithium, 57, 149, 211–221
Lonsdaleite, 233, 235, 237
Lunar Dust, 10, 33–34, 39–42, 116–117, 122–123

M

Maghemite, 74, 131
Magneli Phases, 236
Magnesiowüstite, 132, 134
Magnesite, 235
Magnesium, 23–24, 29, 49, 53–83, 93–94, 97, 147, 149–151, 174, 203–204, 208, 256, 299, 314–315
Magnetic Field, 33–43
Magnetite (including Rims), 47–48, 53–54, 74–75, 80, 82, 90, 119, 131–132, 134, 171, 227, 237–238, 261
Manganese, 53–54, 57, 60, 62, 63, 66, 100, 147, 149–152, 299
Mass of IDPs, 305–306, 310
Mass Distribution, 306, 310
Mass Spectrometry, 66, 94, 173–184, 193–200, 205–206, 208
Melilite, 232
Mercury (planet), 13, 34, 36, 40, 42
Metal, 6, 52, 106, 115, 236, 301–302
Metamorphism, 66, 105, 108, 236, 302
Meteor Streams, 270
Methanol, 24
Microcrystalline Aggregates, see Granular Units
Micrometeorites, 12, 48, 51–84, 89, 205, 214, 237, 242, 277–287, 290, 315, 326–327
Microtomy, 52–84, 105–106, 161, 165–171, 210, 234, 299, 301
Mineralogy (see also specific minerals), 3, 11, 46–84, 89–101, 105, 119, 127–142, 162–163, 168, 171, 227, 229, 306
Mir, 243, 310, 313–314, 317
Molecular Clouds, 206–207
Molybdenum, 147
Moon, 34, 36, 40–41
Morphology, IDPs, 255–273
Multiple Foil Detectors, 313–315

N

Neoformation Products, 112
Neon, 115–125, 133, 149
Neptune, 42
Nickel, 57, 71, 134, 139, 147, 149–152, 203, 222, 256, 278, 283, 299, 301, 313
Nitrogen, 23, 49, 136, 162, 194, 197,

Subject Index   355

204, 206–207, 211–213
Non-equilibrium, 238
Nucleation, Grain, 210, 237, 276
Nucleosynthesis, 159, 211

O

Olivine, 7, 24, 29–30, 47, 49, 53–83, 97, 100, 105–113, 130–132, 227, 229, 231–237, 278, 302–303, 312
Oort Cloud, 24
Optical Microscopy, 310, 332
Orbital Debris (including Space Debris), 12, 144, 290–303, 305, 307–308, 310
Orbital Elements, of Dust, 11, 13–20
Orbital Evolution, of Dust, 11, 13–20
Organics, 3, 7, 50, 97, 159–163, 173–184, 185–191, 193–200, 206–207
Oxides, 74, 198, 233, 281
Oxygen, 7, 23, 165–166, 174, 190, 194, 196–198, 200, 203–204, 211–213, 254, 315
Oxygen Fugacity, 6–7, 70, 74

P

Palapa Satellite, 292
Parent Body Processes, 113, 236
Periclase, 132
Perovskite, 46, 232
Petrofabrics, 105
Petrography, 47, 89–101, 231
Petrologic Classes, 236–237
pH, Solution, 111, 236
Phosphorus, 57, 147, 149
Phyllosilicates (see also specific minerals), 48–49, 53, 72, 74–75, 106, 110, 131–132, 161, 174, 231, 235, 237
PIA Instrument, 198
Plasma, Dust, xi, 33–34, 38, 41–42
Pluto, 13
Point Counting, 166, 170–171, 232
Polycyclic Aromatic Hydrocarbons, 161–163, 185–186, 189–191, 207
Polymers, 196–197

Polyoxymethylene, 23
Polyphase Units (PU) (see also Granular Units), 47, 49, 231–237
Porosity, 47, 90–92, 97, 106, 112–113, 119, 146, 231
Potassium, 54, 57, 71, 147, 149–150, 203, 278, 314
Poynting-Robertson Drag, 6, 13–14, 35, 118, 224
Proton Induced X-ray Emission (PIXE), 145, 147, 149, 153, 161
Protoplanets, 211, 231, 235–236, 238, 270
Pulse-Heating, 115–125
PUMA Instrument, 193–200
Pyrometamorphism, 237–238
Pyroxene
  Augite, 108, 110, 300
  Diopside, 76, 105–112
  Enstatite, xiv, 30, 53, 60, 97, 99, 107–108, 110, 276
  General, 29–30, 47, 49, 66–69, 71–74, 80, 82, 90, 97, 100, 105–113, 132, 227, 231, 233–234, 237, 278, 299–303, 312
  Pigeonite, 108, 110
Pyrrhotite, see Sulfides

R

Recrystallization, 66, 301–302
Regolith, 42, 47
Rubidium, 54, 57, 149

S

Salinity, Fluid, 236
Saponite, 49, 92, 106, 108–109, 210, 235, 237
Saturn, 42
Scandium, 57, 71, 149, 153, 155
Scanning Auger Microscopy, 148
Scanning Electron Microscopy (SEM), xiv, 10, 46, 50, 88, 91, 177, 192, 248, 250–251, 254, 260, 276, 284, 286, 294–300, 302, 306, 310, 332, 337

Subject Index

Secondary Ion Mass Spectrometry (SIMS), 137, 145, 148–149, 153, 161, 203–208, 211–221
Selenium, 149, 152
Serpentine, 49, 72, 74, 92, 106, 131
Shock Effects, 5–6, 243, 299–302
SiC, see Carbides
Silicon, 10, 24, 49, 57, 93–94, 147, 149–151, 154, 174, 198, 211–221, 254, 256, 299, 314–315
Silicone Oil, 3, 48, 138, 154, 161, 166, 175–176, 243, 246, 248, 250
Skylab, 22
Smectite (see also Saponite), 72, 74, 76–79, 210, 278
Sodium, 54, 57, 71, 134–135, 147, 149, 174, 194, 198, 314
Solar Energetic Particles, 56, 116, 122–123
Solar Flare Tracks, 48, 54, 56, 66, 69, 89–90, 100–101, 116–117, 133, 146, 162, 229
Solar Maximum Satellite, 66, 158, 254, 292, 330, 332, 338, 342
Solar Nebula, 5, 7, 23, 47, 145, 185, 193, 207
Solar System (primitive), 5, 23, 145
Solar Wind, 36–37, 43, 122, 211–214
Space Debris, see Orbital Debris
Space Shuttle (STS), 158, 243, 290, 296, 299–300
Space Station, 243, 298
Spacecraft Debris, see Orbital Debris
Spectroscopy
　Energy Dispersive X-ray, 94, 105, 145, 147, 149, 151, 161, 165–171, 175, 231, 238, 248, 250–251, 256, 260, 284, 296–302, 313–327, 337
　Laser-Induced Luminescence, 185–191
　Optical and IR, 3, 7, 11, 23–30, 52, 89–90, 92, 94–98, 100, 131, 163, 185–191, 197, 231, 310
　Raman, 162, 186, 189, 207
Spherules, 222, 281–286
Spinel, 236
Step-Heating, 115–125
Stokes Law, 245
Stratosphere, see Collection
Strontium, 57, 149
Sulfates, 178, 235, 277–278, 283, 285
Sulfides, 42, 52–78, 88, 90, 106–107, 132, 176, 178, 197–198, 210, 222, 232–233, 235, 301–302, 312
Sulfur, 54, 57, 71, 127, 132, 134, 137, 147–151, 174–178, 198, 203, 256, 269–270, 278, 283, 299, 314–315
Sulfur Dioxide, 42
Sun, 37, 185, 191
Synchrotron X-ray Fluorescence (SXRF), 145, 147, 149–150, 152

T

T-Tauri, 185
Talc, 235
Tar Balls (see also Granular Units), 3, 47, 100, 106
Teflon, 329–342
Thermometers, 227–228
TiC, 205
Titania, 307
Titanium, 57, 71, 147, 149, 174, 203, 256, 314
Tochilinite, 92, 278
Trajectories of IDPs, 11, 13–20, 37, 39, 130, 243, 245, 291–303, 305–310
Transmission Electron Microscopy (TEM), 52–84, 90, 92, 99–100, 131, 133, 137, 155, 161, 166–167, 169, 205, 207, 234, 238, 299–302
Tridymite, 137
Troilite, see Sulfides
Tropopause, 48
Tunguska Event, 287

U

Ulysses Spacecraft, xi, 11, 308
Uranium, 54, 149
Uranus, 42

V

Vanadium, 57, 315
VEGA Spacecraft, xi, 61, 66, 94, 193–200
Velocities of IDPs, 11, 30, 47, 118, 130, 223–226, 243, 245, 291–303, 305–306, 310
Velocity Scaling, 329–342
Venus, 42
Volcanic Dust, 33–43, 137–138, 287
Volcanic Plume, 34–35
Voyager Spacecraft, xi, 34–35, 42

W

Water, 176–178, 181
Water/Rock Ratios, 236
Weathering Rate, 111

X

X-ray Element Mapping, 205
Xenon, 133, 149, 205

Y

Yttrium, 57

Z

Zinc, 47–48, 54, 57, 61, 71, 119, 132–134, 137, 139, 149, 152–153, 174, 261, 270, 314
Zirconium, 57
Zodiacal Cloud (Light), 2, 11, 14, 115, 127–128, 223

AIP Conference Proceedings

		L.C. Number	ISBN
No. 92	The State of Particle Accelerators and High Energy Physics (Fermilab, 1981)	82-73861	0-88318-191-6
No. 93	Novel Results in Particle Physics (Vanderbilt, 1982)	82-73954	0-88318-192-4
No. 94	X-Ray and Atomic Inner-Shell Physics – 1982 (International Conference, U. of Oregon)	82-74075	0-88318-193-2
No. 95	High Energy Spin Physics – 1982 (Brookhaven National Laboratory)	83-70154	0-88318-194-0
No. 96	Science Underground (Los Alamos, NM, 1982)	83-70377	0-88318-195-9
No. 97	The Interaction Between Medium Energy Nucleons in Nuclei – 1982 (Indiana University)	83-70649	0-88318-196-7
No. 98	Particles and Fields – 1982 (APS/DPF University of Maryland)	83-70807	0-88318-197-5
No. 99	Neutrino Mass and Gauge Structure of Weak Interactions (Telemark, 1982)	83-71072	0-88318-198-3
No. 100	Excimer Lasers – 1983 (OSA, Lake Tahoe, NV)	83-71437	0-88318-199-1
No. 101	Positron-Electron Pairs in Astrophysics (Goddard Space Flight Center, 1983)	83-71926	0-88318-200-9
No. 102	Intense Medium Energy Sources of Strangeness (UC-Santa Cruz, CA, 1983)	83-72261	0-88318-201-7
No. 103	Quantum Fluids and Solids – 1983 (Sanibel Island, FL)	83-72440	0-88318-202-5
No. 104	Physics, Technology and the Nuclear Arms Race (APS, Baltimore, MD, 1983)	83-72533	0-88318-203-3
No. 105	Physics of High Energy Particle Accelerators (SLAC Summer School, 1982)	83-72986	0-88318-304-8
No. 106	Predictability of Fluid Motions (La Jolla Institute, 1983)	83-73641	0-88318-305-6
No. 107	Physics and Chemistry of Porous Media (Schlumberger-Doll Research, 1983)	83-73640	0-88318-306-4
No. 108	The Time Projection Chamber (TRIUMF, Vancouver, 1983)	83-83445	0-88318-307-2
No. 109	Random Walks and Their Applications in the Physical and Biological Sciences (NBS/La Jolla Institute, 1982)	84-70208	0-88318-308-0

No.	Title		
No. 110	Hadron Substructure in Nuclear Physics (Indiana University, 1983)	84-70165	0-88318-309-9
No. 111	Production and Neutralization of Negative Ions and Beams (3rd Int'l Symposium) (Brookhaven, NY, 1983)	84-70379	0-88318-310-2
No. 112	Particles and Fields – 1983 (APS/DPF, Blacksburg, VA)	84-70378	0-88318-311-0
No. 113	Experimental Meson Spectroscopy – 1983 (7th International Conference, Brookhaven, NY)	84-70910	0-88318-312-9
No. 114	Low Energy Tests of Conservation Laws in Particle Physics (Blacksburg, VA, 1983)	84-71157	0-88318-313-7
No. 115	High Energy Transients in Astrophysics (Santa Cruz, CA, 1983)	84-71205	0-88318-314-5
No. 116	Problems in Unification and Supergravity (La Jolla Institute, 1983)	84-71246	0-88318-315-3
No. 117	Polarized Proton Ion Sources (TRIUMF, Vancouver, 1983)	84-71235	0-88318-316-1
No. 118	Free Electron Generation of Extreme Ultraviolet Coherent Radiation (Brookhaven/OSA, 1983)	84-71539	0-88318-317-X
No. 119	Laser Techniques in the Extreme Ultraviolet (OSA, Boulder, CO, 1984)	84-72128	0-88318-318-8
No. 120	Optical Effects in Amorphous Semiconductors (Snowbird, UT, 1984)	84-72419	0-88318-319-6
No. 121	High Energy e^+e^- Interactions (Vanderbilt, 1984)	84-72632	0-88318-320-X
No. 122	The Physics of VLSI (Xerox, Palo Alto, CA, 1984)	84-72729	0-88318-321-8
No. 123	Intersections Between Particle and Nuclear Physics (Steamboat Springs, CO, 1984)	84-72790	0-88318-322-6
No. 124	Neutron-Nucleus Collisions: A Probe of Nuclear Structure (Burr Oak State Park, 1984)	84-73216	0-88318-323-4
No. 125	Capture Gamma-Ray Spectroscopy and Related Topics – 1984 (Int'l Symposium, Knoxville, TN)	84-73303	0-88318-324-2
No. 126	Solar Neutrinos and Neutrino Astronomy (Homestake, 1984)	84-63143	0-88318-325-0
No. 127	Physics of High Energy Particle Accelerators (BNL/SUNY Summer School, 1983)	85-70057	0-88318-326-9
No. 128	Nuclear Physics with Stored, Cooled Beams (McCormick's Creek State Park, IN, 1984)	85-71167	0-88318-327-7

No. 129	Radiofrequency Plasma Heating (Sixth Topical Conference) (Callaway Gardens, GA, 1985)	85-48027	0-88318-328-5
No. 130	Laser Acceleration of Particles (Malibu, CA, 1985)	85-48028	0-88318-329-3
No. 131	Workshop on Polarized ^{3}He Beams and Targets (Princeton, NJ, 1984)	85-48026	0-88318-330-7
No. 132	Hadron Spectroscopy – 1985 (International Conference, Univ. of Maryland)	85-72337	0-88318-331-5
No. 133	Hadronic Probes and Nuclear Interactions (Arizona State University, 1985)	85-72638	0-88318-332-3
No. 134	The State of High Energy Physics (BNL/SUNY Summer School, 1983)	85-73170	0-88318-333-1
No. 135	Energy Sources: Conservation and Renewables (APS, Washington, DC, 1985)	85-73019	0-88318-334-X
No. 136	Atomic Theory Workshop on Relativistic and QED Effects in Heavy Atoms (Gaithersburg, MD, 1985)	85-73790	0-88318-335-8
No. 137	Polymer-Flow Interaction (La Jolla Institute, 1985)	85-73915	0-88318-336-6
No. 138	Frontiers in Electronic Materials and Processing (Houston, TX, 1985)	86-70108	0-88318-337-4
No. 139	High-Current, High-Brightness, and High-Duty Factor Ion Injectors (La Jolla Institute, 1985)	86-70245	0-88318-338-2
No. 140	Boron-Rich Solids (Albuquerque, NM, 1985)	86-70246	0-88318-339-0
No. 141	Gamma-Ray Bursts (Stanford, CA, 1984)	86-70761	0-88318-340-4
No. 142	Nuclear Structure at High Spin, Excitation, and Momentum Transfer (Indiana University, 1985)	86-70837	0-88318-341-2
No. 143	Mexican School of Particles and Fields (Oaxtepec, México, 1984)	86-81187	0-88318-342-0
No. 144	Magnetospheric Phenomena in Astrophysics (Los Alamos, NM, 1984)	86-71149	0-88318-343-9
No. 145	Polarized Beams at SSC & Polarized Antiprotons (Ann Arbor, MI & Bodega Bay, CA, 1985)	86-71343	0-88318-344-7
No. 146	Advances in Laser Science—I (Dallas, TX, 1985)	86-71536	0-88318-345-5
No. 147	Short Wavelength Coherent Radiation: Generation and Applications (Monterey, CA, 1986)	86-71674	0-88318-346-3
No. 148	Space Colonization: Technology and The Liberal Arts (Geneva, NY, 1985)	86-71675	0-88318-347-1

No. 149	Physics and Chemistry of Protective Coatings (Universal City, CA, 1985)	86-72019	0-88318-348-X
No. 150	Intersections Between Particle and Nuclear Physics (Lake Louise, Canada, 1986)	86-72018	0-88318-349-8
No. 151	Neural Networks for Computing (Snowbird, UT, 1986)	86-72481	0-88318-351-X
No. 152	Heavy Ion Inertial Fusion (Washington, DC, 1986)	86-73185	0-88318-352-8
No. 153	Physics of Particle Accelerators (SLAC Summer School, 1985) (Fermilab Summer School, 1984)	87-70103	0-88318-353-6
No. 154	Physics and Chemistry of Porous Media—II (Ridgefield, CT, 1986)	83-73640	0-88318-354-4
No. 155	The Galactic Center: Proceedings of the Symposium Honoring C. H. Townes (Berkeley, CA, 1986)	86-73186	0-88318-355-2
No. 156	Advanced Accelerator Concepts (Madison, WI, 1986)	87-70635	0-88318-358-0
No. 157	Stability of Amorphous Silicon Alloy Materials and Devices (Palo Alto, CA, 1987)	87-70990	0-88318-359-9
No. 158	Production and Neutralization of Negative Ions and Beams (Brookhaven, NY, 1986)	87-71695	0-88318-358-7
No. 159	Applications of Radio-Frequency Power to Plasma: Seventh Topical Conference (Kissimmee, FL, 1987)	87-71812	0-88318-359-5
No. 160	Advances in Laser Science—II (Seattle, WA, 1986)	87-71962	0-88318-360-9
No. 161	Electron Scattering in Nuclear and Particle Science: In Commemoration of the 35th Anniversary of the Lyman-Hanson-Scott Experiment (Urbana, IL, 1986)	87-72403	0-88318-361-7
No. 162	Few-Body Systems and Multiparticle Dynamics (Crystal City, VA, 1987)	87-72594	0-88318-362-5
No. 163	Pion–Nucleus Physics: Future Directions and New Facilities at LAMPF (Los Alamos, NM, 1987)	87-72961	0-88318-363-3
No. 164	Nuclei Far from Stability: Fifth International Conference (Rosseau Lake, ON, 1987)	87-73214	0-88318-364-1
No. 165	Thin Film Processing and Characterization of High-Temperature Superconductors (Anaheim, CA, 1987)	87-73420	0-88318-365-X
No. 166	Photovoltaic Safety (Denver, CO, 1988)	88-42854	0-88318-366-8

No. 167	Deposition and Growth: Limits for Microelectronics (Anaheim, CA, 1987)	88-71432	0-88318-367-6
No. 168	Atomic Processes in Plasmas (Santa Fe, NM, 1987)	88-71273	0-88318-368-4
No. 169	Modern Physics in America: A Michelson-Morley Centennial Symposium (Cleveland, OH, 1987)	88-71348	0-88318-369-2
No. 170	Nuclear Spectroscopy of Astrophysical Sources (Washington, DC, 1987)	88-71625	0-88318-370-6
No. 171	Vacuum Design of Advanced and Compact Synchrotron Light Sources (Upton, NY, 1988)	88-71824	0-88318-371-4
No. 172	Advances in Laser Science—III: Proceedings of the International Laser Science Conference (Atlantic City, NJ, 1987)	88-71879	0-88318-372-2
No. 173	Cooperative Networks in Physics Education (Oaxtepec, Mexico, 1987)	88-72091	0-88318-373-0
No. 174	Radio Wave Scattering in the Interstellar Medium (San Diego, CA, 1988)	88-72092	0-88318-374-9
No. 175	Non-neutral Plasma Physics (Washington, DC, 1988)	88-72275	0-88318-375-7
No. 176	Intersections Between Particle and Nuclear Physics (Third International Conference) (Rockport, ME, 1988)	88-62535	0-88318-376-5
No. 177	Linear Accelerator and Beam Optics Codes (La Jolla, CA, 1988)	88-46074	0-88318-377-3
No. 178	Nuclear Arms Technologies in the 1990s (Washington, DC, 1988)	88-83262	0-88318-378-1
No. 179	The Michelson Era in American Science: 1870–1930 (Cleveland, OH, 1987)	88-83369	0-88318-379-X
No. 180	Frontiers in Science: International Symposium (Urbana, IL, 1987)	88-83526	0-88318-380-3
No. 181	Muon-Catalyzed Fusion (Sanibel Island, FL, 1988)	88-83636	0-88318-381-1
No. 182	High T_c Superconducting Thin Films, Devices, and Applications (Atlanta, GA, 1988)	88-03947	0-88318-382-X
No. 183	Cosmic Abundances of Matter (Minneapolis, MN, 1988)	89-80147	0-88318-383-8
No. 184	Physics of Particle Accelerators (Ithaca, NY, 1988)	89-83575	0-88318-384-6
No. 185	Glueballs, Hybrids, and Exotic Hadrons (Upton, NY, 1988)	89-83513	0-88318-385-4

No. 186	High-Energy Radiation Background in Space (Sanibel Island, FL, 1987)	89-83833	0-88318-386-2
No. 187	High-Energy Spin Physics (Minneapolis, MN, 1988)	89-83948	0-88318-387-0
No. 188	International Symposium on Electron Beam Ion Sources and their Applications (Upton, NY, 1988)	89-84343	0-88318-388-9
No. 189	Relativistic, Quantum Electrodynamic, and Weak Interaction Effects in Atoms (Santa Barbara, CA, 1988)	89-84431	0-88318-389-7
No. 190	Radio-frequency Power in Plasmas (Irvine, CA, 1989)	89-45805	0-88318-397-8
No. 191	Advances in Laser Science—IV (Atlanta, GA, 1988)	89-85595	0-88318-391-9
No. 192	Vacuum Mechatronics (First International Workshop) (Santa Barbara, CA, 1989)	89-45905	0-88318-394-3
No. 193	Advanced Accelerator Concepts (Lake Arrowhead, CA, 1989)	89-45914	0-88318-393-5
No. 194	Quantum Fluids and Solids—1989 (Gainesville, FL, 1989)	89-81079	0-88318-395-1
No. 195	Dense Z-Pinches (Laguna Beach, CA, 1989)	89-46212	0-88318-396-X
No. 196	Heavy Quark Physics (Ithaca, NY, 1989)	89-81583	0-88318-644-6
No. 197	Drops and Bubbles (Monterey, CA, 1988)	89-46360	0-88318-392-7
No. 198	Astrophysics in Antarctica (Newark, DE, 1989)	89-46421	0-88318-398-6
No. 199	Surface Conditioning of Vacuum Systems (Los Angeles, CA, 1989)	89-82542	0-88318-756-6
No. 200	High T_c Superconducting Thin Films: Processing, Characterization, and Applications (Boston, MA, 1989)	90-80006	0-88318-759-0
No. 201	QED Structure Functions (Ann Arbor, MI, 1989)	90-80229	0-88318-671-3
No. 202	NASA Workshop on Physics From a Lunar Base (Stanford, CA, 1989)	90-55073	0-88318-646-2
No. 203	Particle Astrophysics: The NASA Cosmic Ray Program for the 1990s and Beyond (Greenbelt, MD, 1989)	90-55077	0-88318-763-9
No. 204	Aspects of Electron-Molecule Scattering and Photoionization (New Haven, CT, 1989)	90-55175	0-88318-764-7
No. 205	The Physics of Electronic and Atomic Collisions (XVI International Conference) (New York, NY, 1989)	90-53183	0-88318-390-0

No. 206	Atomic Processes in Plasmas (Gaithersburg, MD, 1989)	90-55265	0-88318-769-8
No. 207	Astrophysics from the Moon (Annapolis, MD, 1990)	90-55582	0-88318-770-1
No. 208	Current Topics in Shock Waves (Bethlehem, PA, 1989)	90-55617	0-88318-776-0
No. 209	Computing for High Luminosity and High Intensity Facilities (Santa Fe, NM, 1990)	90-55634	0-88318-786-8
No. 210	Production and Neutralization of Negative Ions and Beams (Brookhaven, NY, 1990)	90-55316	0-88318-786-8
No. 211	High-Energy Astrophysics in the 21st Century (Taos, NM, 1989)	90-55644	0-88318-803-1
No. 212	Accelerator Instrumentation (Brookhaven, NY, 1989)	90-55838	0-88318-645-4
No. 213	Frontiers in Condensed Matter Theory (New York, NY, 1989)	90-6421	0-88318-771-X 0-88318-772-8 (pbk.)
No. 214	Beam Dynamics Issues of High-Luminosity Asymmetric Collider Rings (Berkeley, CA, 1990)	90-55857	0-88318-767-1
No. 215	X-Ray and Inner-Shell Processes (Knoxville, TN, 1990)	90-84700	0-88318-790-6
No. 216	Spectral Line Shapes, Vol. 6 (Austin, TX, 1990)	90-06278	0-88318-791-4
No. 217	Space Nuclear Power Systems (Albuquerque, NM, 1991)	90-56220	0-88318-838-4
No. 218	Positron Beams for Solids and Surfaces (London, Canada, 1990)	90-56407	0-88318-842-2
No. 219	Superconductivity and Its Applications (Buffalo, NY, 1990)	91-55020	0-88318-835-X
No. 220	High Energy Gamma-Ray Astronomy (Ann Arbor, MI, 1990)	91-70876	0-88318-812-0
No. 221	Particle Production Near Threshold (Nashville, IN, 1990)	91-55134	0-88318-829-5
No. 222	After the First Three Minutes (College Park, MD, 1990)	91-55214	0-88318-828-7
No. 223	Polarized Collider Workshop (University Park, PA, 1990)	91-71303	0-88318-826-0
No. 224	LAMPF Workshop on (π, K) Physics (Los Alamos, NM, 1990)	91-71304	0-88318-825-2

No. 225	Half Collision Resonance Phenomena in Molecules (Caracas, Venezuela, 1990)	91-55210	0-88318-840-6
No. 226	The Living Cell in Four Dimensions (Gif sur Yvette, France, 1990)	91-55209	0-88318-794-9
No. 227	Advanced Processing and Characterization Technologies (Clearwater, FL, 1991)	91-55194	0-88318-910-0
No. 228	Anomalous Nuclear Effects in Deuterium/Solid Systems (Provo, UT, 1990)	91-55245	0-88318-833-3
No. 229	Accelerator Instrumentation (Batavia, IL, 1990)	91-55347	0-88318-832-1
No. 230	Nonlinear Dynamics and Particle Acceleration (Tsukuba, Japan, 1990)	91-55348	0-88318-824-4
No. 231	Boron-Rich Solids (Albuquerque, NM, 1990)	91-53024	0-88318-793-4
No. 232	Gamma-Ray Line Astrophysics (Paris-Saclay, France, 1990)	91-55492	0-88318-875-9
No. 233	Atomic Physics 12 (Ann Arbor, MI, 1990)	91-55595	088318-811-2
No. 234	Amorphous Silicon Materials and Solar Cells (Denver, CO, 1991)	91-55575	088318-831-7
No. 235	Physics and Chemistry of MCT and Novel IR Detector Materials (San Francisco, CA, 1990)	91-55493	0-88318-931-3
No. 236	Vacuum Design of Synchrotron Light Sources (Argonne, IL, 1990)	91-55527	0-88318-873-2
No. 237	Kent M. Terwilliger Memorial Symposium (Ann Arbor, MI, 1989)	91-55576	0-88318-788-4
No. 238	Capture Gamma-Ray Spectroscopy (Pacific Grove, CA, 1990)	91-57923	0-88318-830-9
No. 239	Advances in Biomolecular Simulations (Obernai, France, 1991)	91-58106	0-88318-940-2
No. 240	Joint Soviet-American Workshop on the Physics of Semiconductor Lasers (Leningrad, USSR, 1991)	91-58537	0-88318-936-4
No. 241	Scanned Probe Microscopy (Santa Barbara, CA, 1991)	91-76758	0-88318-816-3
No. 242	Strong, Weak, and Electromagnetic Interactions in Nuclei, Atoms, and Astrophysics: A Workshop in Honor of Stewart D. Bloom's Retirement (Livermore, CA, 1991)	91-76876	0-88318-943-7

No. 243	Intersections Between Particle and Nuclear Physics (Tucson, AZ, 1991)	91-77580	0-88318-950-X
No. 244	Radio Frequency Power in Plasmas (Charleston, SC, 1991)	91-77853	0-88318-937-2
No. 245	Basic Space Science (Bangalore, India, 1991)	91-78379	0-88318-951-8
No. 246	Space Nuclear Power Systems (Albuquerque, NM, 1992)	91-58793	1-56396-027-3 1-56396-026-5 (pbk.)
No. 247	Global Warming: Physics and Facts (Washington, DC, 1991)	91-78423	0-88318-932-1
No. 248	Computer-Aided Statistical Physics (Taipei, Taiwan, 1991)	91-78378	0-88318-942-9
No. 249	The Physics of Particle Accelerators (Upton, NY, 1989, 1990)	92-52843	0-88318-789-2
No. 250	Towards a Unified Picture of Nuclear Dynamics (Nikko, Japan, 1991)	92-70143	0-88318-951-8
No. 251	Superconductivity and its Applications (Buffalo, NY, 1991)	92-52726	1-56396-016-8
No. 252	Accelerator Instrumentation (Newport News, VA, 1991)	92-70356	0-88318-934-8
No. 253	High-Brightness Beams for Advanced Accelerator Applications (College Park, MD, 1991)	92-52705	0-88318-947-X
No. 254	Testing the AGN Paradigm (College Park, MD, 1991)	92-52780	1-56396-009-5
No. 255	Advanced Beam Dynamics Workshop on Effects of Errors in Accelerators, Their Diagnosis and Corrections (Corpus Christi, TX, 1991)	92-52842	1-56396-006-0
No. 256	Slow Dynamics in Condensed Matter (Fukuoka, Japan, 1991)	92-53120	0-88318-938-0
No. 257	Atomic Processes in Plasmas (Portland, ME, 1991)	91-08105	0-88318-939-9
No. 258	Synchrotron Radiation and Dynamic Phenomena (Grenoble, France, 1991)	92-53790	1-56396-008-7
No. 259	Future Directions in Nuclear Physics with 4π Gamma Detection Systems of the New Generation (Strasbourg, France, 1991)	92-53222	0-88318-952-6
No. 260	Computational Quantum Physics (Nashville, TN, 1991)	92-71777	0-88318-933-X

No. 261	Rare and Exclusive B&K Decays and Novel Flavor Factories (Santa Monica, CA, 1991)	92-71873	1-56396-055-9
No. 262	Molecular Electronics—Science and Technology (St. Thomas, Virgin Islands, 1991)	92-72210	1-56396-041-9
No. 263	Stress-Induced Phenomena in Metallization: First International Workshop (Ithaca, NY, 1991)	92-72292	1-56396-082-6
No. 264	Particle Acceleration in Cosmic Plasmas (Newark, DE, 1991)	92-73316	0-88318-948-8
No. 265	Gamma-Ray Bursts (Huntsville, AL, 1991)	92-73456	1-56396-018-4
No. 266	Group Theory in Physics (Cocoyoc, Morelos, Mexico, 1991)	92-73457	1-56396-101-6
No. 267	Electromechanical Coupling of the Solar Atmosphere (Capri, Italy, 1991)	92-82717	1-56396-110-5
No. 268	Photovoltaic Advanced Research & Development Project (Denver, CO, 1992)	92-74159	1-56396-056-7
No. 269	CEBAF 1992 Summer Workshop (Newport News, VA, 1992)	92-75403	1-56396-067-2
No. 270	Time Reversal—The Arthur Rich Memorial Symposium (Ann Arbor, MI, 1991)	92-83852	1-56396-105-9
No. 271	Tenth Symposium Space Nuclear Power and Propulsion (Vols. I–III) (Albuquerque, NM, 1993)	92-75162	1-56396-137-7 (set)
No. 272	Proceedings of the XXVI International Conference on High Energy Physics (Vols. I and II) (Dallas, TX, 1992)	93-70412	1-56396-127-X (set)
No. 273	Superconductivity and Its Applications (Buffalo, NY, 1992)	93-70502	1-56396-189-X
No. 274	VIth International Conference on the Physics of Highly Charged Ions (Manhattan, KS, 1992)	93-70577	1-56396-102-4
No. 275	Atomic Physics 13 (Munich, Germany, 1992)	93-70826	1-56396-057-5

No. 276	Very High Energy Cosmic-Ray Interactions: VIIth International Symposium (Ann Arbor, MI, 1992)	93-71342	1-56396-038-9
No. 277	The World at Risk: Natural Hazards and Climate Change (Cambridge, MA, 1992)	93-71333	1-56396-066-4
No. 278	Back to the Galaxy (College Park, MD, 1992)	93-71543	1-56396-227-6
No. 279	Advanced Accelerator Concepts (Port Jefferson, NY, 1992)	93-71773	1-56396-191-1
No. 280	Compton Gamma-Ray Observatory (St. Louis, MO, 1992)	93-71830	1-56396-104-0
No. 281	Accelerator Instrumentation Fourth Annual Workshop (Berkeley, CA, 1992)	93-072110	1-56396-190-3
No. 282	Quantum 1/f Noise & Other Low Frequency Fluctuations in Electronic Devices (St. Louis, MO, 1992)	93-072366	1-56396-252-7
No. 283	Earth and Space Science Information Systems (Pasadena, CA, 1992)	93-072360	1-56396-094-X
No. 284	US-Japan Workshop on Ion Temperature Gradient-Driven Turbulent Transport (Austin, TX, 1993)	93-72460	1-56396-221-7
No. 285	Noise in Physical Systems and 1/f Fluctuations (St. Louis, MO, 1993)	93-72575	1-56396-270-5
No. 286	Ordering Disorder: Prospect and Retrospect in Condensed Matter Physics: Proceedings of the Indo-U.S. Workshop (Hyderabad, India, 1993)	93-072549	1-56396-255-1
No. 287	Production and Neutralization of Negative Ions and Beams: Sixth International Symposium (Upton, NY, 1992)	93-72821	1-56396-103-2
No. 288	Laser Ablation: Mechanismas and Applications-II: Second International Conference (Knoxville, TN, 1993)	93-73040	1-56396-226-8
No. 289	Radio Frequency Power in Plasmas: Tenth Topical Conference (Boston, MA, 1993)	93-72964	1-56396-264-0
No. 290	Laser Spectroscopy: XIth International Conference (Hot Springs, VA, 1993)	93-73050	1-56396-262-4

No. 291	Prairie View Summer Science Academy (Prairie View, TX, 1992)	93-73081	1-56396-133-4
No. 292	Stability of Particle Motion in Storage Rings (Upton, NY, 1992)	93-73534	1-56396-225-X
No. 293	Polarized Ion Sources and Polarized Gas Targets (Madison, WI, 1993)	93-74102	1-56396-220-9
No. 294	High-Energy Solar Phenomena A New Era of Spacecraft Measurements (Waterville Valley, NH, 1993)	93-74147	1-56396-291-8
No. 295	The Physics of Electronic and Atomic Collisions: XVIII International Conference (Aarhus, Denmark, 1993)	93-74103	1-56396-290-X
No. 296	The Chaos Paradigm: Developments an Applications in Engineering and Science (Mystic, CT, 1993)	93-74146	1-56396-254-3
No. 297	Computational Accelerator Physics (Los Alamos, NM, 1993)	93-74205	1-56396-222-5
No. 298	Ultrafast Reaction Dynamics and Solvent Effects (Royaumont, France, 1993)	93-074354	1-56396-280-2
No. 299	Dense Z-Pinches: Third International Conference (London, 1993)	93-074569	1-56396-297-7
No. 300	Discovery of Weak Neutral Currents: The Weak Interaction Before and After (Santa Monica, CA, 1993)	94-70515	1-56396-306-X
No. 301	Eleventh Symposium Space Nuclear Power and Propulsion (3 Vols.) (Albuquerque, NM, 1994)	92-75162	1-56396-305-1 (Set) 156396-301-9 (pbk. set)
No. 302	Lepton and Photon Interactions/ XVI International Symposium (Ithaca, NY, 1993)	94-70079	1-56396-106-7
No. 304	The Second Compton Symposium (College Park, MD, 1993)	94-70742	1-56396-261-6
No. 305	Stress-Induced Phenomena in Metallization Second International Workshop (Austin, TX, 1993)	94-70650	1-56396-251-9
No. 306	12th NREL Photovoltaic Program Review (Denver, CO, 1993)	94-70748	1-56396-315-9